Geophysical Monograph Series

Geophysical Monograph Series

Geophysical Monograph 212

The Early Earth

Accretion and Differentiation

James Badro
Michael Walter
Editors

This Work is a co-publication between the American Geophysical Union and John Wiley and Sons, Inc.

WILEY

This Work is a co-publication between the American Geophysical Union and John Wiley & Sons, Inc.

Published under the aegis of the AGU Publications Committee

Brooks Hanson, Director of Publications
Robert van der Hilst, Chair, Publications Committee

Published by John Wiley & Sons, Inc., Hoboken, New Jersey
Published simultaneously in Canada

Library of Congress Cataloging-in-Publication Data

ISBN: 978-1-118-86057-1

Cover images: An artist's view of Earth's formation by accretion of planetesimals in the solar nebula. The accretion of large objects and giants impacts melt the outermost part of the planet to produce a magma ocean. The material separates chemically in the magma ocean and undergoes gravitational differentiation to form the core and the bulk silicate Earth. The magma ocean cools and crystallizes and gives rise to the earliest crust. (Copyright: IPGP - Joël Dyon)

Printed in the United States of America

10 9 8 7 6 5 4 3 2

CONTENTS

CONTRIBUTORS

James Badro
Research Director, CNRS
Institut de Physique du Globe de Paris
Université Sorbonne-Paris-Cité
Paris, France
and
Blaustein Visiting Professor
Department of Geological and Environmental Sciences
Stanford University
Stanford, California, USA

Maud Boyet
CNRS Researcher
Laboratoire Magmas et Volcans
Université Blaise Pascal
CNRS UMR 6524, Clermont-Ferrand, France

Richard W. Carlson
Director
Department of Terrestrial Magnetism
Carnegie Institution for Science
Washington, DC, USA

Marc Chaussidon
Directeur de Recherches, CNRS
Institut de Physique du Globe de Paris (IPGP)
CNRS UMR 7154, Université Sorbonne-Paris-Cité
Paris, France

Bruce Fegley, Jr.
Professor
Planetary Chemistry Laboratory
Department of Earth and Planetary Sciences and
McDonnell Center for the Space Sciences,
Washington University,
St. Louis, Missouri, USA

John W. Hernlund
Professor
Earth-Life Science Institute
Tokyo Institute of Technology
Meguro, Tokyo, Japan

Kei Hirose
Professor
Earth-Life Science Institute
Tokyo Institute of Technology
Meguro, Tokyo, Japan

Seth A. Jacobson
Postdoctoral Researcher
Bayerisches Geoinstitut
Universtät Bayreuth
Bayreuth, Germany
and
Laboratoire Lagrange,
Observatoire de la Côte d'Azur
Nice, France

Thorsten Kleine
Professor
Institut für Planetologie
Westfälische Wilhelms-Universität Münster
Münster, Germany

Stéphane Labrosse
Professor
Laboratoire de géologie de Lyon
ENS de Lyon, Université Lyon-1, CNRS
Lyon, France

Ming-Chang Liu
Assistant Research Fellow
Institute of Astronomy and Astrophysics,
Academia Sinica (ASIAA), Taipei, Taiwan
Current Affiliation: Ion Microprobe Specialist/
Researcher Department of Earth,
Planetary and Space Sciences
UCLA, Los Angeles
California, USA

Alessandro Morbidelli
Directeur de Recherches, CNRS
Laboratoire Lagrange
Observatoire de la Côte d'Azur
Université Cote d'Azur, CNRS
Nice, France

Frédéric Moynier
Professor
Institut de Physique du Globe de Paris
Insitut Universitaire de France
Université Paris Diderot
Sorbonne Paris Cité
CNRS UMR 7154, Paris, France

Francis Nimmo
Professor
Department of Earth and Planetary Sciences
University of California, Santa Cruz
Santa Cruz, California, USA

Jonathan O'Neil
Assistant Professor
Department of Earth and Environmental Sciences
University of Ottawa
Ottawa, Ontario, Canada

Hanika Rizo
Postdoctoral Scientist
Department of Terrestrial Magnetism
Carnegie Institution for Science
Washington, DC, USA
and
Department of Geology
University of Maryland
College Park, Maryland, USA
Current Affiliation: Assistant Professor
Geotop, Department des sciences
de la Terre et de l'atmosphere
Universite du Quebec à Montreal
Montreal, Canada

Anat Shahar
Staff Scientist
Geophysical Laboratory
Carnegie Institution for Science
Washington, DC, USA

Julien Siebert
Associate Professor
Institut de Physique du Globe de Paris
Université Paris Diderot
Paris, France

Richard J. Walker
Professor
Department of Geology
University of Maryland
College Park, Maryland, USA

Kevin J. Walsh
Research Scientist
Planetary Science Directorate
Southwest Research Institute
Boulder, Colorado, USA

Michael Walter
Professor and Head of School of Earth Sciences
University of Bristol
Bristol, UK

Bernard J. Wood
Professor
Department of Earth Sciences
University of Oxford
Oxford, UK

PREFACE

The processes that formed the infant Earth set the stage for its subsequent evolution into the dynamic and habitable planet we know today. Probing these earliest events is problematic in the extreme, as over four billion years of subsequent evolution have obscured most of Earth's primary features. The challenge is to unravel the series of complicated and bewildering events that began with condensation in the solar nebula, proceeded through the cataclysm of planetary accretion from which emerged a hot and molten proto-Earth, and ended with the large-scale differentiation including core formation, mantle differentiation, and proto-crust formation. Through a multi-disciplinary effort involving geochemistry, experimental petrology and mineral physics, geophysics, planetary physics, and geodynamics, we have seen a transformation in our understanding of these processes in recent years, and are coming to grips with many of the basic questions involving the details of planetary growth and the timescales of accretion. We have placed new constraints on the composition of nebular materials and the bulk Earth, and developed ever more realistic models of the internal differentiation that led to its current dynamic state. But many fundamental questions remain.

The last decade has witnessed a tremendous interest in early Earth processes, and a series of sessions dedicated to this topic were held at the American Geophysical Union Fall Meeting from 2010 to 2013. In this volume we bring together eight review papers that emerged from these sessions, and which cover a range of topics from the very earliest nebular processes to the formation of the earliest proto-crust. These papers provide a look at the state-of-the-art across this range of topics that can help to inform future research by practitioners, and will be of great use to students who are captivated by early Earth evolution. The volume contains papers falling in two general themes: (1) nebular processes, treated in the first four papers, and (2) early differentiation, covered in the last four papers.

The volume appropriately begins with a paper by *Chaussidon and Liu* that deals with the first few millions years of nebular evolution, when most of the materials that constitute the terrestrial planets formed. The authors provide a review of the high-temperature components of primitive meteorites, summarize constraints on timing from short-lived radionuclides, and compare astrophysical and meteoritic chronologies. Next is a paper by *Moynier and Fegley* in which an updated compositional model for Earth is presented based on the latest estimates for the composition of the core. The authors use a range of calculations and data to explore combinations of meteorite types that can best explain bulk Earth composition. The third in the theme is by *Jacobson and Walsh,* and it provides a summary of how Earth is modeled in a series of growth stages. The authors explain why pebble growth processes may have been important in the early stages of planet formation, revisit popular models for late planetary accretion, and describe how the recent Grand Tack model can explain many of the features of the inner terrestrial planets. Rounding out the nebular processes theme is a paper by *Morbidelli and Wood* in which they tackle the bemusing topic of the late addition of material to the planet. They aim to disentangle chemical concepts of the 'late veneer' from modeling concepts of 'late accretion,' and discuss these processes with respect to internal differentiation occurring over similar timescales, which lead naturally to the second theme.

The second theme on early differentiation begins with a paper by *Nimmo and Kleine* in which core formation is considered in terms of the dynamics of metal segregation during giant impacts among differentiated objects, and the timescales of these events as constrained by short-lived isotopes and dynamical models. *Siebert and Shahar* follow on with a work summarizing the modern view of core formation as deduced from geochemical and isotopic arguments. They review experimental siderophile element partitioning data, and discuss how geochemical models, in combination with models for the fractionation of stable isotopes during metal/silicate separation, can be used to deduce the nature of light elements in the core and the conditions of core formation. Next up, *Labrosse, Hernlund, and Hirose* discuss how seismically deduced structures in the very deep mantle, such as large low shear velocity provinces and ultra-low velocity zones atop the core-mantle boundary, may be explained by early magma ocean crystallization that results in a basal magma ocean. They review experimental and mineral physical data and discuss models for how a basal magma ocean may have formed and evolved to the features we observe today. The volume concludes with a review article by *Carslon, Boyet, O'Neil, Rizo, and Walker* that deals with much of the material presented in the first seven articles and brings it all into sharp focus. The paper provides a narrative of the processes occurring in the first few hundred million years from accretion to core formation to early differentiation, summarizes much of the geochemical and isotopic data that constrain these processes, and describes the subtle signals that inform us about processes extending throughout the first 500 million years of evolution and leading to primitive continental crust.

James Badro, *Institut de Physique du Globe de Paris, France*
Michael Walter, *University of Bristol, UK*

ACKNOWLEDGMENTS

This substance of this book stems from four sessions on the "Early Earth" held at American Geophysical Union (AGU) Fall Meetings. We are grateful to the 2009, 2010, 2012, and 2013 AGU Fall Meeting chairs and committees, without whom this project would have never materialized. We acknowledge support from the staff at the AGU for guiding us through the book proposal process, and extend our recognition to the staff at John Wiley & Sons for handling the book production process. We kindly thank Joel Dyon from the Institut de physique du globe de Paris for designing the book cover.

1

Timing of Nebula Processes That Shaped the Precursors of the Terrestrial Planets

Marc Chaussidon[1] and Ming-Chang Liu[2,3]

ABSTRACT

Two key questions in Solar System formation concern the timescales of high-temperature processing (e.g., formations of solids, temperature fluctuations in the disk, etc.) in the early evolutionary stages, and the astrophysical environment in which the solar protoplanetary disk resided. Astrophysical theories of stellar evolution and astronomical observations of young stellar objects, analog to the forming Sun, provide some constraints on the lifetimes of their different evolutionary stages. Finer scale chronologies of the formation of the first solids and planetary objects in the solar accretion disk can be established from analyses of daughter isotopes from now-extinct, short-lived radionuclides in meteorites and in their components. In this review, we describe the high-temperature components of primitive meteorites, namely Ca-Al-rich Inclusions (CAI) and chondrules, we summarize the current knowledge on the origin of short-lived radionuclides, and we compare the two types of chronologies, the "astrophysical one" (derived from observations of young stellar objects and their disks) and the "meteoritical one" (derived from isotopic analyses of meteorites). Within the first few millions years, most of the mass of the solids, which will be at the origin of the terrestrial planets, were formed.

1.1. INTRODUCTION

Recent developments in astrophysics and cosmochemistry are changing our view of the formation and early evolution of the Solar System. A major recent result, coming independently from astrophysical modeling of accretion processes and from dating of meteorites and their components, is that the first planetesimals and the first planets formed very early, in the first few million years (Myr). Gravitational instabilities due to turbulent concentration of solids and further drag of gas can make objects of a few hundred kilometer-size in a few orbital periods in the inner disk [*Johansen et al.*, 2007]. Refinements of our understanding of the $^{182}Hf/^{182}W$ chronometer and of the most appropriate way to correct for W isotopic modifications due to the capture of cosmic ray-induced neutrons in the parent bodies of meteorites, show that a class of differentiated meteorites (the magmatic iron meteorites) were formed by metal-silicate differentiation ~1 Myr after the start of the Solar System [*Kruijer et al.*, 2014]. Thermal modeling of a planetesimal heated by the decay of short-lived ^{26}Al shows that in order to reach internal temperatures that were high enough for complete metal-silicate differentiation at 1 Myr, this

[1]*Institut de Physique du Globe de Paris (IPGP), CNRS UMR 7154, Université Sorbonne-Paris-Cité, Paris, France*

[2]*Institute of Astronomy and Astrophysics, Academia Sinica (ASIAA), Taipei, Taiwan*

[3]*Now at Department of Earth, Planetary and Space Sciences, UCLA, Los Angeles, California, USA*

The Early Earth: Accretion and Differentiation, Geophysical Monograph 212, First Edition.
Edited by James Badro and Michael Walter.

object must have been accreted at ~0.1–0.3 Myr [*Kruijer et al.*, 2014]. Similarly, Hf/W chronometry shows that Mars is most likely a planetary embryo that had reached half of its present size at ~1.8 Myr, and that it escaped later giant impacts [*Dauphas and Pourmand*, 2011; *Tang and Dauphas*, 2014]. Thus, it can be expected that processes, which occurred early in the accretion disk when the nebular gas and dust were still present, left their traces in the composition of the building blocks of planets.

A key step, to which chemical and isotopic fractionations are linked, is the high-temperature processing of material in the nebula, such as evaporation of preexisting (presolar) dust, (re)condensation of (evaporated) solids, and (re)melting and (re)crystallization of solid material. Part of (re)condensed/(re)crystallized solids further agglomerate to form the first "rocks" of the Solar System. Spectroscopic observations of accretion disks around forming stars reveal that dust is processed in the disk and redistributed between the inner and the outer zones [*van Boeckel et al.*, 2004]. A refractory inclusion, presumably formed within a few tenths of an astronomical unit (AU) of the Sun, is present in the cometary matter returned to Earth by the Stardust NASA mission [*Brownlee et al.*, 2006; *Zolensky et al.*, 2006]. Modeling shows that viscous dissipation in the accretion disk, within the first 50–100 kyrs, is able to transport refractory solids formed close to the star to asteroidal or even cometary distances [*Ciesla*, 2010; *Charnoz et al.*, 2011; *Jacquet et al.*, 2011].

Thus, the first few million years are a key period for the evolution of the Solar System. High-temperature solids (refractory inclusions and chondrules, the major components of chondritic meteorites) are formed at that time. They can be considered to be "fossils" of this period. It is even conceivable that solid precursors of some chondrules are fragments of a first generation of planets that would have formed and been destroyed very early [*Libourel and Krot*, 2007], collisions between planetesimals being the rule in the first few million years of the disk [*Bottke et al.*, 2006]. In this review, we concentrate on the timescales for the formation of the high-temperature components, which are presumably the first-generation solids formed in the Solar System, of chondrites. We first present the astrophysical timescales for the different steps in the formation of a solar mass star as derived from observations of young stars and their accretion disks, and then summarize our understanding of the nature of chondrites and their high-temperature components (chondrules and refractory inclusions) followed by a review of the latest developments with the study of short-lived radioactive nuclides in refractory inclusions and chondrules. We finish the paper by summarizing the cosmochemical timescales derived from the study of short-lived ^{26}Al in meteorites.

1.2. YOUNG STELLAR OBJECTS AND THEIR DISKS: ANALOGS OF THE EARLY SOLAR SYSTEM

1.2.1. From the Interstellar Medium to a Protostellar Core

Stars form in the interstellar medium from gravitational collapse of dense fragments of giant molecular clouds. Such molecular clouds are the densest (typically more than 200 H_2 molecules cm^{-3} and up to 10^4–10^6 H_2 molecules cm^{-3} in their hot dense cores) and coldest (T ~10 K) of all interstellar clouds. They are dominantly made of gas (~95% of this gas is molecular H_2, the remaining being mostly He) and of a small dust fraction (~1%). Their density is much higher than that of HII regions (T ~8000 K, density ~0.5 cm^{-3}) where hydrogen is ionized by radiation emitted from young massive stars, or that of HI regions (T ~100 K, density ~50 cm^{-3}), which contain neutral atomic hydrogen [*Bless*, 1996; *Tielens*, 2010 and references therein]. Spectroscopic studies of starlight absorbed and scattered by the interstellar dust grains show that the grain sizes range from ~5 nm to ~2.5 μm and that they have a large range of composition from carbonaceous dust to amorphous silicates [*Tielens*, 2010; *Henning et al.*, 2010]. Dust is initially produced by high-temperature condensation in the envelopes of stars and their ejecta (where other phases such as diamonds, graphite, or refractory oxides are observed), and then is transported to and mixed with interstellar gas and dust. Thus, most of the refractory elements (C, Si, Mg, Fe, Ca, Al, Ti) in the interstellar space are in the dust and not in the gas. Asymptotic Giant Branch (AGB) stars are a major contributor of amorphous carbon dust (for C-rich AGBs) and of silicates (for O-rich AGBs) [*Henning et al.*, 2010]. Once in the interstellar medium and in the intercloud region, dust is exposed to shock waves produced by supernova explosions and to cosmic rays, which could result in amorphization, evaporation, annealing, and/or shattering of the dust grains [e.g., *Hirashita et al.*, 2014]. The broadening of the 10 μm line, i.e. stretching of the Si-O bond in silicates, due to amorphization, is a characteristic of the infrared spectra of interstellar dust. When in cold clouds, the dust can be coated by ices and low condensation temperature species. Typically the lifecycle of dust [*Tielens et al.*, 2005] is such that dust is cycling many times between the intercloud regions and the clouds with a typical timescale of 3×10^7 yr and that the total cycle from condensation in stellar ejecta to incorporation into a new forming star in the core of a dense molecular cloud is about 2×10^9 yr long.

Gas (and associated dust) in molecular clouds is in equilibrium between contraction due to gravitational attraction and thermal (and magnetic) dilatation. For a given temperature there is a critical mass of gas, known as the Jeans mass, exceeding which the gravitational potential energy within the cloud overcomes the kinetic energy of the gas, according to the virial theorem. After some perturbation (e.g., supernova shock waves) compresses the gas, the Jeans mass can be locally reached and a region of the cloud collapses toward its center of mass. Typically a dense core has to be more massive than $10M_\odot$, the Jeans mass calculated with the average density $\rho = 10^5$ H_2 cm^{-3} and temperature $T = 10$ K, for collapse to take place. During the collapse, any particle of gas is pulled toward the center of mass of the collapsing cloud and gets accelerated. Under the assumption of constant acceleration, a first order estimate of the duration of collapse can be obtained by calculating the so-called free-fall time, which corresponds to the time required for a particle initially at distance r to reach the center due to the attraction of mass within this radius r. For typical dense molecular clouds, this free-fall time, which depends only on density, is on the order of 100 kyrs (for $\rho = 10^8$ H_2 cm^{-3}) to 1 Myr (for $\rho = 10^6$ H_2 cm^{-3}).

1.2.2. From a Protostar to a Pre-main Sequence Star

The formation of a star, initiated by gravitational collapse, can be characterized by three physical "stages" on theoretical grounds [*Shu et al.*, 1987]: (i) that of the in-falling envelope, (ii) that of the accretion disk, and (iii) that of the dissipation of the disk. The gravitational collapse of a fragment of a cloud, having initially some weak movement of rotation, leads to the formation of an optically thick rotating core supported by the thermal pressure of hydrogen gas. The core is enclosed by an in-falling envelope, and a disk develops in a plane perpendicular to the axis of rotation of the core, where the centrifugal force can balance the in-fall. Redistribution of angular momentum in the accretion disk from the inside out occurs because of viscosity enhanced by turbulence, which allows transport of material inward to the star [*Dullemond and Monnier*, 2010]. As accretion of material onto the central star continues, the envelope becomes more diffuse and less dense and then eventually disappears. This marks the onset of the second stage, where a large accretion disk surrounds the young star. The third stage is characterized by the dissipation of the disk essentially due to the winds emitted from the forming star and, in case of the formation of giant planets, to the consumption of the gas to form the planets.

This evolution predicts changes in the luminosity, a measure of the total energy emitted by a star per unit time, and surface temperature of young stellar objects (YSO), both of which can be followed in observations. Initially the cloud is transparent to IR and the gravitational energy released by the collapse is radiated away so that the collapse is isothermal. At some point, the density in the core increases to a level where the gas becomes opaque to IR,

and the gravitational energy begins to heat the gas until hydrostatic equilibrium is attained, thus slowing down the collapse. At this point the object is considered to be a protostar. With the temperature of the core increasing to around 2000 K, collisions between the gas molecules will start to dissociate molecular H_2, thus consuming energy and re-initiating contraction. At higher temperatures, H and He will further consume gravitational energy by being successively ionized. The star will then reach again hydrostatic equilibrium, and its contraction will slow down. At this point, the star is called a pre-main sequence star. During all the pre-main sequence stages several nuclear reactions will take place with D burning at around 10^6 K and Li burning at higher temperatures. This pre-main sequence stage ends when the temperature of the core has reached $\sim 10^7$ K and the hydrogen burning starts. At this point, contraction stops again, and the star is named a zero-age main sequence star (ZAMS).

A classification scheme (see Fig. 1.1) has been established for these protostellar and pre-main sequence stages of YSOs [e.g., *André et al.*, 1993, 2000] based on the observed energy output as a function of wavelength (the so-called spectral energy distribution or SED) and its spectral index α_{IR} ($\alpha_{IR} = d\log(\lambda F_\lambda)/d\log(\lambda)$, F_λ is the infrared (IR) flux at the wavelength λ). Because of the low temperature (100—1000 K range) of the gas in the early stages, the light is emitted in the IR or sub-millimeter range. Class 0 corresponds to a protostar deeply embedded in the circumstellar material and envelope: it is bright in the far-infrared, sub-millimeter, and millimeter regimes but is extremely faint in shorter wavelengths. Class I corresponds to a protostar, which has a more diffuse circumstellar envelope and a slower in-fall rate, and is visible in the mid- to near-infrared observations. Objects of this class are characterized by a slope of the SED (α_{IR}) steeper than 0 for wavelengths between 2 μm and ~ 20 μm [e.g., *Lada*, 1987]. Class II objects are pre-main sequence stars, also called T-Tauri stars for objects with mass < ~ 1.5 $M_\odot$, or Herbig Ae/Be stars for objects of intermediate mass (from 1.5 to 3 $M_\odot$). They have a characteristic α_{IR} ranging between -1.5 and 0 and a large protoplanetary disk. Stars with α_{IR} smaller than -1.5 are "Class III" objects, which are also pre-main sequence stars but no longer accrete considerable amounts of material because the disk has largely dissipated.

The existence of an accretion disk around pre-main sequence stars of low to intermediate mass is ascertained by the presence in the SED of an excess of IR emission that cannot be explained by emission from the stellar photosphere alone (see review by *Dullemond and Monnier*, 2010). Although for T-Tauri stars the bump due to the IR emission from the disk is not totally resolved, it is in the case of Herbig Ae/Be stars because they have a higher stellar surface temperature due to their higher masses.

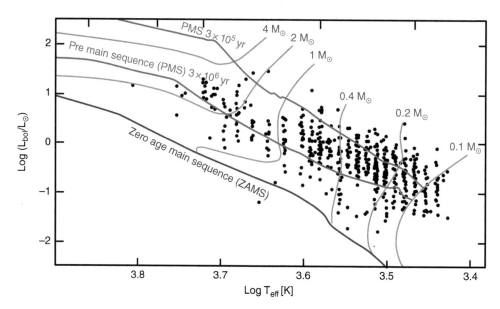

Figure 1.1 An example of HR diagram where observations of X-ray emitting YSOs (dots) are compared to theoretical evolution tracks computed for pre-main sequence stars of low to intermediate masses [redrawn from *Preibisch et al., 2005*]. Color-coded curves are "isochrons" corresponding to theoretically modeled ages of 0.3 Myr, 3 Myr and zero-age main sequence stage (ZAMS). Blue curves are pre-main sequence tracks for stellar masses of 0.1, 0.2, 0.4, 1, 2, and 4 $M_\odot$ according to the evolutionary models of *Siess et al.* [2000].

The presence of this separated bump and its modeled temperature always around 1500 K indicate that it is most likely due to emission from dust that sublimes at higher temperatures closer to the star.

In addition to the different amounts of IR excesses in the SED, other phenomena have also been observed in young stars in other wavelengths, pointing to different physical processes in star formation. For example, bipolar jets and outflows originating from near the central star and the surrounding disk can be observed in Class 0 to Class II sources, and this phenomenon is believed to be a manifestation of the accretion of materials onto the central star and the outward transfer of angular momentum. During the Class I and II stages where active accretion takes place, variations in the accretion rate and the strength of stellar magnetic field can cause the disk to fluctuate, resulting in wrapping of magnetic field lines and magnetic reconnection. These reconnection events liberate excess magnetic field energy and generate intense solar flares and X-rays [*Goldstein et al.*, 1986], which have been observed by the Chandra X-ray Space Telescope [e.g., *Feigelson et al.*, 2002].

1.2.3. Duration of Protostellar and Pre-main Sequence Stages

Just as the free-fall time which characterizes the collapse of the cloud, the protostellar evolution can be described by two characteristic timescales: the Kelvin-Helmholtz timescale (t_{KH}), which is the ratio of the available gravitational energy to the luminosity, and the accretion timescale, which is the ratio of the mass of the protostar core to its accretion rate. Using present day luminosity, $t_{KH} \sim 10$–30 Myr for solar type stars [e.g., *Stahler and Palla*, 2005], but t_{KH} are of course much shorter when taking into account that the luminosity of YSOs is typically enhanced by several orders of magnitude relative to main sequence stars of similar mass [*Feigelson and Montmerle*, 1999].

The age of a pre-main sequence star can be estimated by comparing its measured bolometric luminosity (L_{bol}, i.e., the rate of emission of energy across all wavelengths from a star) and effective temperature (T_{eff}, i.e., the temperature of a black body that would emit the same amount of energy as the star) with theoretical modeling of stellar evolution tracks in the Hertzsprung-Russel diagram (Fig. 1.1). A typical portion of this track is, for instance, the so-called Hayashi track, which is subvertical and corresponds to a nearly adiabatic stage where the star is fully convective, the surface temperature is nearly constant, and the luminosity is decreasing because of contraction. However, unlike main-sequence or more evolved stars whose L_{bol} and T_{eff} can be directly obtained with optical observations, a YSO that is still embedded in the circumstellar envelope or surrounded by a protoplanetary disk is only "visible" in the sub-mm and infrared wavelengths. The observed starlight largely reflects the properties of the ambient dust and gas, rather than those of the stellar surface. Ages can only be

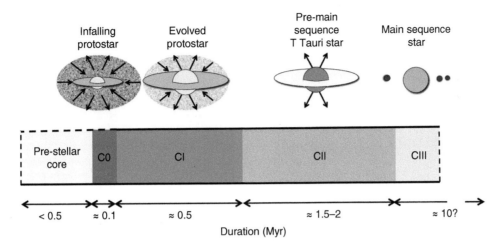

Figure 1.2 Classification of YSOs showing schematically the different evolution phases [*André*, 2002; *Feigelson and Montmerle*, 1999] and an average duration of the corresponding class [*Evans et al.*, 2009], from 0 to III as defined from the spectral energy distribution (see text).

estimated for YSOs whose photospheres can be (partially) seen in the optical wavelengths, such as some of the Class II objects and Class III ones. The lifetimes of earlier phases are then back-calculated by plotting their observed bolometric luminosities against the inferred bolometric temperatures (T_{bol}, defined as the temperature of a blackbody having the same frequency as the observed continuum spectrum; for a main-sequence star, $T_{eff} = T_{bol}$ and for a pre-main-sequence star, $T_{eff} > T_{bol}$) [e.g., *Chen et al.*, 1995]. A L_{bol}-T_{bol} plot has many features of the H-R diagram, including the main sequence, with an extension to cover early (Class 0 and I) phases. However, it should be noted that due to incomplete sampling of the SED and the effects of extinction and reddening, corrections for those complications could result in large systematic uncertainties in the age estimate [e.g., *Gullbring et al.*, 1998]. In addition, the evolutionary tracks of pre-main sequence stars in the L_{bol}-T_{bol} diagram remain relatively poorly calibrated compared to those of main sequence stars.

Based on the observational data of star formation regions, durations have been inferred for each of the SED classes, but these vary substantially from one region to another [e.g., *Evan et al.*, 2009; *Winston et al.*, 2009]. *Evans et al.* [2009] performed a statistical study on the basis of 1024 objects from five clouds and more than 100 sources in each SED class and proposed that the Class 0, Class I, and Class II phases last roughly 0.04−0.3 Myr, 0.2−0.6 Myr, and 0.6−3 Myr, respectively. The duration of Class III phases were not constrained in that study as the sources were missing [*Evans et al.*, 2009]. Other studies, such as *Winston et al.* [2009], showed that the age distributions of Class II and Class III sources are statistically indistinguishable, implying that the duration of the Class III phase is on the order of a few Myr. It should be

pointed out that the inferred lifetimes from observations have an underlying assumption of star formation at steady state [*Evans et al.*, 2009], which may be too simplistic, especially for earlier phases (Class 0 and I), since the star formation rate is highly controlled by the local environment [*Visser et al.*, 2002]. The lifetimes of the inner accretion disks around low to intermediate mass pre-main sequence stars can be studied from focusing on stars having clear IR excess in their SED. This shows that the disk fraction decreases exponentially with age with an approximate mean age for disks of a few Myrs, no inner disk being observable after more than ~10 Myr [*Haisch, Lada, and Lada*, 2001; *Hillenbrand*, 2005; *Fedele et al.*, 2010].

Figure 1.2 shows a schematic timescale for the evolution of YSOs where average values are shown for each class. It is most likely that the early Solar System went through the same evolution. Whether the early evolution of the Sun (as a protosar and a pre-main sequence star) could also be characterized by similar timescales remains an open question as we cannot go back in time to witness the Solar System's birth. However, we have in hand samples of this early epoch: the primitive meteorites and their components, from which a chronology can be established for the first few Myr, as described in the following.

1.3. THE SAMPLES OF THE SOLAR PROTOPLANETARY DISK

1.3.1. Chondrites and Their Putative Parent Bodies

Meteorites are extra-terrestrial rocks, which are debris resulting from collisions or disruptions of various Solar System objects (asteroids, comets, satellites such as the

Moon and planets such as Mars). To first order, they can be divided into two groups according to their petrography and chemistry: "undifferentiated meteorites" and "differentiated meteorites." The term "differentiated meteorites" covers all meteorites, including achondrites and iron meteorites, that have chemical compositions and petrographic characteristics indicating that they are rocks formed after planetary-scale melting and differentiation, either metal-silicate or silicate-silicate. There is a fraction of iron meteorites, the so-called non-magmatic, that are probably formed from impact melts in which silicate-metal segregation took place on a smaller scale. During these planetary processes, these rocks have lost most pre-planetary signatures so that they cannot really be used to reconstruct early Solar System nebular processes. Thus, differentiated meteorites are not the subjects of the present review. The reader might refer to *Mittlefehldt* [2014] and *Benedix et al.* [2014] and references therein for detailed descriptions and studies of differentiated meteorites.

The structure of chondrites implies that they are sediments made from components with very different origins (from high-temperature minerals to water-rich low-temperature minerals) and their high iron content (used as a means of identification in the field) demonstrates that they did not undergo melting and differentiation since their formation. Chondrules (see below) are their characteristic feature (except for CI chondrites, which are dominated by matrix), hence their name. Chondrites are understood as having accumulated in the accretion disk from materials formed very early in the solar nebula or inherited from the presolar molecular cloud. The observation that the bulk compositions of chondrites, except for a few volatile elements (e.g., H, He, noble gases, ...) are close or identical to that of the solar photosphere [*Lodders*, 2003; *Palme and Jones*, 2005; *Asplund et al.*, 2009; and references therein], and by inference that of the Solar nebula, is generally considered as indicating that chondrites are chemically "primitive."

The classification of chondrites relies on the study of a growing number of specimens using more and more elaborate petrographic, chemical, and isotopic criteria. The initial classification [*Van Schmus and Wood*, 1967] identified three groups of different chemical compositions (enstatite chondrites, carbonaceous chondrites, and ordinary chondrites), which were further subdivided into six petrographic types corresponding to metamorphic or alteration grades (not all observed in the three groups), from type 1 to type 6. Based on the amount of iron (high, low, or very low) the enstatite chondrites are subdivided into EH and EL and the ordinary chondrites into H, L, and LL. The six metamorphic grades are now understood as representing two very different types of processes [*Huss et al.*, 2006; *McSween and Huss*, 2010]: from type 3

to type 6 it corresponds to high-temperature transformations that took place at increasing temperatures from ~500°C to 900°C in the depths of the parent bodies, while type 2 and type 1 correspond to an increasing degree of water-assisted low temperature (< ~150°C) alteration, probably throughout the parent bodies (pre-accretion alteration in a nebular setting of chondritic components has also been inferred [*Krot et al.*, 1995; *Brearley*, 2006, 2007]). Other criteria can be used, such as the preservation of refractory presolar grains [*Huss and Lewis*, 1995]. A finer-scaled estimation of the degree of metamorphism (from 0 to 9) has been obtained for type 3 chondrites, with 3.0 being least metamorphosed, from detailed studies of mineralogical changes and compositional zoning in minerals [*Brearley and Jones*, 1998]. Additional constraints on the petrologic type have been obtained from Raman spectroscopic studies of the maturity of organic matter in the matrix of chondrites [*Bonal et al.*, 2006].

An important discovery, post-dating the earliest petrological classification scheme of chondrites, was the presence of non mass-dependent oxygen isotopic variations (expressed as $\Delta^{17}O$ values and calculated from the $\delta^{18}O$ and $\delta^{17}O^*$ values according to $\Delta^{17}O = \delta^{17}O - 0.52 \times \delta^{18}O$) in chondrites and their components [*Clayton et al.*, 1973] and the fact that the different chemical groups have different $\Delta^{17}O$ values [*Clayton and Mayeda*, 1999; *Franchi*, 2008; *Yurimoto et al.*, 2008; *Krot et al.*, 2014]. This makes the oxygen isotopic composition a very powerful criterion to refine the chemical classification. Since then, systematic differences in isotopic compositions have been found for other elements (e.g., Ti, Ca, Cr, Ni, Sr, Zr, ...) within chondrites [*Moynier and Fegley*, this volume, Chapter 2]. At present 15 groups are identified based on chemical and isotopic criteria [*Wasson*, 1985; *Wood et al.*, 1988; *Krot et al.*, 2005; *Scott and Krot*, 2014; *McSween and Huss*, 2010]: carbonaceous chondrites (CI, CM, CV, CO, CR, CK, CB, and CH), ordinary chondrites (H, L, and LL), enstatite chondrites (EH and EL), K chondrites, and R chondrites. There are also a number of carbonaceous chondrites (≈10) that are not classified into any category because of the too limited number of specimens known. Of all groups, the CI chondrites group has a bulk chemical composition nearly identical to that of the Solar photosphere, except for the most volatile elements [*Lodders*, 2003; *Palme and Jones*, 2005; *Asplund et al.*, 2009]. Systematic gross differences exist in bulk composition within the different chondrite groups relative to CI chondrites [*Wasson and Kallemeyn*, 1988; *Lodders and Fegley*, 1998]: (i) carbonaceous chondrites (except CI) are enriched in refractory lithophile elements (Al, Ca, ...) relative to CI chondrites (by ~10 to 40% normalized to

---

* $\delta^{17,18}O_{SMOW} = [(^{17,18}O/^{16}O)_{sample} / (^{17,18}O/^{16}O)_{SMOW} - 1] \times 1000$

Si), while ordinary chondrites and enstatite chondrites are depleted (by ~10 to 20% and ~30 to 50%, respectively), and (ii) all chondrites show decreases of concentration of moderately volatile elements (Cr, Mn, K, Na, ...) with increasing volatility (or decrease of 50% condensation temperature [see *Lodders*, 2003]). These systematics are understood to first order as resulting from formation of chondrites from components having various nebular histories and origins at variable temperatures [*Wasson and Kallemeyn*, 1988; *Scott and Krot*, 2005].

The fact that variations in bulk compositions of chondrites are associated with variations in their bulk $\Delta^{17}O$ values gives support to the idea that each chondrite group is likely to correspond to a different parent body. The bulk chemical and oxygen isotopic composition of a parent body is established during its accretion from the mixing in different proportions of materials with various origins (from very refractory, such as CAIs, to water) and to carry variable $\Delta^{17}O$ values ($\Delta^{17}O$ is conserved during mass fractionation planetary processes). In this context, the existence of different metamorphic grades from 3 to 6 in a given chondrite group (the existence of the four grades from 3 to 6 is observed for H, L, LL, EH, EL, and CK groups) has been used to suggest an "onion shell" structure for the parent bodies of these chondrites. This was proposed from U-Pb ages of metamorphic phosphate minerals [*Göpel et al.*, 1994] and from thermochronometric studies [*Trieloff et al.*, 2003], both showing that the change of peak metamorphic temperature with age along increasing metamorphic grades was following thermal histories modeled at various depths in a parent body of size >100 km heated by the decay of short-lived ^{26}Al. However, recent studies [*Ganguly et al.*, 2013] of interdiffusion profiles between adjacent orthopyroxenes and clinopyroxenes and of metallographic cooling rates [*Scott et al.*, 2014] in H4-H5-H6 chondrites show that samples from different metamorphic grades have similar cooling histories, implying a more complex model with fragmentation and re-accretion for the H chondrite parent body. A similar change in paradigm has been recently provided by the observation that some chondritic meteorites have traces of primary internally generated magnetic fields [*Weiss and Elkins-Tanton*, 2013]. Geophysical modeling shows that bodies having accreted sufficient amounts of mass before ~2 Myr after the start of the Solar System would have enough ^{26}Al to melt in their interiors to form metallic cores that may be capable of generating a magnetic field for more than 10 Myr, while keeping at their surface an undifferentiated crust that could continue to accrete material for as long as it is still available in the disk [*Elkins-Tanton et al.*, 2011; *Weiss and Elkins-Tanton*, 2013]. This introduces a modification of the "onion shell" model, in which the chondritic parent bodies could have metallic cores and differentiated mantles, and in which

the isotopic compositions of the interior of the body (e.g., its $\Delta^{17}O$) would not have been homogenized with that of its crust. The Allende CV3 chondrite, which recorded temperatures up to ~590°C in some of its components [*Krot et al.*, 1995, 1998; *Busemann et al.*, 2007], could have originated from some depths in the undifferentiated crust of the CV parent body, while less metamorphosed CV3 chondrites like Vigarano and Efremovka could have originated from shallower depths. Thus chondrites have retained primitive compositions but might not have originated from primitive undifferentiated bodies sensu-stricto. One can even speculate that parent bodies of chondrites had a protracted evolution with an undifferentiated crust added late, >3–4 Myr after the start of the Solar System, as indicated by the oldest ages of chondritic components dated with ^{26}Al (see next section). In addition, recent developments of spectroscopic studies [*DeMeo and Carry*, 2013] allow for a better link between asteroids and chondrites. In the case for instance of the S-type asteroids, the putative parent bodies of ordinary chondrites, there are several objects (which are not fragments of a single one) having identical compositions, suggesting that one group of chondrites does not necessarily originate from a single parent body [*Vernazza et al.*, 2014]. One of these objects (asteroid 25143 Itokawa) has been recently sampled by the Hayabusa spacecraft, and the isotopic composition of the grains returned to Earth demonstrate that this S-type asteroid is a possible parent body of the L and LL group of equilibrated ordinary chondrites [*Yurimoto et al.*, 2011].

1.3.2. The Major High-Temperature Components of Chondrites

Most chondrites are composed of three main constituents, best preserved in type 3 chondrites: refractory inclusions (mostly Ca-Al-rich Inclusions [CAI], though a few other types of refractory inclusions exist, see below), chondrules (by far the dominant constituent, ~60–80%, in ordinary and enstatite chondrites, but much lower in CK and CI chondrites for instance [*Scott*, 2007]), and fine-grained dust particles (also called matrix materials). The bulk composition of chondrites (chemical or isotopic) depends of course on the fraction of these three constituents, which accreted together in the "feeding zone" of the parent bodies. Systematic isotopic differences (^{17}O, ^{50}Ti, ^{54}Cr...) seem present between material accreted either beyond the snow line (carbonaceous chondrites) or within it (non-carbonaceous materials) [*Warren*, 2011]. In addition, significant variations in composition do exist within chondrules and matrix, requiring some specific mixing fractions (the origin of which remains enigmatic) of these two components (also called complementarity between chondrules and matrix) to match a bulk chondritic composition [*Palme et al.*, 2015]. In the following discussion, we concentrate on the two high-temperature phases, CAIs and chondrules, because most, if not all, of the age information that can be obtained is from them. They reflect conditions and processes in the inner Solar System (within a few AU and even perhaps closer than 0.1 AU from the forming Sun) early in its history. CAIs are the oldest Solar System objects that have been dated. Their Pb-Pb age is of ~4568 Myr, but it is not yet totally clear if all the range reported from 4567.30±0.16 Myr [*Connelly et al.*, 2012] to 4568.22±0.17 Myr [*Bouvier and Wadhwa*, 2010] is only due to the use of an improper U isotopic ratio for the age calculations [*Brennecka et al.*, 2010; *Amelin et al.*, 2010; *Amelin and Ireland*, 2013]. Regardless, there is no reason to consider that when CAIs formed, low-temperature phases (that were subsequently transported to the inner region) were not forming at the same time in the outer region of the disk. Key questions pertaining to the origin of CAIs and chondrules are presented in *Connolly* [2005].

Thermodynamic calculations simulating the cooling of a gas of solar composition heated at T >2000 K predict the formation of solids by condensation (or liquids for higher non-canonical pressures and/or dust enrichments) with initially refractory compositions (e.g., Al- and Ca-rich), and upon cooling, the solid condensates become less refractory by further reaction with the gas [*Grossman*, 1972; *Grossman et al.*, 2000; *Ebel*, 2006]. Refractory solids are observed in the dust around forming Herbig Ae stars from the bump of infrared emission at 2–3 μm corresponding to a Planck temperature of ~1600 K [*Dullemond and Monnier*, 2010]. Grain coagulation in disks is predicted to be fast with time scales of 10^4–10^5 years [*Dullemond and Dominik*, 2005] and observations show that grain growth to mm sizes occurs already in the collapsing envelopes of Class 0 and I protostars [*Miotello et al.*, 2014]. The rapid formation of refractory condensates is in line with the constraints coming from the ^{26}Al systematics of refractory inclusions in CV chondrites (see below).

Most CAIs consist mainly of refractory silicates and oxides, including Al-spinel ($MgAl_2O_4$), melilite [$Ca(Al,Mg)(Si,Al)_2O_7$], Ti-rich pyroxene (a solid solution of four end-member components: $CaMgSi_2O_6$ – $CaAl_2SiO_6$ – $CaTi^{3+}AlSiO_6$ – $CaTi^{4+}Al_2O_6$) and anorthite ($CaAl_2Si_2O_8$) as major components, and corundum (Al_2O_3), hibonite ($CaAl_{12}O_{19}$), and perovskite ($CaTiO_3$) as minor components. CAIs are classified by using different criteria (Fig. 1.3 and its caption) which, for instance, allows us to identify (based on petrography) which of them are the closest to be direct aggregates of condensation products (e.g., fluffy type A or fine-grained CAIs), or (based on isotopic composition) were formed in an isotopically heterogeneous reservoir possibly dating back to the earliest epoch in the disk (e.g., PLAC hibonites or FUN inclusions), or underwent melting/evaporation after their accretion (e.g., type B CAIs or FUN inclusions). In addition, many CAIs have experienced complex thermal and alteration (either in the nebula or in the parent bodies) histories that complicated their morphology, mineralogy, texture, composition, and isotopic chemistry.

The bulk compositions of CAIs correspond to the first ~5% of chemical elements that are predicted to condense from a gas of solar composition [*Davis and Richter*, 2007]. In a CMAS system (composed of only Ca, Mg, Al, and Si) equilibrium condensation predicts the successive formation with decreasing temperature of corundum, hibonite, type A CAI compositions, type B CAI composition, type C CAI composition, and Al-rich chondrule composition [e.g., *Yoneda and Grossman*, 1995; *MacPherson*, 2005]. However in detail, the composition of type B, type C, and of FUN CAIs differ significantly from those predicted from the condensation trends, whatever the total pressure considered is. This has been attributed to alteration of CAIs before their re-melting [*MacPherson*, 2005], to chemical fractionation due to evaporation, or to non-representative sampling during bulk analysis [*Grossman et al.*, 2000]. A key observation is that large enrichments in heavy Si and Mg isotopes are present in type B CAIs (~5‰ in δ^{25}Mg and ~3‰ in δ^{29}Si) and FUN CAIs (~30‰ in δ^{25}Mg and ~15‰ in δ^{29}Si) and that they can be explained by evaporative loss of Mg (~30% for type B and 80% for FUN) and Si (~15% for type B and ~50% for FUN). Correcting the bulk composition of CAIs for this evaporative loss makes their original composition nearly identical (within the uncertainties of the approach) to that predicted from the condensation sequence [*Richter et al.*, 2007; *Mendybaev et al.*, 2013]. Other mechanisms capable of reproducing the observed Mg and Si isotopic fractionations in type B and FUN CAIs have also been proposed [e.g., *Alexander*, 2004].

Chondrules are sub-millimeter sized silicate objects that experienced melting, partial crystallization, and quench in a gas, as indicated by their sub-rounded shapes and the presence of glass. In contrast to theories for condensation of refractory phases in a nebula gas, chondrules are not a direct prediction of astrophysical models. However they are characteristic of chondrites, and in some cases the major component (e.g., up to 80% by volume in ordinary and enstatite chondrites). Despite many analytical and experimental studies, the origin(s) of chondrules remains a key open question [see for instance *Grossman*, 1988; *Wasson*, 1993; *Hewins*, 1997; *Zanda*, 2004; *Jones et al.*, 2005]. Chondrules are typically composed of low-Ca pyroxene and olivine (with various Fe/Mg ratio) along with minor components such as Ca-rich pyroxene, feldspathic glass, metal, and mesostasis. They are classified

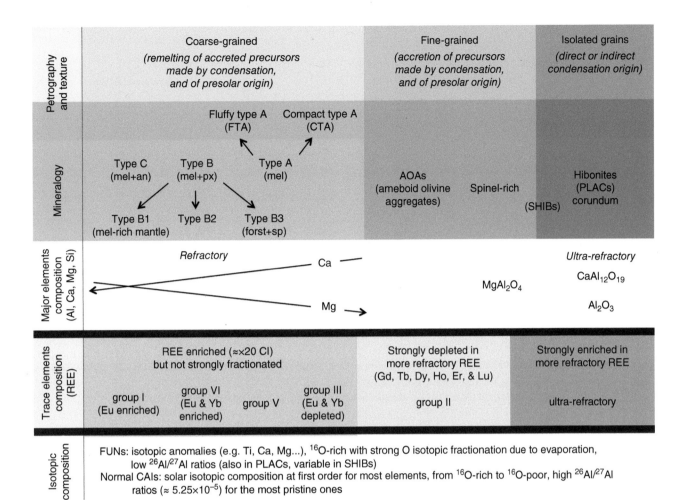

Figure 1.3 Overview of the classification of refractory inclusions.

Five criteria are shown from top to bottom (more or less in the order of their recognition as a classification criterion): petrography and texture, mineralogy, major elements chemistry, trace elements chemistry, and isotopic composition. A thick line separates two criteria for which no direct link has been identified (though some relationships may exist). Different colors underline different criteria or groups, the superposition of colors indicating when two different criteria concur to define a special group.

Coarse-grained CAIs have generally rounded shapes, as expected in case they were once melted droplets in a gas, and they show igneous textures and mineral chemistry consistent with cooling from the outside in (e.g., zoned melilite chemistry in a certain type of CAIs [*MacPherson et al.*, 1988]). In contrast, irregular-shaped CAIs tend to be fine-grained, albeit with exceptions, suggesting that they are aggregates of nebular condensates that have never been significantly melted. There is a general trend of chemical composition (see text), and corresponding mineralogy, from the ultra-refractory objects (isolated grains) to refractory objects (coarse-grained CAIs). Coarse-grained CAIs are further divided based on their mineralogy [*Grossman*, 1973; *MacPherson et al.*, 1988; *MacPherson*, 2005] into types A, B, and C (and further into B1, B2, and B3 for types B). Experiments have reproduced the crystallization sequence observed in type B CAIs, for liquids of appropriate composition cooled from ≈1400°C at a few tenths to tens °C/hour [*Stolper*, 1982; *Stolpe and Paque*, 1986]. CAIs are enriched in the refractory lithophile REE elements by a factor of ≈20 relative to CI, in line with their refractory composition. Six groups have been defined on the basis of REE distribution [*Martin and Mason*, 1974; *MacPherson*, 1988] ranging from ultra-refractory (strongly enriched in more refractory HREE relative to LREE) and group II (strongly depleted in HREE relative to LREE) to the four other groups (I, III, IV, and V, all enriched in REE but with variable anomalies in the more volatile REE Eu and Yb). These variations are understood as reflecting fractionations due to the relative volatility of REE during condensation/evaporation [*Ireland and Fegley*, 2000]. CAIs are generally ^{16}O-rich [*Yurimoto et al.*, 2008], as expected if their precursors condensed from a gas of solar composition [*McKeegan et al.*, 2011], even though evidence for isotopic exchanges with a ^{16}O-poor reservoir very early in their history has been found in some of them [*Krot et al.*, 2005, 2010]. They contain nuclear isotopic anomalies in several elements (e.g. Ti, Ca, Mg, …) which are more pronounced in ultra-refractory and FUNs inclusions [e.g., *Ireland*, 1988; *Liu et al.*, 2009], as expected if they formed very early in the disk when nuclear isotopic anomalies carried by presolar grains were not homogenized or largely diluted within condensation products from the nebula gas.

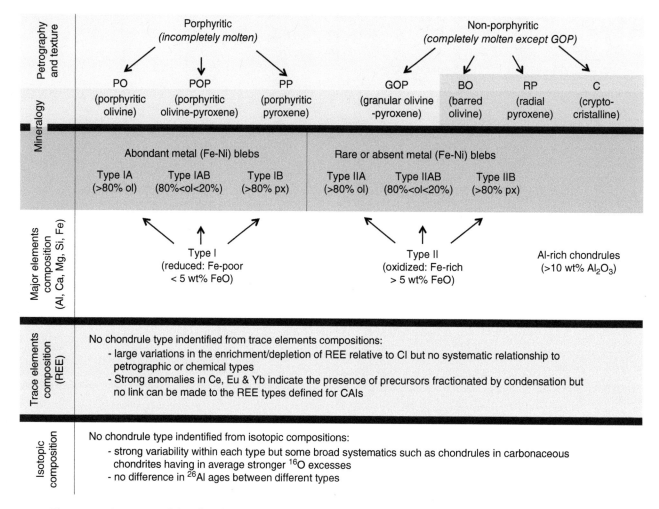

Figure 1.4 Overview of the classification of chondrules (same code and presentation than in Figure 1.3 for the classification of refractory inclusions).

Chondrules are either classified based on their petrographic textures [*Gooding and Keil*, 1981] or on their chemistry [*McSween*, 1977]. The major divide in the petrographic/textural classification is between porphyritic (either PO, POP, or PP depending on the abundance of olivine and pyroxene), which have never been totally melted, and non-porphyritic, which have been fully melted (for BO, RP, and C types). GOP type is a rare subset [*Wasson*, 1993]. Porphytitic chondrules dominate nearly in all chondrite groups [*Grossman et al.*, 1988]. For the chemical classification, the major divide is between type I (or reduced) chondrules with Fe-poor olivine (Fo#>95) and type II (or oxidized) chondrules with Fe-rich olivines (Fo#<95). These two types are further divided in three sub-types (A, AB, and B) based on the relative amounts of olivine and pyroxene. A third chemical type is Al-rich chondrules, which are intermediate in composition between refractory inclusions and "normal" ferro-magnesian chondrules. There is no direct relationship between the chemical types and the petrographic types, but of course there are links between the chemical composition and the textures and mineralogy (see text). Contrary to refractory inclusions, chondrules cannot be classified further based on their REE contents or on their isotopic compositions. Strong variations in REE contents exist in chondrules, e.g., strong depletion in Eu and Yb and to a lesser extent Sm and Gd [*Misawa and Nakamura*, 1988; *Pack et al.*, 2004] and are most likely due to the fact that some chondrule precursors underwent condensation/evaporation under varying fO_2. Within a "normal" chondrule, REE contents are distributed among the different phases following classical magmatic partitioning [e.g., *Jacquet et al.*, 2012]. Chondrules have variable oxygen isotopic compositions, and are generally depleted in ^{16}O relative to refractory inclusions [see review in Yurimoto et al., 2008]. Some of them, but not all, show clear evidence of oxygen isotopic exchange during their formation between their putative precursors and the nebula gas [*Chaussidon et al.*, 2008].

(see Fig. 1.4 and its caption) on the basis of their petrographic/textural, mineralogic, and chemical compositions [*Gooding and Keil*, 1981; *McSween*, 1977; *Brearley and Jones*, 1998; *Scott and Krot*, 2014], allowing the identification of whether they were totally molten or preserved part of their precursors, or if they formed in a reduced or oxidized environment. The chemistry of chondrules is typically less refractory than that of CAIs as expected if their precursors condensed at lower temperatures than that of CAIs (although their liquidus temperature spans a large range from ~1200 to ~1800°C [*Radomsky and Hewins*, 1990]), Al-rich chondrules are intermediate, but there is no simple chemical path to go from CAIs to chondrules. Despite the fact that, for instance, type I chondrules are much more abundant in carbonaceous chondrites than in ordinary chondrites, chondrites generally contain examples of most chondrule types [*Gooding and Keil*, 1981; *Scott and Taylor*, 1983; *Grossman et al.*, 1988].

Experimental simulations [*Lofgren*, 1996; *Hewins et al.*, 2005, and references therein] have demonstrated that the textures of most chondrules could be explained by melting of their precursors up to peak temperatures of 1500–1600°C followed by subsequent cooling at 10–1000°C/hour, i.e. faster than CAIs (few tenths to tens °C/hour [*Stolper and Paque*, 1986]) but slower than if they were isolated droplets in the nebula gas. Note that a slow cooling rate (1°C/hour between 1000 and 800°C), similar to that of type B CAIs, seems required to explain the growth of plagioclase in plagioclase-bearing type I chondrules from CO chondrites [*Wick and Jones*, 2012]. The broad relationships observed between the chemical composition of chondrules and their textures are likely due to the effect of composition on liquidus temperature (and preservation of seed nuclei during melting) so that for instance non-porphyritic textures are nearly absent for compositions having liquidus temperatures above 1680 °C [*Radomsky and Hewins*, 1990]. The duration of the melting event has been tentatively constrained to be short (i.e., flash heating) in order that volatile elements such as Na were retained in chondrule melts [*Yu and Hewins*, 1998], but these constraints can be relaxed if chondrules were melted in a gas with high partial pressures of Na and other volatiles [*Alexander et al.*, 2008; *Hewins et al.*, 2012; *Fedkin and Grossman*, 2013]. Such high pressures of Na are most easily explained by having very high densities of dust in the chondrule-forming regions. More generally, the interactions between chondrule melts and the ambient gas could exert a strong control on the final chemistry of chondrule [*Tissandier et al.*, 2002; *Libourel et al.*, 2006]. Obviously, determining the P-T-time conditions in which chondrules were melted is a key constraint on their origin [e.g., *Desch et al.*, 2012] as well as other parameters such as their size distribution

[*Jacquet*, 2014]. High pressures of alkalis can, for instance, be obtained in the atmosphere of an outgassing magma ocean on a planetary embryo [*Morris et al.*, 2012; *Lupu et al.*, 2014], which would be the environment to consider if chondrules formed in bow shocks of planetary embryos. While nebular models such as shockwave heating of chondrules' precursor-dust [*Ciesla and Hood*, 2002; *Desch and Connolly*, 2012] predict a flash-heating and reproduce the cooling paths inferred for chondrules from experimental simulations, formation in impact-generated plumes [*Asphaug et al.*, 2011] would explain the gas pressures and dust densities (a few orders of magnitude higher than canonical, i.e., gas of solar composition at a pressure of 10^{-4} bar and with no dust-enrichment) required for Na retention [*Fedkin and Grossman*, 2103]. Other origins are possible for chondrules, such as their formation and transport in the strong winds emitted by the Sun when in its T Tauri phase [*Shu et al.*, 1996; *Salmeron and Ireland*, 2012]. Age determination for chondrule melting and chondrule precursors (see next section) is another strong constraint on the origin of chondrules.

1.4. CHRONOLOGY OF THE FIRST FEW MILLION YEARS: THE PERIOD OF THE DISK

1.4.1. Short-Lived Radionuclides Present in the Accretion Disk

Studies of short-lived radionuclides (SLR) have been of keen interest in cosmochemistry because the results have important implications for the immediate astrophysical environment in which the nascent solar system resided as well as for time constraints on processes occurring in the solar nebula (Fig 1.5). Usually the term "short-lived" in cosmochemistry is defined as radioactivities with half-lives ≤100 Myr; however, to constrain the history in the first few million years of the solar system, radionuclides with shorter half-lives (≤10 Myr) are needed. Such half-lives are short compared to the age of the solar system but are long enough for SLRs to survive in the solar nebula for a period of time and then be incorporated into CAIs and chondrules. The presence of SLRs at the beginning of the Solar System can be deduced through the excesses of daughter isotopes caused by radioactive decay in meteoritic components. ^{129}I ($t_{1/2}$ = 15.7 Myr) is the first short-lived radionuclide whose presence in the early Solar System was inferred via the excesses of its daughter isotope ^{129}Xe in meteorites by *Jeffery and Reynold* [1961]. This discovery was followed by the work of Rowe and *Kuroda* [1965], which revealed the excesses of fission Xenon isotopes, indicative of the prior existence of ^{244}Pu ($t_{1/2}$ = 82 Myr). Later, large excesses of ^{26}Mg in the WA CAI of the Allende meteorite were found, and the degree of such excesses appeared to be correlated with Al/Mg in CAI minerals, implying the in-situ decay of a much shorter-lived radionuclide ^{26}Al ($t_{1/2}$ = 0.7 Myr) [*Gray and Compston*, 1974; *Lee et al.*, 1976]. With more advanced analytical techniques becoming available, such as secondary ion mass spectrometry (SIMS) and inductively coupled plasma mass spectrometry (ICPMS), a number of other extinct radioactivities have also been successfully inferred in various meteoritic components, ranging from minerals formed in the solar nebula or parent bodies to rocks produced during planetary differentiation processes.

The abundance of a short-lived radionuclide in a meteoritic object at the time when this object formed, or when isotopic closure was reached in this object, is expressed as a ratio of that given radionuclide to its stable counterpart, for instance, ^{26}Al/^{27}Al. This is because the abundance is inferred by using an "isochron" (see Fig. 1.6). Because CAIs are known to be the oldest solids in the Solar System that have been dated and are assemblages of refractory minerals enriched in Ca and Al, their fossil records of short-lived ^{41}Ca (decays to ^{41}K, $t_{1/2}$ = 0.1 Myr) and ^{26}Al (if free of loss of radiogenic daughter isotopes due to resetting or secondary alteration) would in principle represent the starting ^{41}Ca/^{40}Ca and ^{26}Al/^{27}Al ratio in the solar system (see discussion in Section 1.4.4). Having a comprehensive understanding of the initial abundances of extinct radionuclides is key to reconstructing the astrophysical environment of the forming Solar System and is still a matter of ongoing work. Improvements on analytical techniques have made revisions of some of the previously inferred values successful, such as ^{41}Ca/^{40}Ca [*Liu et al.*, 2012] and ^{60}Fe/^{56}Fe [e.g., *Moynier et al.*, 2011; *Tang and Dauphas*, 2012]. However, it should be noted that to infer the initial abundances of radioactivities that are volatile in nature, e.g., ^{36}Cl (decays to ^{36}S and ^{36}Ar, $t_{1/2}$ = 0.3 Myr), ^{53}Mn (decays to ^{53}Cr, $t_{1/2}$ = 3.7 Myr) and ^{60}Fe, requires knowledge of formation time relative to CAIs in the objects being measured [*McKeegan and Davis*, 2007]. Moreover, discrepancies exist in inferred abundances obtained with different analytical methods (e.g., bulk vs. in-situ measurements of Ni isotopes for ^{60}Fe), see *Moynier et al.* [2011], *Tang and Dauphas* [2012] and *Mishra and Chaussidon* [2014b], for which explanations are still lacking. In the following section, discussions will be focused on the literature results and implications of ^{26}Al (decays to ^{26}Mg, $t_{1/2}$ = 0.7 Myr) and ^{10}Be (decays to ^{10}B, $t_{1/2}$ = 1.3 Myr), since they have been extensively used to infer the origins of SLRs and Solar System chronology in the first few million years. The origins of these SLRs have important implications for the Solar System's astrophysical environment as well as processes that occurred when the Sun was still forming.

1.4.2. The Dual Origin of SLRs: Presolar Stellar Sources and Solar System Irradiation

Several thorough reviews about the origins of SLRs [e.g., *Meyer*, 2005; *Wasserburg et al.*, 2006; *McKeegan and Davis*, 2007; *Huss et al.*, 2009; *Dauphas and Chaussidon*, 2011] have been published, and no major change in our understanding has occurred since. From theoretical viewpoints, SLRs present 4.5 billion years ago can be produced by stellar nucleosynthesis and then injected into the molecular cloud that ended up forming the Solar System or into the just-formed solar protoplanetary disk. Alternatively, non-thermal nuclear reactions between energetic charged particles and gas or/and solid targets in the interstellar medium or in the early solar system could also account for the presence of some SLRs (Fig. 1.5). Of all aforementioned radionuclides, ^{60}Fe requires a stellar source (whether it was a last minute event or inherited from the parental molecular cloud is still a matter of discussion), ^{10}Be is primarily produced by irradiation, and the rest can be made by either nucleosynthetic pathway [e.g., *Wasserburg et al.*, 2006]. Disentangling different sources accounting for different SLRs is key to constraining the immediate environment of the forming Solar System, but this requires a detailed and quantitative understanding of the distributions and initial abundances of different SLRs. For example, ^{26}Al and ^{41}Ca were found to have been correlated with one another in terms of the presence or absence in CV CAIs and CM hibonite grains, indicative of a common origin for the two radionuclides [*Sahijpal and Goswami*, 1998]. On the contrary, isotopic analysis of the same groups of inclusions revealed that ^{26}Al and ^{10}Be abundances are decoupled (see below), implying whatever mechanism accounting for ^{10}Be nucleosynthesis did not contribute much to the ^{26}Al inventory in the early Solar System. It should however be noted that, except ^{26}Al, there are usually significant uncertainties (as much as 50%) [e.g., *Liu et al.*, 2012] associated with the inferred initial abundances of shortest-lived radionuclides, which sometimes would not allow for a statistically meaningful comparison between them. More work is certainly needed to better address this issue.

1.4.3. Timing of Irradiation Processes in the Early Solar System

Star formation is not a peaceful process. Observations of solar-type YSOs have shown enhanced X-ray luminosities, ~5 orders of magnitude stronger than what the contemporary Sun emits, suggesting high-energy magnetic activities around the newborn stars [e.g., *Feigelson and Montmerle*, 1999; *Feigelson et al.*, 2002; *Wolk et al.*, 2005 and references therein]. Such activities include wrapping of magnetic field lines around the star and flares caused

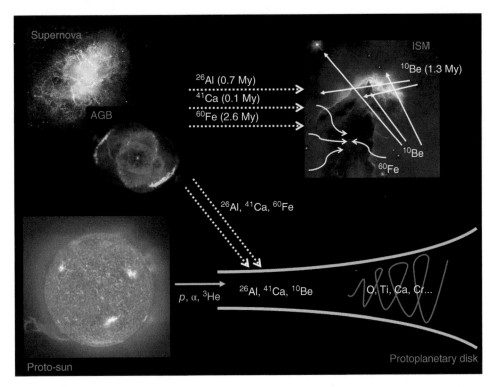

Figure 1.5 Origins of short-lived radionuclides as a proxy for understanding the Solar System nascent environment and astrophysical processes in the solar nebula. Extinct radionuclides can in theory be synthesized in various nucleosynthetic pathways (see text). Here ^{10}Be, ^{26}Al, ^{41}Ca and ^{60}Fe are taken as an example because they are among shortest-lived radionuclides of interest and are of importance for deducing the astrophysical environment of the forming Solar System and processes that occurred in the solar nebula. ^{10}Be requires an irradiation origin, which could be the in situ production near the proto-Sun [e.g., *Gounelle et al.*, 2001] or inheritance from the molecular cloud prior to the formation of the Solar System [*Desch et al.*, 2004]. In contrast, ^{60}Fe could either be injected into the forming solar nebula from a stellar source [e.g., *Tachibana et al.*, 2006; *Wasserburg et al.*, 2006] or came in as an inherited isotope from the background molecular cloud [*Tang and Dauphas*, 2012]. ^{26}Al and ^{41}Ca can be produced by either irradiation or stellar nucleosynthesis. That ^{10}Be and ^{60}Fe both were present in the solar nebula suggests that irradiation and stellar nucleosynthesis followed by injection both took place in the beginning of the Solar System history. An alternative scenario is that the existence of ^{60}Fe did not involve injection from a stellar source but instead was a result of inheritance of the background molecular cloud [*Tang and Dauphas*, 2012]. To constrain which source made up the inventory of ^{26}Al and ^{41}Ca in the early Solar System and thus to infer the associated environment and processes, a coordinated understanding of their abundances and those of ^{60}Fe and ^{10}Be needs to be established. If two or more radionuclides were produced by the same means and co-delivered into the solar nebula, their abundances should evolve with time in a concordant manner, that is, synchronocity should exist between radionuclides of the same astrophysical origin. Recent work has shown that the initial abundances of ^{26}Al, ^{41}Ca, and ^{60}Fe could be understood in the context of supernova injection [*Liu*, 2014] or of the contribution of several supernovae to the evolving parent molecular cloud [*Gounelle et al.*, 2009; *Gounelle and Meynet*, 2012]. It should however be pointed out that the true initial abundances of ^{41}Ca and ^{60}Fe are still relatively poorly constrained because of the analytical challenges and small numbers of samples analyzed. More high quality data are needed to obtain a more comprehensive picture of the origin of short-lived radionuclides. In addition to radioisotopes, mixing of stable isotopes (O, Ca, Ti ...) between various reservoirs took place, which in the end homogenized large isotope variations to a small dispersion. The processes of homogenization in stable isotope anomalies are also found in meteorite components (see text).

by magnetic reconnection. These reconnection events liberate excess magnetic field energy and accelerate charged particles [*Goldstein et al.*, 1986], which could induce nonthermal nuclear reactions with ambient gas and solids and leave traces of spallation, such as extinct radioactivities

or stable isotope anomalies, in early Solar System materials [e.g., *Feigelson*, 1982; *Feigelson and Montmerle*, 1999]. ^{10}Be is a radionuclide mainly produced by charged particle irradiation, but convincing evidence for its former presence has only been found in CAIs and refractory hibonite grains so far [*McKeegan et al.*, 2000; *Sugiura et al.*, 2001; *Marhas et al.*, 2002; *MacPherson et al.*, 2003; *Chaussidon et al.*, 2006; *Liu et al.*, 2009, 2010; *Wielandt et al.*, 2012; *Gounelle et al.*, 2013]. This radionuclide, in theory, could be a product of charged particle irradiation in the early Solar System [*Lee et al.*, 1998; *Gounelle et al.*, 2001, 2006] or could come in as trapped galactic cosmic rays by the Solar System's parental molecular cloud core [*Desch et al.*, 2004]. Given that the molecular cloud is so much larger than the Solar System, trapped-GCR-^{10}Be should be uniformly distributed within the solar nebula [*Desch et al.*, 2004]. However, there is no single initial ratio that could characterize refractory inclusions that presumably formed at the same time. The overall range of ^{10}Be/^{9}Be from 4.5×10^{-4} to 1×10^{-3} was found in about two dozen CV CAIs that contained ^{26}Al/^{27}Al ranging from 3.7×10^{-5} to 5.2×10^{-5} [e.g., *MacPherson et al.*, 2003; *Srinivasan and Chaussidon*, 2013]. The time difference inferred from the ^{26}Al records, which is on the order of 300,000 years, does not account for the observed variation in ^{10}Be/^{9}Be ratios, implying that the two radionuclides were not coupled in terms of the distribution in the solar nebula. An even stronger decoupling between the two radionuclides is that refractory samples devoid of ^{26}Al also contained radiogenic excesses of ^{10}B that can be attributed to the decay of ^{10}Be at the level of ^{10}Be/^{9}Be $\sim (3-5) \times 10^{-4}$. The two observations indicate that the spread in the ^{10}Be/^{9}Be ratios of CAIs is not time-dependent, but instead represent an initial heterogeneity, most likely due to a protosolar irradiation origin of ^{10}Be [*Liu et al.*, 2009, 2010].

Theoretical modeling of irradiation production of ^{10}Be has shown that the observed level of ^{10}Be/^{9}Be can be accounted for by exposing refractory inclusions to energetic protons with a flux of $\sim 10^{10}$ cm^{-2} s^{-1} for tens of years [e.g., *McKeegan et al.*, 2000; *Gounelle et al.*, 2001, 2006; *Chaussidon et al.*, 2006; *Liu et al.*, 2010; *Gounelle et al.*, 2013]. Such a high flux is available in the pre-main sequence, solar-type stellar environments, according to the X-ray luminosities observed by *Chandra* [e.g., *Feigelson et al.*, 2002; *Wolk et al.*, 2005], and the tens-of-year timescale is compatible with that of the disk fluctuation in the Class I to II stages [e.g., *Shu et al.*, 2001]. The stars surveyed in the Chandra observations have an age distribution from 0.3 to ~10 Myr, which is consistent with the hypothesis that ^{10}Be was produced in refractory inclusions by charged particle irradiation in an early epoch of stellar evolution. Very recently, indirect astronomical evidence for young stars emitting a high flux of energetic (≥ 10 MeV) particles has been found via Herschel observations of a Class 0, intermediate-mass protostar OMC-2 FIR 4 [*Ceccarelli et al.*, 2014]. In this study, the authors discovered an enhanced abundance ratio of HCO$^+$ to N$_2$H$^+$ in the protostellar envelope, which is derived from collisions between energetic particles and envelope

material, and estimated the particle flux at 1 AU from the central star to be $\sim (1-3) \times 10^{19}$ cm^{-2} yr^{-1}. This new result provides another line of support for intense irradiation in the very early history of the Solar System. The in-situ production of ^{10}Be in CAIs by solar cosmic ray irradiation could have started as early as in the Class 0 phase and perhaps lasted hundreds of thousands of years according to the ^{26}Al records in CAIs.

In addition to ^{10}Be, analysis of isotopic compositions of rare light elements (e.g., Li and B) and noble gases (Ne and Ar) in meteoritic inclusions has also revealed the records of charged particle irradiation in the beginning of the Solar System history (see review by *Chaussidon and Gounelle* [2006]). *Caffee et al.* [1987] analyzed mineral grains extracted from meteorites for ^{21}Ne and ^{38}Ar abundances and found that grains with tracks produced by solar flares were much more enriched in the two noble gas isotopes compared to the grains without tracks. After correcting for exposure ages of meteorites, the amount of ^{21}Ne and ^{38}Ar excesses could only be understood in the context of enhanced solar flare activity in the early epoch of the solar system, such as the T-Tauri stage [*Caffee et al.*, 1987]. Similar inferences have been drawn based on the stable (non-radiogenic) Li and B isotopic compositions in meteoritic refractory inclusions [*Liu et al.*, 2010; *Srinivasan and Chaussidon*, 2013]. Li and B in the universe are mainly produced by galactic cosmic ray spallation as the two elements are destroyed during stellar nucleosynthesis [*Burbidge et al.*, 1957; *Fowler et al.*, 1961]. The production ratios of ^{7}Li/^{6}Li and ^{10}B/^{11}B are energy-dependent [*Ramaty et al.*, 1996]. In the energy regime of protosolar cosmic rays, spallogenic ^{7}Li/^{6}Li and ^{10}B/^{11}B ratios are expected to be ~1 and 0.44, respectively, considerably different from the average solar ratios of 12.02 and 0.2481 (from *Seitz et al.*, 2007 and *Zhai et al.*, 1996, respectively). If spallogenic Li and B mix with the solar components, "subchondritic" ^{7}Li/^{6}Li and "superchondritic" ^{10}B/^{11}B ratios can be expected. Such ratios have been observed in CAIs and hibonites, indicating that the solids have been irradiated by energetic charged particles and later were contaminated with solar Li and B. Variations of Li and B isotopic ratios have also been found in chondrules [e.g., *Chaussidon and Robert*, 1995, 1998]. The overall range ($11.86 < ^7$Li/^{6}Li < 12.44 and $0.236 < ^{10}$B/^{11}B < 0.259) appeared to require multiple (irradiation) sources (galactic cosmic rays, solar cosmic rays, and big bang) to be accounted for, but nevertheless chondrules still recorded traces of early Solar System irradiation. Given the age difference between the majority of chondrules and CAIs (~3 Myr [e.g., *Villeneuve et al.*, 2009; *Connelly et al.*, 2012; *Kita et al.*, 2013]), it is conceivable that intense charged particle irradiation could have lasted a few Myr. It should also be pointed out that the Li and B isotopic variations seen in refractory inclusions and chondrules could not have been inherited from the nebular gas as extremely high proton fluences ($>10^{21}$ cm^{-2}) would be required to produce the observed deviations from the chondritic ^{7}Li/^{6}Li and ^{10}B/^{11}B ratios [*Liu et al.*, 2010].

Potentially charged particle irradiation could also modify the isotopic compositions of other elements in the irradiated targets, such as O in CAIs [*Lee*, 1978] and Mg in gas [*Heymann et al.*, 1978] and in hibonite [*Liu and McKeegan*, 2009]. However, those models required some ad-hoc irradiation conditions or very unphysically high proton fluences to make the isotopic anomalies compatible with the observations in meteorites. More searches in other elements for traces of irradiation are needed to better address this issue.

Another type of irradiation involves photochemical reactions induced by photons, rather than energetic charged particles, and molecules. Most relevant to the early Solar System is the self-shielding of CO, an isotopically selective photodissociation process that occurs in the far ultraviolet (FUV) wavelengths [*Thiemens and Heidenreich*, 1983]. This mechanism has been hypothesized to have taken place in the inner part of the solar nebula [e.g., *Clayton*, 2002], outer regions of the solar nebula [e.g., *Lyons and Young*, 2005], or the solar parental molecular cloud [*Yurimoto and Kuramoto*, 2004] and is one of the possible explanations for the oxygen isotopic anomalies seen in the refractory grains, CAIs and chondrules. Generally, refractory hibonite/corundum grains and CAIs are ^{16}O-rich relative to the Standard Mean Ocean Water (SMOW), with $\delta^{17}O_{SMOW} \approx \delta^{18}O_{SMOW}$ from -5 to -50 ‰, depending on the phases analyzed. Chondrule components can be either ^{16}O-rich or ^{16}O-poor relative to SMOW and span a smaller range, from ~ -5‰ to 5‰ in both $\delta^{17}O_{SMOW}$ and $\delta^{18}O_{SMOW}$ (for detailed descriptions about oxygen anomalies in meteoritic and planetary material, see *Clayton*, 2005). The fact that components of CAIs and chondrules broadly plot on a slope-1 line has been a long-standing problem in cosmochemistry. In the self-shielding hypothesis, UV, which originated from the central and nearby stars, dissociates CO molecules to C + ^{x}O, where x = 16, 17, and 18, and the UV penetration depth of the respective photons is inversely proportional to the abundances of CO isotopologues. The most abundant isotopologue C^{16}O saturates during photodissociation, and it reduces the rate of C^{16}O dissociation compared to that of the less abundant isotopologues C^{17}O and C^{18}O. As a result, this self-shielding process produces a zone enriched in ^{17}O and ^{18}O relative to ^{16}O. The free ^{17}O and ^{18}O then recombine with H in the cold regions of the solar nebula or the molecular cloud to become isotopically heavy (^{16}O-poor) water ice. Transport of such water ice into the inner, hot region of the solar nebula to mix with initially ^{16}O-enriched dust and gas would result in a slope-1 mixing line in a three-oxygen-isotope ($\delta^{17}O_{SMOW}$–$\delta^{18}O_{SMOW}$) plot. The overall scatter along this line that CAIs and chondrules display implies that the CO photodissociation processes occurred very early in the Solar System history, and the product, ^{16}O-depleted water ice, kept being brought into the inner Solar System perhaps for at least a few million years.

1.4.4. A Chronology for the Formation of the First Solar System Minerals and Rocks Based on ^{26}Al

There are three reasons why ^{26}Al is the most extensively studied SLR to establish a relative chronology of high-temperature processes that took place in the accretion disk: (1) it has an appropriate half-life of 0.72 Myr, (2) Al and Mg are refractory lithophile and major elements, thus abundant in CAIs and chondrules, and (3) strong parent/daughter fractionations are produced due to the different condensation temperatures and magmatic partitioning of Al and Mg. In addition, significant advances have been made over the last ~5–10 years in the developments of high (10–20 ppm level) to ultra-high precision (1–2 ppm level) techniques, often also with high spatial resolution, for Mg isotope measurements. These techniques include bulk measurements by chemistry and multi-collector inductively coupled plasma mass spectrometry (MC-ICPMS) [*Galy et al.*, 2000; *Jacobsen et al.*, 2008; *Bizzaro et al.*, 2011; *Paton et al.*, 2012], in-situ measurements by laser ablation and MC-ICPMS [*Young et al.*, 2002], and in-situ measurements by multi-collector secondary ion mass spectrometry (MC-SIMS) [*Villeneuve et al.*, 2009; *Luu et al.*, 2012; *MacPherson et al.*, 2012; *Kita and Ushikubo*, 2012; *Kita et al.*, 2012].

Because ^{26}Al is now extinct, only relative ages can be calculated for a given object by comparing the initial abundance of ^{26}Al (given by the initial ^{26}Al/^{27}Al ratio inferred from the isochron, see Fig. 1.6a) in this object to the initial abundance of ^{26}Al in the Solar System (relative ages can also be calculated between two objects). A major assumption in ^{26}Al-dating is that the inner Solar System (in which CAIs, chondrules, and parent bodies of meteorites formed) was early in its evolution homogeneous for its ^{26}Al/^{27}Al ratio. Recent high-precision ^{26}Al data on CAIs and chondrules, and comparison of ^{26}Al relative ages with relative ^{182}Hf-^{182}W ages and with absolute Pb-Pb ages bring new arguments for and against this assumption (see below).

In an isochron diagram (Fig. 1.6a), as predicted by mass balance of Al and Mg isotopes, a linear correlation must exist between the ^{26}Mg excesses (noted δ^{26}Mg*) and the ^{27}Al/^{24}Mg concentration ratio according to:

$$\left(\delta^{26}Mg*\right) = \left(\delta^{26}Mg*\right)_{i} + \frac{\left(^{26}Al/^{27}Al\right)_{i} \times \left(^{27}Al/^{26}Mg\right) \times 1000}{\left(^{26}Mg/^{24}Mg\right)_{ref}}$$

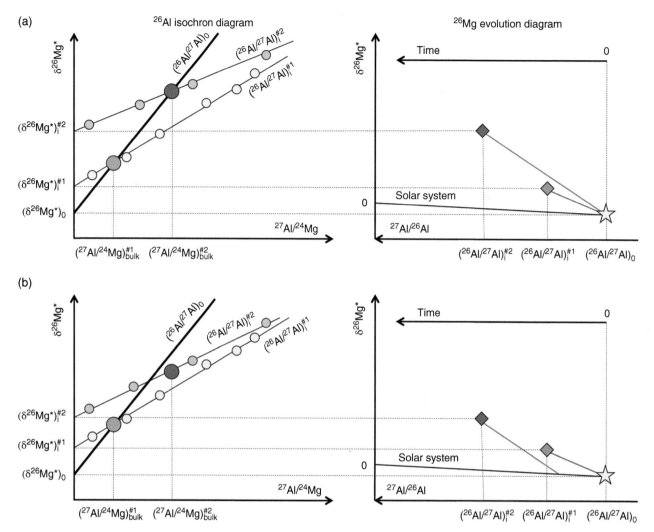

Figure 1.6 Principles of ^{26}Al-^{26}Mg chronology. This figure shows the two diagrams pertinent to the description of the variations of Mg isotopic compositions due to the decay of short-lived ^{26}Al (two objects, which could be CAIs or chondrules, are shown in blue for object #1 and in red for object #2). It is assumed here that the system was homogenized at some "time zero" in Al and Mg isotopic compositions to $\delta^{26}Mg_0$ and $(^{26}Al/^{27}Al)_0$ (see text). The left diagram is the "Isochron diagram," which shows how the excesses in ^{26}Mg, if of radiogenic origin, must be positively correlated to the ^{27}Al/^{24}Mg ratios. The right diagram is the "Evolution diagram," which shows how the radiogenic ^{26}Mg excesses in a given reservoir must develop with time, i.e., with decreasing ^{26}Al/^{27}Al ratios. The ^{26}Mg excesses (here shown in delta notation) are calculated from the measured ^{25}Mg/^{24}Mg and ^{26}Mg/^{24}Mg isotopic ratios relative to a mass fractionation law characterized by an exponent β ($\alpha^{26/24} = (\alpha^{25/24})^\beta$, $\alpha^{26/24}$ being the isotopic fractionation of the ^{26}Mg/^{24}Mg ratio relative to a reference value [see *Young and Galy*, 2004; *Davis et al.*, 2005; *Wasserburg et al.*, 2012; *Luu et al.*, 2013; *Mishra and Chaussidon*, 2014 and references therein for discussion on the approach to calculate at best the ^{26}Mg excesses and the values of β to use]. Two types of ^{26}Al isochrons, either mineral- or bulk-^{26}Al isochrons are shown (see text for details). The initial δ^{26}Mg* of a given object is related to its bulk ^{27}Al/^{24}Mg ratio and to the duration of its evolution with this bulk ^{27}Al/^{24}Mg ratio before its last melting/crystallization event assumed to correspond to isotopic closure for Mg isotopes: in the top diagrams (panel a) the two objects are made from precursors having the same age, while in the bottom diagrams (panel b) the precursors of the two objects were formed at different times (see text for details).

where $(^{26}Mg/^{24}Mg)_{ref}$ is the reference Mg isotopic ratio. Most studies use the reference ratios ^{26}Mg/^{24}Mg = 0.13932 and ^{25}Mg/^{24}Mg = 0.12663 from *Catanzaro et al.* [1966], but the more precise ratios determined recently by *Bizzarro et al.* [2011] can be used without significantly changing δ^{26}Mg* values, which by definition are relative variations to an initial [see also discussion in *Luu et al.*, 2013]. The isochron allows for the determinations of the initial Mg and Al isotopic compositions $(\delta^{26}Mg^*)_i$ and $(^{26}Al/^{27}Al)_i$, at the time of isotopic closure for the system considered. Two types of ischrons exist: a "mineral isochron" is made by the analysis of different minerals that crystallized rapidly from the same parent melt, and a "bulk isochron" is made up of analyses of different objects, e.g. different CAIs, which originated from the same nebular reservoir. Here we use CAIs as an example to show how Al-Mg dating is done. In the case of a mineral isochron, the system is the object being considered (e.g., a CAI), and the inferred $^{26}Al/^{27}Al$ ratio marks the time of Al/Mg fractionation between the different minerals, i.e., magmatic partitioning during crystallization of the different minerals. If the parent melt of the object was made by rapid and complete melting in a closed system of solid precursors, the initial $(\delta^{26}Mg^*)_i$ value and $(^{26}Al/^{27}Al)_i$ ratio are those of the precursors at the time of melting. In the case of a "bulk isochron," the system is this nebular reservoir in which CAI precursors formed. The $^{26}Al/^{27}Al$ ratio given by the "bulk isochron" dates the process that fractionated the $^{27}Al/^{24}Mg$ ratio between the different objects, e.g., condensation/evaporation that produced the precursors of the CAIs. The initial Mg and Al isotopic compositions of the bulk CAIs isochron are denoted in the following by $(\delta^{26}Mg^*)_0$ and $(^{26}Al/^{27}Al)_0$ because they are the oldest Mg and Al isotopic compositions that can be determined for the Solar System; they are generally considered as the Solar System initial.

An "evolution diagram" or a "growth curve" of $\delta^{26}Mg^*$ (Fig. 1.6) shows how the $\delta^{26}Mg^*$ of an object increases with time (i.e., with decreasing $^{26}Al/^{27}Al$ ratio), under the assumptions listed above. If the formation of an object, for example, a CAI, follows shortly after (within the time resolution of the ^{26}Al-^{26}Mg system) that of its precursors in the solar nebula, then the initial $(\delta^{26}Mg^*)_i$ and $(^{26}Al/^{27}Al)_i$ of this object can be calculated by the following equation:

$$\left(\delta^{26}Mg^*\right)_i = \frac{\left(^{27}Al/^{24}Mg\right)_{solar}}{\left(^{26}Mg/^{24}Mg\right)_{ref}}$$
$$\times\left[\left(^{26}Al/^{27}Al\right)_0 - \left(^{26}Al/^{27}Al\right)_i\right]\times1000$$

If there is some evolution of the precursors (or of the proto-CAI or proto-chondrule) in a closed system before the final melting event dated by the mineral isochron, then the initial $(\delta^{26}Mg^*)_i$ at the time of last melting and $(^{26}Al/^{27}Al)_i$, of the object are related by:

$$\left(\delta^{26}Mg^*\right)_i = \left(\delta^{26}Mg^*\right)_0 + \frac{1000}{\left(^{26}Mg/^{24}Mg\right)_{ref}}$$
$$\times\left[\left(\left(^{26}Al/^{27}Al\right)_0 - \left(^{26}Al/^{27}Al\right)_{precursors}\right)\right.$$
$$\times\left(^{27}Al/^{24}Mg\right)_{solar} - \left(\left(^{26}Al/^{27}Al\right)_{precursors}\right.$$
$$\left.\left. - \left(^{26}Al/^{27}Al\right)_i\right)\times\left(^{27}Al/^{24}Mg\right)_{precursors}\right]$$

In this equation, it is assumed that the precursors were isolated from the solar nebula when the Al isotopic ratio is $(^{26}Al/^{27}Al)_{precursors}$ and kept subsequently their bulk $^{27}Al/^{24}Mg$ ratio [noted $(^{27}Al/^{24}Mg)_{precursors}$]. These two scenarios (single-step and two-step evolution) are the simplest ones to consider and can be complicated by adding open system evolution (mixing, evaporation, …). Thus, combining high-precision bulk and in-situ measurements allows one in theory to determine not only the age of a CAI or of a chondrule, but also the age of its precursors, and to reconstruct its evolution (Al/Mg fractionations, closed or open system evolution, …) by comparing its composition to a simple one- or two-step scenario (Fig. 1.6).

A recent major finding was the demonstration that CAIs from the CV chondrite Allende define a very tight bulk ^{26}Al isochron [Thrane et al., 2006; Jacobsen et al., 2008; Larsen et al., 2011], which implies that they were all formed within a very narrow time window that is defined by the uncertainties of ^{26}Al measurements and that their precursors (if any) also formed within that time interval (otherwise, because of their high $^{27}Al/^{24}Mg$ ratios, the CAIs would not share a common $(\delta^{26}Mg^*)_0$ initial). The CAIs' initial $(^{26}Al/^{27}Al)_0$ and $\delta^{26}Mg^*_0$ inferred from all the CAIs analyzed by Jacobsen et al. [2008] and Larsen et al. [2011] are $(5.25\pm0.12)\times10^{-5}$ and $-0.034\pm0.032‰$ (2 sigma errors), respectively. The small errors on $(^{26}Al/^{27}Al)_0$ and $(\delta^{26}Mg^*)_0$ imply that the precursors of these CAIs had a closed system evolution of less than 30 kyr (or even less than a few kyr according to the data by Larsen et al. [2011] only). However, because these data are bulk data, they cannot track the last melting events that the CAIs underwent. Mineral isochrons of CAIs show that in fact while some of them did not undergo any remelting since their formation at $(^{26}Al/^{27}Al)_0$, others were remelted later on [MacPherson et al., 2010, 2013; Mishra and Chaussidon, 2014a]. The analysis of different refractory objects (CAIs of different types, AOAs, and Al-rich chondrules) in the CV chondrites Efremovka and Vigarano shows that, despite the fact that they were last melted at different times as evidenced from their mineral isochrons (for some of them up to 2 Myr after "time zero"), their initial compositions can be understood in a simple one-step scenario in which all their precursors would have been extracted from a solar nebula having $\delta^{26}Mg^*_{SSI} = -0.052\pm0.013‰$ and $(^{26}Al/^{27}Al)_{SSI} = (5.62\pm0.42)\times10^{-5}$ ("SSI" standing for Solar System Initial [Mishra and Chaussidon, 2014a]). These Al and Mg isotopic compositions are within errors comparable to those derived from the bulk CV CAI's isochron, but they point to the possibility of a short (<100 kyr) evolution of the CAIs' precursors at high-temperature in the nebula gas, with continuous isotopic exchange between solids and gas.

A similar approach (high-precision mineral isochron and comparison between bulk- and mineral-isochron)

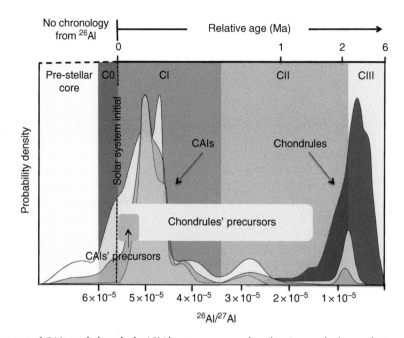

Figure 1.7 Summary of CAIs and chondrules' ^{26}Al ages compared to the timescale for evolution of YSOs of Solar mass derived from astrophysical observations [adapted from *Mishra and Chaussidon*, 2014]. The ^{26}Al ages are calculated by assuming ^{26}Al and Mg isotopic homogeneity in the inner Solar System at "time zero." To anchor the astrophysical scale to the meteoritic scale, it is assumed that this "time zero" corresponds to the end of the Class 0 phase during which the protostar is infalling and fully convective. ^{26}Al/^{27}Al ratios higher than that assumed for "time zero" (see text) have no chronological meaning; they can either be a result of the uncertainty of ^{26}Al-^{26}Mg measurements in CAIs or of ^{26}Al or Mg isotopic heterogeneities in the forming Solar System. The formation of CAIs is simultaneous (within at max a few 10 kyr, see text) to that of their precursors solids (formed by condensation) and is taking place at the transition between C0 and C1 protostellar phases. At variance chondrules precursors are presumably forming over a much longer timescale (<1.5 Myr, i.e., from C0 to C1 and C2 phases) [*Luu et al.*, in press]. There appears to be a peak of formation of chondrule at ~2–4 Myr, i.e., at the transition between C2 and C3 phases, but some chondrules (the fraction of which cannot yet be determined because of the too restricted data set) did form very early, nearly simultaneously with CAIs.

has been applied to chondrules [*Villeneuve et al.*, 2009; *Luu et al.*, in press]. These studies show that chondrules and their precursors formed over a much longer period than CAIs and their precursors (Fig. 1.7). All existing bulk data for chondrules (at present ~50 chondrules from only two chondrites, the CV chondrite Allende and the LL unequilibrated ordinary chondrite Semarkona, data from *Galy et al.*, 2000; *Bizzarro et al.*, 2004, *Villeneuve et al.*, 2009; *Luu et al.*, 2015) show ^{26}Al/^{27}Al ratios between the CAIs' bulk isochron and an isochron that can be called the "minimum bulk-chondrule isochron" having $\delta^{26}Mg^* = -0.009\pm0.012‰$ and $(^{26}Al/^{27}Al)_{SSI} = (1.23\pm0.21)\times10^{-5}$ [*Luu et al.*, 2015]. The implication is that no chondrule (at least the ones studied so far) has precursors formed later than 1.5 Myr after "time zero" and that some chondrules' precursors formed at the same time as CAIs' precursors [*Luu et al.*, 2015]. In the case of Al-rich chondrules that have a high ^{27}Al/^{24}Mg, precise model ^{26}Al ages can be calculated for their precursors between 0±0.12 Myr and 1.51±0.17 Myr after CAIs [*Villeneuve et al.*, 2009;

Luu et al., 2015]. Melting ages determined from mineral ^{26}Al isochrons for the chondrules studied in *Luu et al.* [2015], range from ~0 to ~2 Myr after "time zero." This shows that some chondrules have ^{26}Al melting ages, which within errors, are undisinguishable from CAIs, in agreement with recent high-precision Pb-Pb dating of CAIs and chondrules [*Connelly et al.*, 2012]. These high-precision ^{26}Al-^{26}Mg data do not modify the view established previously from ^{26}Al studies of chondrules in ordinary chondrites [lower precision data with no precise measurement of the $(\delta^{26}Mg^*)_i$] that chondrules underwent their last melting event when $3\times10^{-6} <(^{26}Al/^{27}Al)_i <1\times10^{-5}$, i.e. ~1.5 to ~3 Myr after CAIs [*Hutcheon et al.*, 1989; *Kita et al.*, 2000; *Kunihiro et al.*, 2004; *Rudraswami et al.*, 2008; *Kurahashi et al.*, 2008; *Kita et al.*, 2005]. There is no clear relationship between chondrule ages and chondrite class [*Kita and Ushikubo*, 2012], nor between chondrule age and chondrule type or chemistry [*Villeneuve et al.*, 2013]. These recent findings on the age range for the formation of chondrules' precursors and of chondrules themselves, bring strong new constraints on the origin of chondrules and on condensation and transport processes in the accretion disk. Either the temperature in the condensation zone of the inner disk decreased below the condensation temperature of chondrules' precursors 1.5 Myr after time zero, or the transport of chondrules's precursors to the region where chondrules were melted by energetic events stopped after 1.5 Myr, or a combination of both [*Luu et al.*, 2015]. Also important to mention is the fact that these results could also be understood in the framework of the formation of chondrules during impacts between planetary embryos (see Section 1.3.2). In this hypothesis, the ^{26}Al age of the precursors could tentatively be considered as the age of differentiation of the planetary object parent of the chondrule precursors and the chondrule age as the age of the impact.

Finally, a major assumption underlying ^{26}Al chronology (as described above) is that of the existence of a "time zero" when Al and Mg isotopes were homogenized in the accretion disk, at least in the inner regions where chondrites and their components formed. Discordant views exist on this topic. The observation that the Earth, ferromagnesian chondrules, Al-rich chondrules, AOAs, and CAIs of different types have $(\delta^{26}Mg^*)_i$ values and $(^{26}Al/^{27}Al)_i$ ratios that can all be reconciled with a single Solar System initial [*Villeneuve et al.*, 2009; *Mishra and Chaussidon*, 2014] is a strong argument supporting homogeneity at a level of ~10% relative, allowing for ^{26}Al chronology. The agreement between Al/Mg, Hf/W and Pb/Pb ages of objects that formed late, such as the differentiated meteorites angrites, which were formed ~5 Myr after CAIs [*Kruijer et al.*, 2014], is also a strong argument in favor of a homogeneous distribution of ^{26}Al in the accretion disk. A discordant view is given in *Larsen et al.* [2011] from an ultra-high precision Mg isotopic study of CAIs and AOAs, which appear to define an ^{26}Al isochron with an initial Mg isotopic composition different from that proposed in the case of a homogeneous distribution of Mg isotopes. However, if CAIs and AOAs have slightly different ages, they cannot be combined to define an ^{26}Al isochron. Correlations between initials $\delta^{26}Mg^*$ and nucleosynthetic isotopic anomalies, such as ^{54}Cr variations, is another important argument put forward by *Larsen et al.* [2011] for isotopic heterogeneity of Al and Mg in the accretion disk when CAIs, chondrules, and their precursors formed. There is no question about the fact that large Al and Mg isotopic heterogeneities existed early in the forming Solar System as shown by the range of $\delta^{26}Mg^*$ values and $^{26}Al/^{27}Al$ ratios (from ~3×10^{-6} up to ~6×10^{-5}) found in ultra-refractory objects such as FUN inclusions, hibonites, corundum [*Wasserburg et al.*, 1977; *Ireland et al.*, 1988; *Sahijpal and Goswami*, 1998; *Sahijpal et al.*, 2000; *Makide et al.*, 2011; *Krot et al.*, 2012; *Liu et al.*, 2009, 2012; *Holst et al.*, 2013]. However, data on

CAIs and chondrules show that such rare inclusions had mostly disappeared when these objects formed. This "grand homogenization" of the Solar System could have taken place when the Sun was in Class 0 (Fig. 1.7). Anchoring in such a way, the astrophysical timescale of YSOs to the meteoritic timescale defined by ^{26}Al in CAIs and chondrules, shows that the peak of chondrule ages is at the transition between Class II and Class III, at the beginning of the dissipation of the accretion disk.

ACKNOWLEDGMENTS

We acknowledge the financial support of the UnivEarthS Labex program at Sorbonne Paris Cité (ANR-10-LABX-0023 and ANR-11-IDEX-0005-02) and of Institut de Physique du Globe de Paris (invitation program). Conel Alexander and Trevor Ireland are thanked for their in-depth reviews.

REFERENCES

Alexander, C. M. O.' D., A. P. Boss, and R. W. Carlson (2001), The early evolution of the inner Solar System: a meteoritic perspective, *Science*, *293*, 64–68.

Alexander, C. M. O.' D. (2004), Chemical equilibrium and kinetic constraints for chondrule and CAI formation conditions, *Geochim. Cosmochim. Acta.*, *68*, 3943–3969.

Alexander C. M. O.' D., J. N. Grossman, D. S. Ebel, and F. J. Ciesla (2008), The formation conditions of chondrules and chondrites, *Science*, *320*, 1617–1619.

Alexander, C. M. O.' D., and D. S. Ebel (2012), Questions, questions: Can the contradictions between the petrologic, isotopic, thermodynamic, and astrophysical constraints on chondrule formation be resolved?, *Meteor. Planet. Sci.*, *47*, 1157–1175.

Amelin, Y., A. Kaltenbach, T. Iizuka, C. H. Stirling, T. R. Ireland, M. Petaev, and S. B. Jacobsen (2010), U–Pb chronology of the Solar System's oldest solids with variable ^{238}U/^{235}U, Earth Planet. *Sci. Lett.*, *300*, 343–350.

Amelin, Y. and T. R. Ireland (2013), Dating the oldest rocks and minerals in the Solar System, *Elements*, *9*, 39–44.

André, P. (2002), The initial conditions for protostellar collapse: observational constraints, in Star Formation and the Physics of Young Stars, edited by. J Bouvier and J. P. Zahn, pp. 1–38, *EDP Science*, Les Ulis, France.

André, P., D. Ward-Thompson, and M. Barsony (1993), Submillimeter continuum observations of Rho Ophiuchi A — The candidate protostar VLA 1623 and prestellar clumps, *Astrophys. J.*, *406*, 122–141.

André, P., D. Ward-Thompson, and M. Barsony (2000), From Prestellar Cores to Protostars: the initial conditions of star formation, *Protostars and Planets IV*, 59–96.

Asphaug, E., M. Jutzi, and N. Movshovitz (2011), Chondrule formation during planetesimal accretion, *Earth Planet. Sci. Lett.*, *308*, 369–379.

Asplund, M., N. Grevesse, A. J. Sauval, and P. Scott (2009), The chemical composition of the Sun, *Annu. Rev. Astron. Astrophys.*, *47*, 481–522.

Benedix, G. K., H. Haack, and T. McCoy (2014), Iron and stony-iron meteorites, in Meteorites and Cosmochemical Processes, *Treatise on Geochemistry Second Edition vol. 1*, edited by H. D. Holland and K. K. Turekian, pp. 267–285, Elsevier.

Bizzarro, M., J. A. Baker, and H. Haack (2004), Mg isotope evidence for contemporaneous formation of chondrules and refractory inclusions, *Nature*, *431*, 275–278.

Bizzarro, M., C. Paton, K. K. Larsen, M. Schiller, A. Trinquier, and D. Ulfbeck (2011), High-precision Mg-isotope measurements of terrestrial and extraterrestrial material by HR-MC-ICPMS-implications for the relative and absolute Mg-isotope composition of the bulk silicate Earth, *J. Anal. At. Spectrom.*, *26*, 565–577.

Bless, R. C. (2009), *Discovering the Cosmos*, University Science Books, Mill Valley, California, U.S.A.

Bonal, L, E. Quirico, M. Bourot-Denise, and G. Montagnac (2006), Determination of the petrologic type of CV3 chondrites by Raman spectroscopy of included organic matter, *Geochim. Cosmochim. Acta.*, *70*, 1849–1863.

Bottke, W. F., D. Nesvorny, R. E. Grimm, A. Morbidelli, and D. O'Brien (2006), Iron meteorites as remnants of planetesimals formed in the terrestrial planet region, *Nature*, *439*, 821–824.

Bouvier, A. and M. Wadhwa, M. (2010), The age of the Solar System redefined by the oldest Pb–Pb age of a meteoritic inclusion, *Nature Geoscience*, *3*, 637–641.

Brearley, A. J. (2006), The action of water, in *Meteorites and the early Solar System II*, edited by D. S. Lauretta and H. Y. McSween Jr., pp. 587–624, University of Arizona Press, Tucson.

Brearley, A. J. (2007), Nebular versus parent-body processing, in Meteorites, Planets, and Comets, Treatise on Geochemistry vol. 1, edited by A. M. Davis, H. D. Holland, and K. K. Turekian, pp. 247–268. Elsevier-Pergamon, Oxford.

Brearley, A. J. and R. H. Jones (1998), Chondritic meteorites, in Planetary Materials, Reviews in Mineralogy, Vol. 36, edited by J. J. Papike, pp. 313–398, Mineralogical Society of America, Washington, DC.

Brennecka, G. A., S. Weyer, M. Wadhwa, P. E. Janney, J. Zipfel, J., and A. D. Anbar (2010), $^{238}U/^{235}U$ variations in meteorites: extant ^{247}Cm and implications for Pb-Pb dating, *Science*, *327*, 449–451.

Brownlee, D. and 182 coauthors (2006), Comet 81P/Wild 2 Under a Microscope, *Science*, *314*, 1711–1716.

Burbidge, E. M., G. R. Burbidge, W. A. Fowler, and F. Hoyle (1957), Synthesis of the Elements in Stars, *Rev. Modern Phys.*, *29*, 547–650.

Busemann, H., C. M. O.'D. Alexander, and L. R. Nittler (2007), Characterization of insoluble organic matter in primitive meteorites by microRaman spectroscopy, *Meteor. Planet. Sci.*, *42*, 1387–1416.

Caffee, M. W., C. M. Hohenberg, T. D. Swindle, and J. N. Goswami (1987), Evidence in meteorites for an active early sun, *Astrophys. J.*, *313*, L31–L35.

Catanzaro, E. J., T. J. Murphy, E. L. Garner, and W. R. Shields (1966), Absolute isotopic abundance ratios and atomic weights of magnesium, *J. Res. Nat. Bur. Stand.*, *70A*, 453–458.

Ceccarelli, C., C. Dominik, A. López-Sepulcre, M. Kama, M. Padovani, E. Caux, and P. Caselli (2014), Herschel finds evidence for stellar wind Particles in protostellar envelope: is this what happened to the young Sun?, *Astrophys. J. Lett.*, *790*, L1.

Charnoz, S., L. Fouchet, J. Aléon, and M. Moreira (2011), Three-dimensional lagrangian turbulent diffusion of dust grains in a protoplanetary disk: method and first applications, *Astrophys. J.*, *737*, 33–50.

Chaussidon, M. and F. Robert (1995), Nucleosynthesis of ^{11}B-rich boron in the pre-solar cloud recorded in meteoritic chondrules, *Nature*, *374*, 337–339.

Chaussidon, M. and F. Robert (1998), $^{7}Li/^{6}Li$ and $^{11}B/^{10}B$ variations in chondrules from the Semarkona unequilibrated chondrite, *Earth Planet. Sci. Lett.*, *164*, 577–589.

Chaussidon, M. and M. Gounelle (2006), Irradiation processes in the early Solar System, in Meteorites and the early Solar System II, edited by D. S. Lauretta and H. Y. McSween Jr., pp. 323–339, University of Arizona Press, Tucson.

Chaussidon, M., F. Robert, and K. D. McKeegan (2006), Li and B isotopic variations in an Allende CAI: Evidence for the in situ decay of short-lived ^{10}Be and for the possible presence of the short-lived nuclide ^{7}Be in the early solar system, *Geochim. Cosmochim. Acta.*, *70*, 224–245.

Chaussidon, M., G. Libourel, and A. N. Krot (2008), Oxygen isotopic constraints on the origin of magnesian chondrules and on the gaseous reservoirs in the early Solar System, *Geochim. Cosmochim. Acta.*, *72*, 1924–1938.

Chen, H., P. C. Myers, E. F. Ladd, and D. O. S. Wood (1995), Bolometric temperature and young stars in the Taurus and Ophiuchus complexes, *Astrophys. J.*, *445*, 377–392.

Ciesla, F. J. (2010a), Residence Times of Particles in Diffusive Protoplanetary Disk Environments. I. Vertical Motions, *Astrophys. J.*, *723*, 514–529.

Ciesla, F. J. (2010b), The distributions and ages of refractory objects in the solar nebula, *Icarus*, *208*, 455–467.

Ciesla, F. J. and L. L. Hood (2002), The Nebular shock wave model for chondrule formation: shock processing in a particle-gas suspension, *Icarus*, *158*, 281–293.

Clayton, R. N. (2002), Solar System: Self-shielding in the solar nebula, *Nature*, *415*, 860–861.

Clayton, R. N. (2005), Oxygen isotopes in meteorites, in Meteorites, Planets, and Comets, Treatise on Geochemistry vol. 1, edited by A. M. Davis, H. D. Holland, and K. K. Turekian, pp. 129–142. Elsevier-Pergamon, Oxford.

Clayton, R. N. and T. K. Mayeda (1999), Oxygen isotope studies of carbonaceous chondrites, *Geochim. Cosmochim. Acta.*, *63*, 2089–2104.

Connelly, J. N., M. Bizzarro, A. N. Krot, Å. Nordlund, D. Wielandt, and M. A. Ivanova (2012), The absolute chronology and thermal processing of solids in the solar protoplanetary disk, *Science*, *338*, 651–655.

Connolly, H. C. Jr. (2005), Refractory inclusions and chondrules: insights into protoplanetary disk and planet formation, in Chondrites and the protoplanetary disk, ASP Conf. Ser., vol. 341, edited by A. N. Krot, E. R. D. Scott, and B. Reipurth, pp. 215–224, Astronomical Society of the Pacific, San Francisco.

Dauphas, N. and M. Chaussidon (2011), A perspective from extinct radionuclides on a young stellar object: the Sun and its accretion disk, *Ann. Rev. Earth Planet. Sci.*, *39*, 351–386.

Dauphas, N. and A. Pourmand (2011), Hf-W-Th evidence for rapid growth of Mars and its status as a planetary embryo, *Nature*, *473*, 489–493.

Davis, A. M. and F. M. Richter (2007), Condensation and evaporation of Solar System materials, in Meteorites, Planets, and Comets, Treatise on Geochemistry vol. 1, edited by A. M. Davis, H. D. Holland, and K. K. Turekian, pp. 431–460. Elsevier-Pergamon, Oxford.

DeMeo, F. E., and B. Carry (2013), The taxonomic distribution of asteroids from multi-filter all-sky photometric surveys, *Icarus*, *226*, 723–741.

Desch, S. J. and H. C. Connolly (2002), A model of the thermal processing of particles in solar nebula shocks: application to the cooling rates of chondrules, *Meteor. Planet. Sci.*, *37*, 183–207.

Desch, S. J., M. A. Morris, H. C. Connolly, and A. P. Boss (2012), The importance of experiments: constraints on chondrule formation models, *Meteor. Planet. Sci.*, *47*, 1139–1156.

Desch, S. J., H. C. Connolly, and G. Srinivasan (2004), An interstellar origin for the beryllium-10 in calcium-rich, aluminum-rich inclusions, *Astrophys. J.*, *602*, 528–542.

Dullemond, C. P. and J. D. Monnier (2010), The inner regions of protoplanetary disks, *Ann. Rev. Astro. Astrophys.*, *48*, 205–239.

Ebel, D. S. (2006), Condensation of rocky materials in astrophysical environments, in Meteorites and the early Solar System II, edited by D. S. Lauretta and H. Y. McSween Jr., pp. 253–277, University of Arizona Press, Tucson.

Elkins-Tanton, L. T., B. P. Weiss, and M. T. Zuber (2011), Chondrites as samples of differentiated planetesimals, *Earth Planet. Sci. Lett.*, *305*, 1–10.

Evans II, N. J., M. M. Dunham, J. K. Jørgensen, M. L. Enoch, B. Mern, E. F. van Dishoeck, J. M. Alcalá, P. C. Myers, K. R. Stapelfeldt, T. L. Huard, L. E. Allen, P. M. Harvey, T. van Kempen, G. A. Blake, D. W. Koerner, L. G. Mundy, D. L. Padgett, and A. I. Sargent (2009), The Spitzer c2d Legacy Results: Star-Formation Rates and Efficiencies; Evolution and Lifetimes, *Astrophys. J. Supp.*, *181*, 321–350.

Fedele, D, M. E. van den Ancker, Th. Henning, R. Jayawardhana, and J. M. Oliveira (2010), Timescale of mass accretion in pre-main-sequence stars, *Astron. Astrophys*, *510*, A72.

Fedkin, A.V., and L. Grossman (2013), Vapor saturation of sodium: key to unlocking the origin of chondrules, *Geochim. Cosmochim. Acta.*, *112*, 226–250.

Feigelson, E. D. (1982), X-ray emission from young stars and implications for the early solar system, *Icarus*, *51*, 155–163.

Feigelson, E. D. and T. Montmerle (1999), High-Energy Processes in Young Stellar Objects, *Ann. Rev. Astro. Astrophys*, *37*, 363–408.

Feigelson, E. D., P. Broos, J. A. Gaffney III, G. Garmire, L. A. Hillenbrand, S. H. Pravdo, L. Townsley, and Y. Tsuboi (2002), X-Ray-emitting Young Stars in the Orion Nebula, *Astrophys. J.*, *574*, 258–292.

Fowler, W. A., J. L. Greenstein, and F. Hoyle (1961), Deuteronomy. Synthesis of Deuterons and the Light Nuclei during the Early History of the Solar System, *Am. J. Phys.*, *29*, 393–403.

Franchi, I. A. (2008), Oxygen isotope in asteroidal materials, in Oxygen in the Solar System, *Reviews in Mineralogy and Geochemistry*, *vol. 68*, edited by G. J. MacPherson, pp. 345–390, Mineralogical Society of America, Washington, DC.

Gail, H.-P. (2003), Formation and evolution of minerals in accretion disks and stellar outflows, in *Astromineralogy*, edited by T. Henning, pp. 55–120, Springer.

Galy, A, E. D. Young, R. D. Ash, and R. K. O'Nions (2000), The formation of chondrules at high gas pressures in the solar nebula, *Science*, *290*, 1751–1753.

Ganguly, J., M. Tirone, S. Chakraborty, and K. Domanik (2013), H-chondrite parent asteroid: A multistage cooling, fragmentation and re-accretion history constrained by thermometric studies, diffusion kinetic modeling and geochronological data, *Geochim. Cosmochim. Acta.*, *105*, 206–220.

Goldstein, M. L., W. H. Matthaeus, and J. J. Ambrosiano (1986), Acceleration of charged particles in magnetic reconnection: Solar flares, the magnetosphere, and solar wind, *Geophys. Res. Lett.*, *13*, 205–208.

Gooding, J. L. and K. Keil (1981), Relative abundances of chondrule primary textural types in ordinary chondrites and their bearing on conditions of chondrule formation, *Meteoritics*, *16*, 17–43.

Göpel, C., Manhès G., and C. J. Allègre (1994), U-Pb systematics of phosphates from equilibrated ordinary chondrites, *Earth Planet. Sci. Lett.*, *121*, 153–171.

Gounelle, M., M. Chaussidon, and C. Rollion-Bard (2013), Variable and extreme irradiation conditions in the early Solar System inferred from the initial abundance of [10]Be in Isheyevo CAIs, *Astrophys. J. Lett.*, *763*, L33.

Gounelle, M., F. H. Shu, H. Shang, A. E. Glassgold, K. E. Rehm, and T. Lee (2001), Extinct radioactivities and protosolar cosmic rays: self-shielding and light elements, *Astrophys. J.*, *548*, 1051–1070.

Gounelle, M., F. H. Shu, H. Shang, A. E. Glassgold, K. E. Rehm, and T. Lee (2006), The irradiation origin of beryllium radio-isotopes and other short-lived radionuclides, *Astrophys. J.*, *640*, 1163–1170.

Gounelle, M., A. Meibom, P. Hennebelle, and S.-I. Inutsuka (2009), Supernova propagation and cloud enrichment: a new model for the origin of [60]Fe in the early solar system. *Astrophys. J. Lett.*, *694*, L1–L5.

Gounelle, M. and G. Meynet, (2012), Solar system genealogy revealed by extinct short-lived radionuclides in meteorites, *Astron. Astrophys.*, *545*, A4.

Gray, C. M. and W. Compston (1974), Excess [26]Mg in the Allende Meteorite, *Nature*, *251*, 495–497.

Grossman, J. N. (1988), Formation of chondrules. In Meteorites and the early Solar System, eds. J. F. Kerridge and M. S. Matthews, University of Arizona Press, 680–696.

Grossman, J. N., A. E. Rubin, N. Nagahara, and E. A. King (1988), Properties of chondrules, in Meteorites and the early Solar System, edited by J. F. Kerridge and M. S. Matthews, pp. 619–659, University of Arizona Press.

Grossman, L., D. S. Ebel, S. B. Simon, A. M. Davis, F. M. Richter, and N. M. Parsad (2000), Major element chemical and isotopic compositions of refractory inclusions in C3 chondrites: the separate roles of condensation and evaporation, *Geochim. Cosmochim. Acta.*, *64*, 2879–94.

Grossman, L. (1972), Condensation in primitive solar nebula, *Geochim. Cosmochim. Acta.*, *36*, 597–619.

Grossman, L. (1975), Petrology and mineral chemistry of Ca-rich inclusions in the Allende meteorite, *Geochim. Cosmochim. Acta.*, *39*, 438–451.

Gullbring, E., L. Hartmann, C. Briceño, and N. Calvet (1998), Disk accretion rates for T Tauri stars, *Astrophys. J.*, *492*, 323–341.

Haisch Jr., K. E., E. A. Lada, and C. J. Lada (2001), Disk frequencies and lifetimes in young clusters, *Astrophys. J. Lett.*, *553*, L153–L156.

Henning T. (2010), Cosmic silicates, *Annu. Rev. Astron. Astrophys*, *48*, 21–46.

Hewins, R. H. (1997), Chondrules, *Ann. Rev. Earth Planet. Sci.*, *25*, 61–83.

Hewins, R. H., H. C. Connolly Jr., G. E. Lofgren, and G. Libourel (2005), Experimental constraints on chondrule origins, in Chondrites and the protoplanetary disk, ASP Conf. Ser., vol. 341, edited by A. N. Krot, E. R. D. Scott and B. Reipurth, pp. 286–317, Astronomical Society of the Pacific, San Francisco.

Hewins, R. H., B. Zanda, and C. Bendersky (2012), Evaporation and recondensation of sodium in Semarkona Type II chondrules, *Geochim. Cosmochim. Acta.*, *78*, 1–17.

Heymann, D., M. Dziczkaniec, A. Walker, G. Huss, J. A. Morgan (1978), Effects of proton irradiation on a gas phase in which condensation takes place. I: Negative Mg-26 anomalies and Al-26, *Astrophys. J.*, *225*, 1030–1044.

Hillenbrand, L. A. (2005), Observational constraints on dust disk lifetimes: implications for planet formation. ArXiv Astrophysics e-prints, arXiv:astro-ph/0511083.

Hirashita, H., R. S. Asano, T. Nozawa, T., Z.-Y. Li, and M.-C. Liu (2014), Dense molecular cloud cores as a source of micrometer-sized grains in galaxies, *Planet. Space Sci.*, *100*, 40–45.

Holst, J. C., M. B. Olsen, C. Paton, K. Nagashima, M. Schiller, D. Wielandt, K. K. Larsen, J. N. Connelly, J. K. Jorgensen, A. N. Krot, A. Nordlund, and M. Bizzarro (2013), [182]Hf-[182]W age dating of a [26]Al-poor inclusion and implications for the origin of short-lived radioisotopes in the early Solar System, *Proc. Nat. Acad. of Sci.*, *110*, 8819–8823.

Huss, G. R., and R. S. Lewis (1995), Presolar diamond, SiC, and graphite in primitive chondrites: abundances as a function of meteorite class and petrologic type, *Geochim. Cosmochim. Acta.*, *59*, 115–160.

Huss, G. R., A. E. Rubin, and J. N. Grossman (2006), Thermal metamorphism in chondrites, in Meteorites and the early Solar System II, edited by D. S. Lauretta and H. Y. McSween Jr., pp. 567–586, University of Arizona Press, Tucson.

Huss, G. R., B. S. Meyer, G. Srinivasan, J. N. Goswami, and S. Sahijpal (2009), Stellar sources of the short-lived radionuclides in the early solar system, *Geochim. Cosmochim. Acta.*, *73*, 4922–4945.

Huss, G. R., G. J. MacPherson, G. J. Wasserburg, S. S. Russell, and G. Srinivasan (2001), Aluminum-26 in calcium-aluminum-rich in inclusions and chondrules from unequilibrated ordinary chondrites, *Meteor. Planet. Sci.*, *36*, 975–997.

Hutcheon, I. D. and R. Hutchison (1989), Evidence from the Semarkona ordinary chondrite for Al-26 heating of small planets, *Nature*, *337*, 238–241.

Hutcheon, I. D., K. K. Marhas, A. N. Krot, J. N. Goswami, and R. H. Jones (2009), [26]Al in plagioclase-rich chondrules in carbonaceous chondrites: evidence for an extended duration of chondrule formation, *Geochim Cosmochim. Acta.*, *73*, 5080–5099.

Ireland, T. R. (1988), Correlated morphological and isotopic characteristics of hibonites from the Murchison carbonaceous chondrite, *Geochim. Cosmochim. Acta.*, *52*, 2827–2839.

Ireland, T. R. and B. Jr. Fegley (2000), The solar system's earliest chemistry: systematics of refractory inclusions, *Int. Geol. Rev.*, *42*, 865–894.

Jacobsen, B., Q. Z. Yin, F. Moynier, Y. Amelin, A. N. Krot, K. Nagashima, I. D. Hutcheon, and H. Palme (2008), [26]Al-[26]Mg and [207]Pb-[206]Pb systematics of Allende CAIs: canonical solar initial [26]Al/[27]Al ratio reinstated, *Earth Planet. Sci. Lett.*, *272*, 353–364.

Jacquet, E. (2014), The quasi-universality of chondrule size as a constraint for chondrule formation models, *Icarus*, *232*, 176–186.

Jacquet, E., S. Fromang, and M. Gounelle (2011), Radial transport of refractory inclusions and their preservation in the dead zone, *Astron. Astrophys.*, *526*, L8.

Jacquet, E., O. Alard, and M. Gounelle (2012), Chondrule trace element geochemistry at the mineral scale, *Meteor. Planet. Sci.*, *47*, 1695–1714.

Jeffery, P. M. and J. H. Reynolds (1961), Origin of excess [129]Xe in stone meteorites, *J. Geophys. Res.*, *66*, 3582–3583.

Johansen, A., J. S. Oishi, M.-M. Mac Low, H. Klahr, T. Henning, and A. Youdin (2007), Rapid planetesimal formation in turbulent circumstellar disks, *Nature*, *448*, 1022–1025.

Jones, R. H., J. N. Grossman, and A. E. Rubin (2005), Chemical, mineralogical and isotopic properties of chondrules: clues to their origin, in Chondrites and the protoplanetary disk, *ASP Conf. Ser.*, vol. 341, edited by A. N. Krot, E. R. D. Scott and B. Reipurth, pp. 251–285, Astronomical Society of the Pacific, San Francisco.

Kita, N. T., and T. Ushikubo (2012), Evolution of protoplanetary disk inferred from [26]Al chronology of individual chondrules, *Meteor. Planet. Sci.*, *47*, 1108–1119.

Kita, N. T., G. R. Huss, S. Tachibana, Y. Amelin, L. E. Nyquist, and I. D. Hutcheon, et al. (2005) Constraints on the origin of chondrules and CAIs from short-lived and long-lived radionuclides, in Chondrites and the protoplanetary disk, *ASP Conf. Ser.*, vol. 341, edited by A. N. Krot, E. R. D. Scott and B. Reipurth, pp. 558–587, Astronomical Society of the Pacific, San Francisco.

Kita, N. T., H. Nagahara, S. Togashi, and Y. Morishita (2000), A short duration of chondrule formation in the solar nebula: evidence from [26]Al in Semarkona ferromagnesian chondrules, *Geochim Cosmochim. Acta.*, *64*, 3913–3922.

Kita, N. T., T. Ushikubo, K. B. Knight, R. A. Mendybaev, A. M. Davis, F. M. Richter, and J. H. Fournelle (2012), Internal [26]Al-[26]Mg isotope systematics of a type B CAI: remelting of refractory precursor solids, *Geochim. Cosmochim. Acta.*, *86*, 37–51.

Kita, N. T., Q.-Z. Yin, G. J. MacPherson, T. Ushikubo, B. Jacobsen, K. Nagashima, E. Kurahashi, A. N. Krot, and S. B. Jacobsen (2013), [26]Al-[26]Mg isotope systematics of the first solids in the early Solar System, *Meteor. Planet. Sci.*, *48*, 1383–1400.

Krot A. N., K. Keil, E. R. D. Scott, C. A. Goodrich, and M. K. Weisberg (2014), Classification of meteorites and their genetic relationships, in Meteorites and Cosmochemical Processes, *Treatise on Geochemistry Second Edition vol 1*, edited by H. D. Holland and K. K. Turekian, pp 1–63, Elsevier.

Krot A. N., E. R. D. Scott, and M. E Zolensky (1995), Mineralogical and chemical modification of components in CV3 chondrites: nebular or asteroidal processing, *Meteoritics*, *30*, 748–775.

Krot, A. N., K. Nagashima, F. J. Ciesla, B. J. Meyer, I. D. Hutcheon, A. M. Davis, G. R. Huss, and E. R. D. Scott (2010), Oxygen isotopic composition of the Sun and mean oxygen isotopic composition of the protosolar silicate dust: evidence from refractory inclusions, *Astrophys. J.*, *713*, 1159–1166.

Krot, A. N., I. D. Hutcheon, H. Yurimoto, J. N. Cuzzi, K. D. McKeegan, E. R. D. Scott, G. Libourel, M. Chaussidon, J. Aléon, and M. I. Petaev (2005), Evolution of oxygen isotopic composition in the inner Solar System, *Astrophys. J.*, *622*, 1333–1342.

Krot, A. N., K. Makide, K. Nagashima, G. R. Huss, R. C. Ogliore, F. J. Ciesla, L. Yang, E. Hellebrand, and E. Gaidos (2012), Heterogeneous distribution of ^{26}Al at the birth of the solar system: evidence from refractory grains and inclusions, *Meteorit. Planet. Sci.*, *47*, 1948–1979.

Krot, A. N., M. I. Petaev, E. R. D. Scott, B. G. Choi, M. E. Zolensky, and K. Keil (1998), Progressive alteration in CV3 chondrites: more evidence for asteroidal alteration, *Meteorit. Planet. Sci.*, *33*, 1065–1085.

Kruijer, T. S., T. Kleine, M. Fischer-Gödde, C. Burkhardt, and R. Wieler (2014), Nucleosynthetic W isotope anaomalies and the Hf–W chronomery of Ca-Al-rich inclusions, *Earth Planet. Sci. Lett.*, *403*, 317–327.

Kruijer, T. S., M. Touboul, M. Fischer-Gödde, K. R. Bermingham, R. J. Walker, and T. Kleine (2014), Protracted core formation and rapid accretion of protoplanets, *Science*, *344*, 1150–1154.

Kunihiro, T., A. E. Rubin, K. D. McKeegan, and J. T. Wasson (2004), Initial Al-26 / Al-27 in carbonaceous chondrite chondrules: too little Al-26 to melt asteroids, *Geochim Cosmochim. Acta.*, *68*, 2947–2957.

Kurahashi, E. N., T. Kita, H. Nagahara, and Y. Morishita (2008), Al-26-Mg-26 systematics of chondrules in a primitive CO chondrite, *Geochim Cosmochim. Acta.*, *72*, 3865–3882.

Lada, C. J. (1987), Star formation: from OB associations to protostars, in Star forming regions, edited by M. Peimbert and J. Jugaku, pp. 1–18, International Astronomical Union.

Larsen, K. K., A. Trinquier, C. Paton, M. Schiller, D. Wielandt, M. A. Ivanova, J. N., Connelly, O. Nordlund, A.N. Krot, and M. Bizzaro (2011), Evidence for magnesium isotope heterogeneity in the solar protoplanetary disk. *Astrophys. J.*, *735*, L37–L43.

Lee, T. (1978), A local proton irradiation model for isotopic anomalies in the solar system, *Astrophys. J.*, *224*, 217–226.

Lee, T., D. A. Papanastassiou, and G. J. Wasserburg (1976), Demonstration of Mg-26 excess in Allende and evidence for Al-26, *Geophys. Res. Lett.*, *3*, 41–44.

Lee, T., F. H. Shu, H. Shang, A. E. Glassgold, and K. E. Rehm (1998), Protostellar cosmic rays and extinct radioactivities in meteorites, *Astrophys. J.*, *506*, 898–912.

Libourel, G. and A. N. Krot, (2007), Evidence for the presence of planetesimal material among the precursors of magnesian chondrules of nebular origin, *Earth Planet. Sci. Lett.*, *254*, 1–8,

Libourel, G., A. N. Krot, and L. Tissandier (2006), Role of gas-melt interaction during chondrule formation, *Earth Planet. Sci. Lett.*, *251*, 232–240.

Liu, M.-C (2014), On the injection of short-lived radionuclides from a supernova into the solar nebula: Constraints from the oxygen isotopes. *Astrophys. J.*, *781*, L28 (5 pp.).

Liu, M.-C., M. Chaussidon, C. Goepel, and T. Lee (2012a), A heterogeneous solar nebula as sampled by CM hibonite grains, *Earth Planet. Sci. Lett.*, *327*, 75–83.

Liu, M.-C., M. Chaussidon, G. Srinivasan, and K. D. McKeegan (2012b), A lower initial abundance of short-lived ^{41}Ca in the early Solar System and its implications for Solar System formation, *Astrophys. J.*, *761*, 137–144.

Liu, M.-C., K. D. McKeegan, J. N. Goswami, K. K. Marhas, S. Sahijpal, T. R. Ireland, and A. M. Davis (2009), Isotopic records in CM hibonites: implications for timescales of mixing of isotope reservoirs in the solar nebula, *Geochim. Cosmochim. Acta.*, *73*, 5051–5079.

Liu, M.-C. and K. D. McKeegan (2009), On an irradiation origin for magnesium isotope anomalies in meteoritic hibonite, *Astrophys. J.*, *697*, L145–L148.

Liu, M.-C., L. R. Nittler, C. M. O. Alexander, and T. Lee (2010) Lithium-beryllium-boron isotopic compositions in meteoritic hibonite: implications for origin of ^{10}Be and early Solar System irradiation, *Astrophys. J.*, *719*, L99–L103.

Lodders, K. and B. Fegley Jr. (1998), The planetary scientist's companion. Oxford University Press, New York, 371 p.

Lofgren, G. E. (1996), Dynamic crystallization model for chondrule melts, in Chondrules and the protoplanetary disk, edited by R. H. Hewins, R. H. Jones, and E. R. D. Scott, pp. 187–196, Cambridge University Press, Cambridge.

Lupu, R. E., K. Zanhle, M. S. Marley, L. Schaefer, B. Fegley Jr., C. Morley, C. Kerri, R. Freedman, and J. J. Fortney (2014), The atmospheres of earthlike planets after giant impact events, *Astrophys. J.*, *748*, 27–46.

Luu, T.-H., M. Chaussidon, R. K. Mishra, C. Rollion-Bard, J. Villeneuve, G. Srinivasan, and J. L. Birck (2013), High precision Mg isotope measurements of meteoritic samples by secondary ion mass spectrometry, *J. Anal. Atom. Spec.*, *28*, 67–76.

Luu, T.-H., E. D. Young, M. Gounelle, and M. Chaussidon (2015), A short time interval for condensation of high temperature silicates in the Solar accretion disk, *Proc. Nat. Acad. Sci.*, *112*, 1298–1303.

Lyons, J. R. and E. D. Young (2005), CO self-shielding as the origin of oxygen isotope anomalies in the early solar nebula, *Nature*, *435*, 317–320.

MacPherson G. J. (2005), Calcium-aluminum-rich inclusions in chondritic meteorites, in Meteorites, Planets, and Comets, *Treatise on Geochemistry vol. 1*, edited by edited by A. M. Davis, H. D. Holland, and K. K. Turekian, pp. 201–246. Elsevier-Pergamon, Oxford.

MacPherson, G. J., D. A. Wark, and J. T. Armstrong (1988), Primitive materials surviving in chondrites: refractory inclusions, in Meteorites and the early Solar System, edited by J. F. Kerridge and M. S. Matthews, pp. 746–807, University of Arizona Press, Tucson.

MacPherson, G. J., G. R. Huss, and A. M. Davis (2003), Extinct ^{10}Be in Type A calcium-aluminum-rich inclusions from CV chondrites, Geochim. Cosmochim. Acta., 67, 3165–3179.

MacPherson, G. J., E. S. Bullock, P. E. Janney, N. T. Kita, T. Ushikubo, A. M. Davis, M. Wadhwa, and A. N. Krot (2010), Early Solar nebula condensates with canonical, not supracanonical, initial ^{26}Al/^{27}Al ratios, Astrophys. J., 711, L117–L121.

MacPherson, G. J., N. T. Kita, T. Ushikubo, E. S. Bullock, and A. M. Davis (2012), Well-resolved variations in the formation ages for Ca-Al-rich inclusions in the early Solar System, Earth Planet. Sci. Lett., 331, 43–54.

Makide, K., K. Nagashima, A. N. Krot, G. R. Huss, F. J. Ciesla, E. G. Hellebrand, and L. Yang (2011), Heterogeneous distribution of ^{26}Al at the birth of the Solar System, Astrophys. J., 733, L31–L35.

Marhas, K. K., J. N. Goswami, and A. M. Davis (2002), Short-lived nuclides in hibonite grains from Murchison: evidence for Solar System evolution, Science, 298, 2182–2185.

Martin, P. M., and B. Mason (1974), Major and trace elements in the Allende meteorite, Nature, 249, 333–334.

McKeegan, K. D., A. P. A. Kallio, V. S. Heber, G. Jarzebinski, P. H. Mao, C. D. Coath, T. Kunihiro, R. C. Wiens, J. E. Nordholt, R. W. Moses, D. B. Reisenfeld, A. J. G. Jurewicz, and D. S. Burnett (2011), The oxygen isotopic composition of the Sun inferred from captured Solar wind, Science, 332, 1528–1532.

McKeegan, K. D., M. Chaussidon, and F. Robert (2000), Incorporation of short-Lived ^{10}Be in a Calcium-Aluminum-Rich Inclusion from the Allende meteorite, Science, 289, 1334–1337.

McSween, H. Y. Jr., and G. R. Huss (2010), Cosmochemistry, Cambridge University Press, Cambridge UK.

McSween, H. Y. (1977), Chemical and petrographic constraints on the origin of chondrules and inclusions in carbonaceous chondrites, Geochim. Cosmochim. Acta., 41, 1843–1860.

Mendybaev, R. A., F. M. Richter, R. B. Georg, P. E, Janney, M. J. Spicuzza, A. M. Davis, and J. W. Valley (2013), Experimental evaporation of Mg- and Si-rich melts: Implications for the origin and evolution of FUN CAIs, Geochim. Cosmochim. Acta., 123, 368–384.

Miotello A., L. Testi, G. Lodato, L. Ricci, G. Rosotti, K. Brooks, A. Maury, and A. Natta (2014), Grain growth in the envelopes and disks of Class I protostars, Astron. Astrophys., A32, 1–11.

Misawa, K. and N. Nakamura (1988), Demonstration of REE fractionation among individual chondrules from the Allende (CV3) chondrite, Geochim. Cosmochim. Acta., 52, 1699–1710.

Mishra, R. K. and M. Chaussidon (2014a), Timing and extent of Mg and Al isotopic homogenization in the early inner Solar System, Earth Planet Sci Lett., 390, 318–326.

Mishra, R. K. and M. Chaussidon (2014b), Fossil records of high level of ^{60}Fe in chondrules from unequilibrated chondrites, Earth Planet. Sci. Lett., 398, 90–100.

Mittlefehldt, D. (2014), Achondrites, in Meteorites and Cosmochemical Processes, Treatise on Geochemistry Second Edition vol. 1, edited by H. D. Holland and K. K. Turekian, pp. 235–266, Elsevier.

Molster, F. J., L. B. F. M. Waters, and F. Kemper (2010), The mineralogy of interstellar and circumstellar dust in galaxies, in Astromineralogy, edited by T. Henning, pp. 143–201, Springer.

Morris, M. A., A. C. Boley, S. J. Desch, and T. Athanassiadou (2012), Chondrule formation in bow shocks around eccentric planetary embryos, Astrophys. J., 752, 27–44.

Mostéfaoui, S., N. T. Kita, S. Togashi, S. Tachibana, H. Nagahara, and Y. Morishita (2002), The relative formation ages of ferromagnesian chondrules inferred from their initial ^{26}Al/^{27}Al ratios, Meteorit. Planet. Sci., 37, 421–438.

Moynier, F. and B. Fegley Jr. (2015), The Earth's building blocks, (this volume).

Moynier, F., J. Blichert-Toft, K. Wang, G. F. Herzog, and F. Albarède (2011), The Elusive ^{60}Fe in the Solar Nebula, Astrophys. J., 741, 71–77.

Pack, A., J. M. G. Shelley, and H. Palme (2004), Chondrules with peculiar REE patterns: implications for solar nebula condensation at high C/O, Science, 303, 997–1000.

Palme, H. and A. Jones (2005), Solar system abundances of the elements, in Meteorites, Planets, and Comets, Treatise on Geochemistry vol. 1, edited by edited by A. M. Davis, H. D. Holland and K. K. Turekian, pp. 41–62. Elsevier-Pergamon, Oxford.

Palme, H., D. C. Hezel, and D. S. Ebel (2015), The origin of chondrules: Constraints from matrix composition and matrix-chondrule complementarity, Earth Planet. Sci. Lett., 411, 11–19.

Paton C., M. Schiller, D. Ulfbeck, and M. Bizzarro (2012), High-precision ^{27}Al/^{24}Mg ratio determination using a modified isotope-dilution approach, J. Anal. At. Spectrom., 27, 644–652.

Preibisch, T., Y.-C. Kim, F. Favata, E. D. Feigelson, E. Flaccomio, K. Getman, G. Micela, S. Sciortino, K. Stassun, B. Stelzer, and H. Zinnecker (2005), The origin of T Tauri X-ray emission: new insights from the Chandra Orion ultradeep project, Astrophys. J. Suppl., 160, 401–422.

Radomsky, P.M. and R. H. Hewins (1990), Formation conditions of pyroxene-olivine and magnesian olivine chondrules, Geochim. Cosmochim. Acta., 54, 3475–3490.

Ramaty, R., B. Kozlovsky, and R. E. Lingenfelter (1996), Light isotopes, extinct radioisotopes, and gamma-ray lines from low-energy cosmic-ray interactions, Astrophys. J., 456, 525–540.

Richter, F. M., P. E. Janney, R. A. Mendybaev, A. M. Davis, and M. Wadhwa (2007), Elemental and isotopic fractionation of Type B CAI-like liquids by evaporation, Geochim. Cosmochim. Acta., 71, 5544–5564.

Rowe, M. W. and P. K. Kuroda (1965), Fissiogenic Xenon from the Pasamonte Meteorite, J. Geophys. Res., 70, 709–714.

Rudraswami, N. G., J. N. Goswami, B. Chattopadhyay, S. K. Sengupta, and A. P. Thapliyal (2008), Al-26 records in chondrules from unequilibrated ordinary chondrites: II. duration of chondrule formation and parent body thermal metamorphism, Earth Planet Sci Lett., 274, 93–102.

Russell, S. S., G. Srinivasan, G. R. Huss, G. J. Wasserburg, and G. J. MacPherson (1996), Evidence for widespread ^{26}Al in the solar nebula and constraints for nebula time scales, *Science*, *273*, 757–762.

Sahijpal, S. and J. N. Goswami (1998), Refractory phases in primitive meteorites devoid of ^{26}Al and ^{41}Ca: representative samples of first solar system solids?, *Astrophys. J.*, *509*, 137–140.

Salmeron, R. and T. R. Ireland (2012), Formation of chondrules in magnetic winds blowing through the proto-asteroid belt, *Earth Planet. Sci. Lett.*, *327–328*, 61–67.

Scott, E. R. D. (2007), Chondrites and the protoplanetary disk, *Annu. Rev. Earth Planet. Sci.*, *35*, 577–620.

Scott, E. R. D. and A. N. Krot (2014), Chondrites and their components, in Meteorites and Cosmochemical Processes, *Treatise on Geochemistry Second Edition vol. 1*, edited by H. D. Holland and K. K. Turekian, pp. 65–137, Elsevier.

Scott, E. R. D and G. J. Taylor (1983), Chondrules and other components in C, O, and E chondrites: similarities in their properties and origins, Proc 14th Lunar Planetary Science Conference, 275–286.

Scott, E. R. D. and J. T. Wasson (1975), Classification and properties of iron-meteorites, *Rev. Geophys.*, *13*, 527–546.

Scott, E. R. D., T. V. Krot, J. I. Goldstein, and S. Wakita (2014), Thermal and impact history of the H chondrite parent asteroid during metamorphism: constraints from metallic Fe–Ni, *Geochim. Cosmochim. Acta.*, *136*, 13–37.

Seitz, H.-M., G. P. Brey, J. Zipfel, U. Ott, S. Weyer, S. Durali, and S. Weinbruch (2007), Lithium isotope composition of ordinary and carbonaceous chondrites, and differentiated planetary bodies: bulk solar system and solar reservoirs,. *Earth Planet Sci. Lett.*, *260*, 582–596.

Shu, F. H., F. C. Adams, and S. Lizano (1987), Star formation in molecular clouds: observation and theory, *Ann. Rev. Astro. Astrophys.*, *25*, 23–81.

Shu, F. H., H. Shang, and T. Lee (1996), Toward an astrophysical theory of chondrites, *Science*, *271*, 1545–1552.

Shu, F. H., H. Shang, M. Gounelle, A. E. Glassgold, and T. Lee (2001), The Origin of chondrules and refractory inclusions in chondritic meteorites, *Astrophys. J.*, *548*, 1029–1050.

Siess, L., E. Dufour, and M. Forestini (2000), An internet server for pre-main sequence tracks of low- and intermediate-mass stars, *Astron. Astrophys.*, *358*, 593–599.

Srinivasan, G. and M. Chaussidon (2013), Constraints on ^{10}Be and ^{41}Ca distribution in the early solar system from ^{26}Al and ^{10}Be studies of Efremovka CAIs,. *Earth Planet Sci. Lett.*, *374*, 11–23.

Stahler, S. W. F. and Palla (2005), The formation of stars, Wiley-VCH.

Stopler, E. and J. Paque (1986), Crystallization sequences of Ca-Al-rich inclusions from Allende: the effect of cooling rate and maximum temperature, *Geochim. Cosmochim. Acta.*, *50*, 1785–1806.

Sugiura, N., Y. Shuzou, and A. Ulyanov (2001), Beryllium-boron and aluminum-magnesium chronology of calcium-aluminum-rich inclusions in CV chondrites, *Meteor. Planet. Sci.*, *36*, 1397–1408.

Tachibana, S., H. Nagahara, S. Mostefaoui, and N. T. Kita (2003), Correlation between relative ages inferred from ^{26}Al and bulk compositions of ferromagnesian chondrules in least equilibrated ordinary chondrites, *Meteorit. Planet. Sci.*, *38*, 939–962.

Tachibana, S., G. R. Huss, N. T. Kita, G. Shimoda, and Y. Morishita (2006), ^{60}Fe in chondrites: debris from a nearby supernova in the early Solar System?, *Astrophys. J.*, *639*, L87–L90.

Tang, H. and N. Dauphas (2012), Abundance, distribution, and origin of ^{60}Fe in the solar protoplanetary disk, *Earth Planet. Sci. Lett.*, *359*, 248–263.

Tang, H. and N. Dauphas (2014), ^{60}Fe-^{60}Ni chronology of core formation in Mars, *Earth Planet. Sci. Lett.*, *390*, 264–274.

Thiemens, M. H. and J. E. Heidenreich (1983), The mass-independent fractionation of oxygen: a novel isotope effect and its possible cosmochemical implications, *Science*, *219*, 1073–1075.

Thrane, K. B., M. Bizzarro, and J. A. Baker (2006), Extremely brief formation interval for refractory inclusions and uniform distribution of ^{26}Al in the early solar system, *Astrophys. J.*, *646*, L159–162.

Tielens, A. G. G. M. (2010), The Physics and Chemistry of the Interstellar Medium, Cambridge University Press, Cambridge, UK.

Tielens, A. G. G.M., L. B. F. M. Waters, and T. J. Bernatowicz (2005), Origin and evolution of dust in circumstellar and interstellar environments, in Chondrites and the protoplanetary disk, *ASP Conf. Ser.*, *vol. 341*, edited by A. N. Krot, E. R. D. Scott and B. Reipurth, pp. 605–631, Astronomical Society of the Pacific, San Francisco.

Tissandier, L., G. Libourel, and F. Robert (2002), Gas-melt interactions and their bearing on chondrule formation, *Meteorit. Planet. Sci.*, *37*, 1377–1389.

Trieloff, M., E. K. Jesserberger, I. Herrwerth, J. Hopp, C. Fiéni, M. Ghélis, M. Bourot- Denise, and P. Pellas (2003), Structure and thermal history of the H-chondrite parent asteroid revealed by thermochronometry, *Nature*, *422*, 502–506.

van Boekel, R., M. Min, Ch. Leinert, L. B. F. M. Waters, A. Richichi, O. Chesneau, C. Dominik, W. Jaffe, A. Dutrey, U. Graser, Th. Henning, J. de Jong, R. Köhler, A. de Koter, B. Lopez, F. Malbet, S. Morel, F. Paresce, G. Perrin, Th. Preibisch, F. Przygodda, M. Schöller, and M. Wittkowski (2004), The building blocks of planets within the 'terrestrial' region of protoplanetary disks, *Nature*, *432*, 479–482.

Van Schmus, W. R. and J. A. Wood (1967), A chemical-petrologic classification for the chondritic meteorites, *Geochim. Cosmochim. Acta.*, *31*, 747–65.

Vernazza, P., B. Zanda, R. P. Binzel, T. Hiroi, F. E. DeMeo, M. Birlan, R. Hewins, L. Ricci, P. Barge, and M. Lockhart (2014), Multiple and fast: the accretion of ordinary chondrite parent bodies, *Astrophys. J.*, *791*, 120–142.

Villeneuve, J., M. Chaussidon, and G. Libourel (2009), Homogeneous distribution of ^{26}Al in the solar system from the Mg isotopic composition of chondrules, *Science*, *325*, 985–988.

Villeneuve, J., M. Chaussidon, and G. Libourel (2011), Magnesium isotopes constraints on the origin of Mg-rich olivines from the Allende chondrite: nebula versus planetary?, *Earth Planet. Sci. Lett.*, *301*, 107–116.

Villeneuve, J., M. Chaussidon, and G. Libourel (2013), Lack of relationship between ^{26}Al ages of chondrules and their mineralogical and chemical compositions, *C. R. Geoscience*, *344*, 423–431.

Visser, A. E., J. S. Richer, and C. J. Chandler (2002), Completion of a SCUBA survey of lynds dark clouds and implications for low-mass star formation. *Astrophys. J.*, *124*, 2756–2789.

Warren, P. H. (2011), Stable-isotopic anomalies and the accretionary assemblage of the Earth and Mars: a subordinate role for carbonaceous chondrites, *Earth Planet. Sci. Lett.*, *311*, 93–100.

Wasserburg, G. J., T. Lee, and D. A. Papanastassiou (1977), Correlated O and Mg isotopic anomalies in Allende inclusions: II. *Magnesium, Geophy. Res. Lett.*, *4*, 299–302.

Wasserburg, G. J., M. Busso, R. Gallino, and K. M. Nollett (2006), Short-lived nuclei in the early Solar System: possible AGB sources, *Nucl. Phys. A*, *777*, 5–69.

Wasserburg, G. J., J. Wimpenny, and Q. Z. Yin (2012), Mg isotopic heterogeneity, Al-Mg isochrons, and canonical $^{26}Al/^{27}Al$ in the early solar system, *Meteorit. Planet. Sci.*, *47*, 1980–1997.

Wasson, J. T. (1985), Meteorites: their record of early Solar System history, Freeman and Co., New York.

Wasson, J. T. (1993), Constraints on chondrule origins, *Meteoritics*, *28*, 13–28.

Wasson, J. T. and G. W. Kallemeyn (1988), Compositions of chondrites, *Phil. Trans. Roy. Soc. London A*, *325*, 535–544.

Weiss, B. P. and L. T. Elkins-Tanton (2013), Differentiated planetesimals and the parent bodies of chondrites, *Annu. Rev. Earth Planet. Sci.*, *41*, 529–60.

Wick, M. J. and R. H. Jones (2012), Formation conditions of plagioclase bearing type I chondrules in CO chondrites: a study of natural samples and experimental analogs, *Geochim. Cosmochim. Acta.*, *98*, 140–159.

Wielandt, D., K. Nagashima, A. N. Krot, G. R. Huss, M. A. Ivanova, and M. Bizzarro (2012), Evidence for multiple sources of ^{10}Be in the early Solar System, *Astrophys. J. Lett.*, *748*, L25.

Winston, E., S. T. Megeath, S. J. Wolk, J. Hernandez, R. Gutermuth, J. Muzerolle, J. L. Hora, K. Covey, L. E. Allen, B. Spitzbart, D. Peterson, P. Myers, and G. G. Fazio (2009), A spectroscopic study of young stellar objects in the Serpens cloud core and NGC 1333, *Astrophys. J.*, *137*, 4777–4794.

Wolk, S. J., F. R. Harnden Jr., E. Flaccomio, G. Micela, F. Favata, H. Shang, and E. D. Feigelson (2005), Stellar activity on the young suns of Orion: COUP observations of K5-7 pre-main-sequence stars. *Astrophys. J. Suppl.*, *160*, 423–449.

Wood, J. A. (1988), Chondritic meteorites and the solar nebula, *Annu. Rev. Earth Planet. Sci.*, *16*, 53–72.

Yoneda, S. and L. Grossman (1995), Condensation of CaO–MgO–Al_2O_3–SiO_2 liquids from cosmic gas, Geochim. *Cosmochim. Acta.*, *59*, 3413–3444.

Young, E. D. and A. Galy (2004), The isotope geochemistry and cosmochemistry of magnesium, *Rev. Mineral. Geochem.*, *55*, 197–230.

Young, E. D., R. D. Ash, A. Galy, and N. S. Belshaw (2002), Mg isotope heterogeneity in the Allende meteorite measured by UV laser ablation-MC-ICPMS and comparisons with O isotopes, *Geochim. Cosmochim. Acta.*, *66*, 683–698.

Young, E. D., J. I. Simon, A. Galy, S. S. Russell, E. Tonui, and O. Lovera (2005), Supra-canonical $^{26}Al/^{27}Al$ and the residence time of CAIs in the solar protoplanetary disk, *Science*, *308*, 223–227.

Young, E. D., A. Galy, and H. Nagahara (2002), Kinetic and equilibrium mass-dependent isotope fractionation laws in nature and their geochemical and cosmochemical significance, *Geochim. Cosmochim. Acta.*, *66*, 1095–1104.

Yu, Y. and R. H. Hewins (1998), Transient heating and chondrule formation: evidence from Na loss in flash heating simulation experiments, *Geochim. Cosmochim. Acta.*, *62*, 159–172.

Yurimoto, H. and K. Kuramoto (2004), Molecular cloud origin for the oxygen isotope heterogeneity in the Solar System, *Science*, *305*, 1763–1766.

Yurimoto, H., A. N. Krot, B.-G. Choi, J. Aléon, T. Kunihiro, and A. J. Brearley (2008), Oxygen isotopes in chondritic components, in Oxygen in the Solar System, *Reviews in Mineralogy and Geochemistry*, *vol. 68*, edited by G. J. MacPherson, pp. 141–186, Mineralogical Society of America, Washington, DC.

Yurimoto, H., et al. (2011), Oxygen isotopic compositions of asteroidal materials returned from Itokawa by the Hayabusa mission, *Science*, *333*, 1116–1119.

Zanda, B. (2004), Chondrules, *Earth Planet. Sci. Lett.*, *224*, 1–17.

Zhai, M., E. Nakamura, D. M. Shaw, and T. Nakano (1996), Boron isotope ratios in meteorites and lunar rocks, *Geochim. Cosmochim. Acta.*, *60*, 4877–4881.

Zolensky, M. E., T. J. Zega, H. Yano, S. Wirick, A. J. Westphal, M. K. Weisberg, I. Weber, J. L. Warren, M. A. Velbel, A. Tsuchiyama, P. Tsou, A. Toppani, N. Tomioka, K. Tomeoka, N. Teslich, M. Taheri, J. Susini, R. Stroud, T. Stephan, F. J. Stadermann, C. J. Snead, S. B. Simon, A. Simionovici, T. H. See, F. Robert, F. J. Rietmeijer, W. Rao, M. C. Perronnet, D. A. Papanastassiou, K. Okudaira, K. Ohsumi, I. Ohnishi, K. Nakamura-Messenger, T. Nakamura, S. Mostefaoui, T. Mikouchi, A. Meibom, G. Matrajt, M. A. Marcus, H. Leroux, L. Lemelle, L. Le, A. Lanzirotti, F. Langenhorst, A. N. Krot, L. P. Keller, A. T. Kearsley, D. Joswiak, D. Jacob, H. Ishii, R. Harvey, K. Hagiya, L. Grossman, J. N. Grossman, G. A. Graham, M. Gounelle, P. Gillet, M. J. Genge, G. Flynn, T. Ferroir, S. Fallon, D. S. Ebel, Z. R. Dai, P. Cordier, B. Clark, M. Chi, A. L. Butterworth, D. E. Brownlee, J. C. Bridges, S. Brennan, A. Brearley, J. P. Bradley, P. Bleuet, P. A. Bland, and R. Bastien (2006), Mineralogy and petrology of comet 81P/ Wild 2 nucleus samples, *Science*, *314*, 1735–1739.

2

The Earth's Building Blocks

Frédéric Moynier[1] and Bruce Fegley, Jr.[2]

ABSTRACT

In this chapter we propose an updated bulk chemical composition of the Earth based on new estimates for the composition of the Earth's core. We present the different cosmochemical constraints that allow us to search for the Earth's building blocks. Chemical equilibrium calculations of gas–solid equilibria in solar composition gas show that at first approximation the Earth is chondritic for the refractory lithophile (rock-loving) elements. Elemental and isotopic compositions are then used to discuss the different possible types of meteoritic material (exemplified by the meteorites in our collections) that may represent the bulk Earth composition.

2.1. INTRODUCTION

Consideration of Earth's building blocks requires knowledge of its bulk composition today, chemistry in the solar nebula, and chemical effects during accretion of the Earth [e.g., see pp. 35–38 of *Lewis and Prinn*, 1984]. The Earth is a differentiated planet consisting of five physically distinct geochemical reservoirs (atmosphere, biosphere, hydrosphere, petrosphere, and core) with different chemical compositions. In this chapter, we *geochemically* define petrosphere as the crust plus the entire mantle so it encompasses the entire silicate portion of the Earth. Thus, the petrosphere is similar, but not identical to the bulk silicate Earth, which also includes the atmosphere, biosphere, and hydrosphere (see below). All accessible terrestrial samples have been modified to a greater or lesser extent from Earth's original composition. For example, the atmosphere, biosphere, and hydrosphere originated from Earth's primordial silicate portion. Thus, the mass-weighted chemical composition of the atmosphere, biosphere, hydrosphere, and petrosphere is that of the bulk silicate Earth (BSE), also known as Earth's primitive mantle (PM). We use the term BSE throughout this chapter. Two major processes shaped the composition of the BSE: (1) gas–solid (and/or gas–melt) volatility fractionation of Earth-forming materials in the solar nebula (including volatility effects on planetesimals), and during accretion of the Earth, and (2) silicate, metal, and sulfide elemental partitioning during core formation on the early Earth and its planetesimal precursors.

We could also divide the Earth into geochemically different reservoirs such as undepleted primitive mantle (UPM) and depleted MORB mantle (DMM), but we did not take this approach for two reasons. First, the size of the undepleted and depleted reservoirs may vary for each element and isotope under consideration (e.g., different size reservoirs for a lithophile and a noble gas). Second, the depleted and undepleted reservoirs are model constructs that do not necessarily correspond to physically different reservoirs such as

[1]*Institut de Physique du Globe de Paris, Insitut Universitaire de France, Université Paris Diderot, Sorbonne Paris Cité, CNRS UMR 7154, Paris, France*

[2]*Planetary Chemistry Laboratory, Department of Earth and Planetary Sciences and McDonnell Center for the Space Sciences, Washington University, St. Louis, Missouri, USA*

The Early Earth: Accretion and Differentiation, Geophysical Monograph 212, First Edition.
Edited by James Badro and Michael Walter.

upper and lower mantle. For example, the entire mantle may be a mixture of more and less depleted regions like raisins in a cake [Figure 6 of *Helffrich and Wood*, 2001].

It is more convenient for us to consider the five physically distinct reservoirs mentioned above. In principle, Earth's bulk composition can be reconstructed from chemical analyses of elements in each of these five reservoirs combined together in their relative mass fractions. Thus, the concentration C of an element E in the bulk Earth (Bulk) is given by

$$C_E^{Bulk} = F_A C_E^A + F_B C_E^B + F_H C_E^H + F_P C_E^P + F_C C_E^C$$

The mass fraction of each reservoir R is F_R, and the subscripts denote the atmospheric (A), biospheric (B), hydrospheric (H), petrospheric (P), and core (C) reservoirs. Table 2.1 lists the mass and mass fraction of each of these reservoirs and shows the further subdivision of the petrospheric and core reservoirs. The concentration of an element E in each reservoir is C_E^R. For example, the bulk Earth concentration of Mg is given by

$$C_{Mg}^{Bulk} = F_A C_{Mg}^A + F_B C_{Mg}^B + F_H C_{Mg}^H + F_P C_{Mg}^P + F_C C_{Mg}^C$$

Table 2.1 Earth's Geochemical Reservoirs[a]

Reservoir	Mass (kg)	Mass Fractions of Bulk Silicate Earth[b]	of Total Earth
Total Atmosphere	**5.137×10¹⁸**	**1.27×10⁻⁶**	**8.60×10⁻⁷**
Troposphere	4.22×10¹⁸	1.05×10⁻⁶	7.06×10⁻⁷
Stratosphere	0.906×10¹⁸	2.25×10⁻⁷	1.52×10⁻⁷
Upper Atmosphere	4×10¹⁵	9.9×10⁻¹⁰	6.7×10⁻¹⁰
Biosphere	**1.148×10¹⁶**	**2.85×10⁻⁹**	**1.92×10⁻⁹**
Hydrosphere	**1.664×10²¹**	**4.13×10⁻⁴**	**2.79×10⁻⁴**
Total petrosphere[c]	**4.031×10²⁴**	**0.9995**	**0.6748**
Continental Crust	1.522×10²²	3.77×10⁻³	2.55×10⁻³
Oceanic Crust	0.845×10²²	2.10×10⁻³	1.41×10⁻³
Total Crust	2.367×10²²	5.87×10⁻³	3.96×10⁻³
Upper Mantle	1.068×10²⁴	0.2648	0.1788
Lower Mantle	2.939×10²⁴	0.7287	0.4920
Total Mantle	4.007×10²⁴	0.9936	0.6708
Total Core	**1.941×10²⁴**	**–**	**0.3249**
Outer Core	1.839×10²⁴	–	0.3078
Inner Core	0.102×10²⁴	–	0.0171
Entire Earth	**5.9736×10²⁴**		**1.0000**

[a] Modified from K. Lodders and B. Fegley, "The Planetary Scientist's Companion," Oxford, New York, NY, 1998.
[b] Total mass of the bulk silicate Earth (BSE) is 4.033×10²⁴ kg.
[c] We define petrosphere as the silicate portion of the BSE, i.e., crust + entire mantle.

$$C_{Mg}^{Bulk} = (8.6 \times 10^{-7})(0\%) + (1.9 \times 10^{-9})(0.098\%)$$
$$+ (2.79 \times 10^{-4})(0.128\%) + (0.6748)(22.17\%)$$
$$+ (0.3249)(0\%)$$

$$C_{Mg}^{Bulk} = 14.96\%$$

In comparison, *Morgan and Anders* [1980], *Kargel and Lewis* [1993], and *McDonough* [2003] give 13.90%, 14.86%, and 15.4% Mg, respectively, in the bulk Earth. The variations are due to the different values selected for the Mg percentage in Earth's mantle. The amounts of Mg in the atmosphere (none), biosphere (0.098% in phytomass), hydrosphere (0.128% in seawater) [*Lodders and Fegley*, 1998, 2011], and core (none, *McDonough*, 2003) are very small compared to the petrosphere (22.17%, Table 4 of *Palme and O'Neill*, 2014). Thus, the Mg concentration in the bulk Earth is effectively just the mass-weighted Mg concentration in the petrosphere.

We can simply write the mass balance equations by using elemental concentrations in the BSE; then a two-term equation results (e.g., the equation for Mg) becomes:

$$C_{Mg}^{Bulk} = F_{BSE} C_{Mg}^{BSE} + F_C C_{Mg}^C$$

$$C_{Mg}^{Bulk} = (0.6751)(22.17\%) + (0.3249)(0\%) = 14.96\%$$

Other lithophile elements that are insoluble in water and do not behave as siderophiles or chalcophiles at core-forming conditions (e.g., Al, Ba, Be, Ca, F, Li, REE, Sc, Sr, Th, Ti, U, Y, Zr) have similar mass balance equations comprised of only the petrospheric reservoir term. (We defer discussion of Na, K, Rb, and Cs to later.)

Several other elements are concentrated in one reservoir, but not to the same extent as the lithophiles mentioned above. For example, most of the iodine in the bulk silicate Earth is primarily found in organic-rich marine sediments and marine organisms (i.e., biospheric reservoir) with ~11% residing in the petrosphere. However, the core may contain up to 82% of Earth's total iodine inventory if metal/silicate partition coefficients measured at high pressures and temperatures are relevant to core formation on Earth [*Armytage et al.*, 2013].

About 75% of Cl and Br in the BSE are in the oceans (hydrospheric reservoir) with the remaining 25% in the petrosphere. Neglecting any nitrogen that may be dissolved in the molten outer core, 73.1% of Earth's nitrogen is in the atmosphere, 26.5% is in the crust, and 0.4% in the oceans [*Lodders and Fegley*, 2011]. Geochemical modeling [*Allegre et al.*, 1995; *Albarede*, 1998; *Ballentine and Holland*, 2008] shows only 45–50% of Earth's ⁴⁰Ar is in the atmosphere. This modeling depends on the amount of K in the BSE, which is fairly well constrained at 230–280 ppm by mass [260 ppm, *Jagoutz et al.*, 1979; 240 ppm,

McDonough and Sun, 1995; 232 ppm, *Kargel and Lewis*, 1993; 258 ppm, *Hofmann*, 1988; 231 ppm, *Wänke et al.*, 1984; 258 ppm, *Hart and Zindler*, 1986; 280 ppm, *Allègre et al.*, 1995; and 260 ppm, *Palme and O'Neill*, 2014]. If the degassing efficiency is similar for other noble gases, the atmosphere contains ~50% of Earth's $^{36+38}$Ar, Kr, and Xe [*Fegley and Schaefer*, 2014]. Helium is not considered in this comparison because it is continually escaping to space with an atmospheric lifetime of 0.9–1.8Ma for ^{4}He and 0.4–0.8Ma for ^{3}He [*Torgersen*, 1989]. Neon is not considered because the different isotopic compositions of atmospheric and mantle Ne suggest atmospheric Ne is probably a mixture of Ne degassed from the mantle and from chondritic bolides by impact degassing [*Marty*, 2012; *Moreira*, 2013; *Zhang*, 2014]. An alternative scenario is the Ne has been lost by hydrodynamic escape to space associated with an isotopic mass-fractionation [*Pepin*, 2006].

The atmospheric inventory of Xe is lower than it should be by analogy with Ne, Ar, and Kr, the missing Xe problem. Various solutions to the missing Xe problem have been suggested over the past 50 years, which generally involve sequestering the "missing" Xe inside the Earth (e.g., in glacial ice, sediments, or clathrate hydrates; see *Ozima and Podosek*, 2001). More recently *Sanloup et al.* [2005, 2011] suggested that significant amounts of Xe are in the lower mantle or in the crust as XeO_2 in quartz and olivine, which is supported by the synthesis of XeO_2 [*Brock and Schrobilgen*, 2011]. A last scenario is that Xe has been lost to space through an unknown process associated with an isotopic fractionation [*Pujol et al.*, 2011]. Over the past 30 years geoscientists have speculated that the "missing" Xe is in Earth's core [*Ozima and Podosek*, 2001]. Their reasoning is that Xe becomes a metal with hexagonal close packed (hcp) structure above 120 GPa, and it may alloy with or dissolve into Fe, which is also an hcp metal at core conditions. However, the experimental study of *Nishio-Hamane et al.* [2010] finds no evidence for Xe alloy formation with or dissolution into Fe at P, T conditions similar to those in the core. A more detailed discussion of noble gas geochemistry is beyond our scope, but is given by *Zhang* [2014].

As evident from our discussion above, this simple mass balance approach is difficult in practice because of our incomplete knowledge of reservoir compositions (e.g., the core, which is ~32.5% by mass, and the mantle, which is ~67.5% by mass), of the Earth. For example, recent metal/silicate partitioning experiments indicate that Earth's core may be a significant reservoir of helium [*Bouhifd et al.*, 2013a], iodine [*Armytage et al.*, 2013], and lead [*Bouhifd et al.*, 2013b]. The measured partition coefficients depend upon several variables (e.g., P, T, oxygen fugacity, sulfur content of the molten metal) and the "true" concentrations of these and many other elements in Earth's core will be incredibly difficult to constrain.

Likewise, the concentrations of several elements in the mantle are not well known (e.g., As, Bi, Cd, H, In, S, Sb, Se, Te, Tl). The estimate of H in the mantle varies from 1–2 oceans of water [e.g., *Saal et al.*, 2002, *Hirschmann and Dasgupta*, 2009; *Zhang*, 2014]. Even the crustal concentrations of some elements such as As, Bi, Hg, Sb, and Tl are poorly known [*Palme and O'Neill*, 2014].

Our knowledge of the compositions of the five major geochemical reservoirs comes from four principal types of data: (1) seismological observations, (2) chemical and isotopic analyses of the atmosphere, biosphere, hydrosphere, and petrosphere, (3) experimental studies of elemental partitioning between coexisting phases (e.g., metal, silicate, sulfide, melt) as a function of temperature, pressure, oxygen and/or sulfur fugacity, (4) ratios of elements with very similar geochemical behavior (e.g., Sm/Sn, K/U, K/La, W/Th, P/Nd, Rb/Ba, Nb/Ta) combined with analysis of one element in the pair, and (5) cosmochemical constraints (e.g., elemental and isotopic abundances in meteorites and solar composition material).

Palme and O'Neill [2014] describe in more detail several of these approaches that they use to determine the composition of the bulk silicate Earth. Thus, we do not repeat their discussion here but refer the reader to their chapter. Published values for composition of the BSE are in good agreement. For example, Table 2.3 of *Palme and O'Neill* [2014] compares BSE major element compositions from *Ringwood* [1979], *Jagoutz et al.* [1979], *Wänke et al.* [1984], *Palme and Nickel* [1985], *Hart and Zindler* [1986], *McDonough and Sun* [1995], and *Allègre et al.* [1995]. The range for SiO_2 (the most abundant oxide in the BSE) is only from 45.0% [*McDonough and Sun*, 1995] to 46.2% [*Palme and Nickel*, 1985], and the unweighted mean value of all seven compilations is 45.6 ± 0.5% (1 sigma error), identical within error to the value of 45.40 ± 0.30% recommended by *Palme and O'Neill* [2014]. Likewise the concentration range for MgO, the second most abundant oxide in the BSE, is from 35.5% [*Palme and Nickel*, 1985] to 38.3% [*Jagoutz et al.*, 1979]. The unweighted mean MgO concentration from all seven compilations is 37.4 ± 0.9% (1 sigma error), identical within error to the value of 36.77 ± 0.44% given by *Palme and O'Neill* [2014]. Table 6.9 of *Lodders and Fegley* [1998] compares 10 BSE compositions, and the same conclusion is reached, that the major element composition of the BSE is well constrained. As we discuss later, these BSE compositions assume that the mantle is compositionally uniform. There is good evidence for this as summarized by *Palme and O'Neill* [2014].

More than one line of evidence is often used to constrain the composition of the petrospheric and core reservoirs. For example, seismological observations show the Earth is physically divided (to the first order) into crust, upper mantle, lower mantle, and core, and that the Earth's core is divided into a molten outer core and a solid inner core. Seismic data and analyses of Fe-Ni meteorite metal show that most of the core is Fe-Ni alloy (Fe/Ni ~16.5 by weight = Fe 94.3%, Ni 5.7%). This is the solar composition value that is found in most iron meteorites. However, the molten outer core is ~5–10% less dense and the inner core is ~2–3% less dense than this alloy [*Badro et al.*, 2014]. One or more light elements are present in the core [*Birch*, 1952, 1964]. The identity of the light element(s) in the molten outer core and in the solid inner core is not uniquely constrained by the seismic data alone but also depends upon observed elemental abundances in Fe-Ni alloy in meteorites, Fe – element binary phase diagrams, metal/silicate partition coefficients, and solar elemental abundances; the plausible light elements include C, S, Si, O, N, P, and H. Molecular dynamics simulations of density and sound speed at plausible P & T for the core by *Badro et al.* [2014] show oxygen is the most abundant light element in the core with their best fit model containing 3.7% O and 1.9% Si by weight. The maximum light element contents permitted by their modeling are 5.4% O, 4.5% Si, and 2.4% S.

Seismic tomography shows the penetration of downwelling slabs into the lower mantle, which requires mass exchange between upper and lower mantle. This evidence is important because it implies a generally well-mixed, compositionally uniform mantle [*Palme & O'Neill*, 2014]. *Bina and Helffrich* [2014] computed density and sound velocity profiles along plausible lower mantle adiabats for a variety of compositions and concluded that lower mantle physical properties are consistent with a pyrolite (i.e., upper mantle) composition. *Palme and O'Neill* [2014] review other lines of evidence supporting a homogenous mantle composition.

Thus, most geoscientists accept that analyses of relatively unaltered upper mantle xenoliths (fertile peridotites, spinel lherzolites, garnet lherzolites), petrological models of peridotite–basalt melting (e.g., Ringwood's pyrolite model), and concentration ratios of geochemically similar elements in basalts, can be used to determine the composition of the entire mantle, not just the upper mantle. But see *Anderson* [2002] and *Javoy et al.* [2010] for arguments that the upper and lower mantles have different compositions. The latter group requires this in order for their enstatite chondrite model to match the BSE.

In this review we focus on three lines of evidence, cosmochemical constraints and geochemical and isotopic analyses, that provide information about the bulk composition of the Earth and its building blocks.

2.2. COSMOCHEMICAL CONSTRAINTS

We discuss meteoritic constraints on Earth's composition and origin in a subsequent section. Here we discuss constraints from solar nebula chemistry, but before doing

so we briefly summarize the different stages of solar nebula evolution and planetary accretion based on *Cameron* [1995], *Lodders and Fegley* [2011], and *Chambers* [2014].

First Stage: Molecular cloud collapse. During this stage, the solar nebula disk is built up of infalling material from a collapsing molecular cloud core. This stage lasts for a few times 10^5 years. Most (if not all) of the matter in the disk during this stage ultimately goes into the proto-Sun.

Second Stage: Disk dissipation. The Sun forms during this stage, which lasts for about 50,000 years (i.e., the Sun accretes at approximately 2×10^{-5} solar masses per year). The disk mass is less than that of the proto-Sun. Most matter falling onto the accretion disk is transported through the disk into the proto-Sun, and angular momentum is transported outward in the disk. The major dissipation (i.e., transport) mechanisms include spiral density waves, disk-driven bipolar outflows, and the Balbus-Hawley magnetic instability. *Cameron* [1995] discusses why turbulent viscosity driven by thermal convection is less important than these other dissipative mechanisms, but some other modelers continue to rely on turbulent viscosity as the main dissipative mechanism. The amount of outward mass transport, that would tend to "contaminate" the outer nebula with products from thermochemical processing in the innermost few AU of the nebula, is controversial and uncertain. Near the end of this stage, some disk material survived and is preserved in meteorites (e.g., ^{26}Al-bearing minerals).

Third Stage: Terminal accumulation of the Sun. The final accumulation of the Sun occurs during this stage, which lasts for about $(1–2) \times 10^6$ years. The Sun's accumulation rate decreases from about 10^{-7} to 10^{-8} solar masses per year. The proto-Sun becomes a classical T Tauri star in this phase. Planetary accretion (almost complete for Jupiter and Saturn, and less advanced for the other planets) occurs during this stage.

Fourth Stage: Loss of nebular gas. The Sun becomes a weak line T Tauri star in this stage, which lasts $(3–30) \times 10^6$ years, and is no longer accreting material from the disk. The T Tauri wind removes gas in the inner nebula and photo-evaporation due mainly to UV radiation from the T Tauri wind removes gas in the outer nebula. Ultraviolet photochemistry may be important in the outer nebula, but its importance depends on the poorly constrained and time-dependent nebular column density. Nebular gas–solid thermochemistry ceases sometime (although not necessarily at the same time everywhere) during this stage.

In *Cameron's* [1995] model of nebular evolution, accumulation of the Sun consumed essentially all of the material accreted by the nebular disk during the early stages of its history. Planets and all the other bodies in the solar system were then assembled from material accreted after the Sun had formed (i.e., during the time from the end of stage 2 into stage 4). Accretion of the terrestrial planets proceeded via a multistage process of planetary embryo growth from nebular dust grains (dust grains → chunks → boulders → km size bodies → asteroidal size bodies → planetary embyros → planets). The number of objects decreased at each stage of accretion until roughly a few hundred planetary embryos of lunar to Mars size are produced. The four terrestrial planets formed by accretion of these planetary embryos. For example, the last stage of Earth accretion apparently involved collision of the ~0.9 M_E proto-Earth with an ~0.1 M_E Mars-size body.

The composition of the dust grains that were the starting point for planetesimal growth and accretion was mainly determined by nebular thermochemistry; thus, we discuss constraints on chemistry of nebular solids from chemical equilibrium calculations of gas–solid equilibria in solar composition gas. *Prinn and Fegley* [1989] showed thermochemistry was the dominant chemical process in the inner solar nebula where the terrestrial planets formed. At 1 AU they found thermochemistry is 10 times more important than shock chemistry (caused by nebular lightning), 40,000 times more important than photochemistry, and over 13 million times as important as radiochemistry. For example, absorption of UV radiation by water vapor (the third most abundant gas in the solar nebula after H_2 and He) and scattering by dust dramatically decreases the Ly-α photon flux from the proto-Sun at 1 AU. Hotter, denser conditions inward of 1 AU in the solar nebula made thermochemistry even more important closer to the proto-Sun. This is also due in part to increased abundances at higher temperatures of CO and UV absorbing gases of rock-forming elements (e.g., SiO, SiS, HS, H_2S, AlOH, Al_2O, AlS, AlH, AlO, PO, PN, PS, NaCl, HCl, KCl, KOH, TiO, and TiO_2) that are thermodynamically stable in the inner region of the solar nebula (e.g., see Table 2.2 and Figures 1–17 of *Schaefer and Fegley*, 2010b). Some of these gases absorb UV light at the same wavelengths as H_2O and others absorb at shorter (e.g., CO, SiO) or longer wavelengths (e.g., H_2S). Methane, CO_2, and NH_3 become more abundant and contribute to UV opacity of the solar nebula at a few hundred Kelvin where the CO – CH_4 (625 K at 10^{-4} bar total pressure) and

Table 2.2 Bulk Earth Compositional Models (ppm by mass or mass %)[a]

Z	Element	MF14[b]	KL93[c]	McD03[d]	MA80[e]	CI[f]	H[g]	EH[g]
1	H	81	36.9	260	33	21,015	400	–
6	C	68	44	730	446	35,180	1200	3900
7	N	1	0.59	25	4.1	2940	34	–
8	O	33.62%	31.67%	29.7%	30.12%	45.82%	33.86%	31.00%
9	F	17	15.8	10	13.5	60.6	27	180
11	Na	1748	2450	0.18%	1250	5010	6110	6880
12	Mg	14.96%	14.86%	15.4%	13.90%	9.587%	14.10%	10.73%
13	Al	1.61%	1.433%	1.59%	1.41%	0.850%	1.06%	0.820%
14	Si	16.22%	14.59%	16.1%	15.12%	10.65%	17.10%	16.59%
15	P	59	1180	715	1920	920	1200	2130
16	S	135	0.893%	0.635%	2.92%	5.41%	2.00%	5.60%
17	Cl	20	264	76	19.9	704	77	570
19	K	176	225	160	135	530	780	840
20	Ca	1.76%	1.657%	1.71%	1.54%	0.907%	1.22%	0.850%
22	Ti	854	797	810	820	440	630	460
24	Cr	1701	3423	4700	4120	2590	3500	3320
25	Mn	709	2046	800	750	1910	2340	2170
26	Fe	29.54%	32.04%	32.0%	32.07%	18.28%	27.20%	30.40%
27	Co	808	779	880	840	502	830	870
28	Ni	1.74%	1.72%	1.82%	1.82%	1.064%	1.71%	1.84%
	TOTAL	100.09%	99.99%	100.05%	99.92%	99.75%	99.96%	99.96%
mean μ (amu)		25.19	25.87	26.21	26.26	15.01	24.68	25.91

[a] Values in ppm by mass (μg/g) unless otherwise noted for selected elements.

[b] This work using *Palme and O'Neill* [2014] BSE abundances that are recomputed as 0.675(BSE value), the best fit core model of *Badro et al.* [2014] (3.7% O, 1.9% Si, remainder $Fe_{94}Ni_6$ by mass), and solar Co/Fe in the core (0.0029 by weight).

[c] *Kargel and Lewis* [1993], Table II.

[d] *McDonough* [2003], Table 3.

[e] *Morgan and Anders* [1980], Table 2.

[f] *Lodders* [2003], Table 3, weighted mean for CI chondrites.

[g] *Lodders and Fegley* [2011], Appendix B.

$N_2 - NH_3$ equal abundance lines (345 K at 10^{-4} bar total pressure) are reached (e.g., see Figures 14–15 in *Schaefer and Fegley*, 2010b). With decreasing temperature below the equal abundance lines, CH_4 and NH_3 become more important until they are the major C- and N-bearing gases. The UV opacity due to all of the gases listed above vanishes once the respective elements condense. For example at 10^{-4} bar total pressure Si is 50% in the vapor as SiO and SiS and 50% condensed as forsterite plus enstatite at 1310 K. The UV opacity due to H_2S persists to much lower temperatures because sulfur is 50% condensed as FeS at 664 K. The UV opacity due to water vapor persists until low temperatures of ~180 K (10^{-4} bar total pressure) where 50% of H_2O is in the vapor and 50% is condensed as water ice. At 10^{-4} bar total pressure, ammonia condenses as NH_4OH (131 K) and CH_4 as CH_4 clathrate hydrate at 78 K. Thus, all of the gaseous UV opacity sources except for CO should be absent from the outermost solar nebula.

In contrast, chemical modeling of protoplanetary disks by astronomers [e.g., *Henning and Semenov*, 2013] uses chemical kinetic codes and focuses on lower temperature, lower density disk regions that are "thin" to electromagnetic radiation (e.g., R ≥ 10 AU beyond the condensation fronts for molecules such as H_2O that absorb UV radiation). Indeed, the calculations of *Prinn and Fegley* [1989] showed that stellar UV photochemistry is important in these outer nebular regions (see pp. 85–92 of their paper). For example, the CO photochemical models of *Lyons and Young* [2005] for production of oxygen isotopic anomalies are done at 30 AU and require a far UV flux about 10^3 times larger than the local interstellar medium. But as mentioned earlier, we are interested in the higher temperature, higher density inner disk regions that are opaque (hence convective), where chemical equilibrium is reached within the disk lifetime.

In the next 10 to 20 years we can expect quantitative information about the composition of protoplanetary disks from ALMA (Atacama Large Millimeter/ submillimeter Array), JWST (James Webb Space Telescope), and other telescopes. Ultimately, we may have good constraints on the composition of the dust and gas as a function of radial distance in the solar nebula from astronomical observations of solar metallicity protoplanetary disks, but at the present time we must rely on chemical equilibrium models.

Until the discovery of planets and planetary systems around other stars (exoplanets), it was generally believed that the Earth formed at about its present location from the Sun. However, many of the exoplanets discovered to date are gas giant planets very close to their parent stars (e.g., see the catalog at exoplanet.eu). The accepted model for formation of these extrasolar gas giant planets involves formation at much larger radial distances and radial migration inward to their present positions [*Perryman*, 2011].

Thus, it may be possible that the Earth also formed much farther outward of its present position and migrated inward to one astronomical unit (AU; 1 AU = Earth's average distance from the Sun) (e.g., as in the giant gaseous protoplanet models of Cameron and colleagues) [*DeCampli*, 1978; *DeCampli and Cameron*, 1979; *Slattery*, 1978]. We cannot rule out this model definitively, but we think it is extremely unlikely for two reasons. First, formation of the Earth from a giant gaseous protoplanet that has migrated inward to 1 AU involves the loss of large amounts of H_2, He, other noble gases, water, and C-bearing compounds. For example, a solar composition giant gaseous protoplanetary Earth (with 14.96% Mg in the "rocky + metallic" portion as in the bulk Earth) would have to lose ~180 Earth masses (M_E) of H_2, ~60 M_E of He, and ~2.7 M_E of Ne, H_2O, N-, and C-compounds (~243 M_E total) to get to its present composition. Second, calculations by *DeCampli and Cameron* [1979] show that a giant gaseous protoplanet rapidly collapses to a state where removal of light elements (H, He) is impossible within the lifetime of the solar nebula. As a result the formation of the Earth by sedimentation of condensed material within the giant gaseous protoplanet leads instead to a Jovian planet rich in H and He [*Slattery*, 1978; *Slattery et al.*, 1980]. It may be interesting to explore the chemistry of giant gaseous protoplanets for application to rocky exoplanets with H_2-bearing atmospheres in other planetary systems. However, these models seem irrelevant to formation of the Earth and thus our discussion assumes the Earth formed at 1 AU and accreted material at about this location.

Figures 2.1 and 2.2 show the chemical equilibrium calculations of *Barshay* [1981] given as condensate abundances versus inverse radial distance (1/R) from the proto-Sun in astronomical units. These calculations are done for a convective solar nebula model with an adiabatic temperature–pressure profile with these characteristics: T_o = 600 K, P_o = 10^{-4} bar, R_o = 1 AU. The variation of temperature with radial distance and pressure with temperature in the midplane of the nebula are given by *Lewis* [1974]

$$\frac{T}{T_o} = \left(\frac{R}{R_o}\right)^{-1.1}$$

$$\frac{P}{P_o} = \left(\frac{T}{T_o}\right)^{C_p/R}$$

To first approximation, the details of this model are irrelevant because (1) the condensation sequence is relatively insensitive to total pressure (see Fig. 1 in *Lewis*, 1974; Fig. 3.8 in *Lodders and Fegley*, 2011) and (2) other models of the solar nebula accretion disk give the same basic result, T and P decreasing with increasing radial distance from the proto-Sun (e.g., Fig. 3.5 in *Lodders and Fegley*, 2011).

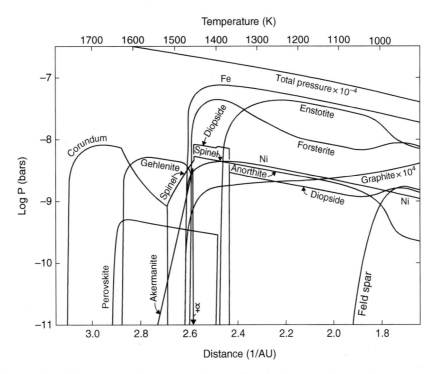

Figure 2.1 Chemical equilibrium abundances of stable condensates in the solar nebula as a function of inverse radial distance (AU^{-1}) along an adiabatic P – T profile. The condensate abundances are the logarithm of each condensate partial pressure (in bar) if the condensate were vaporized to a molecular gas of the same composition. The average orbital distances of the terrestrial planets Mercury, Venus, Earth, and Mars are indicated by their astrological symbols along the bottom of the figure. This figure shows the higher temperature region and Figure 2.2 shows the lower temperature region of the inner solar nebula. Reproduced from *Barshay* [1981] with the permission of Dr. Barshay.

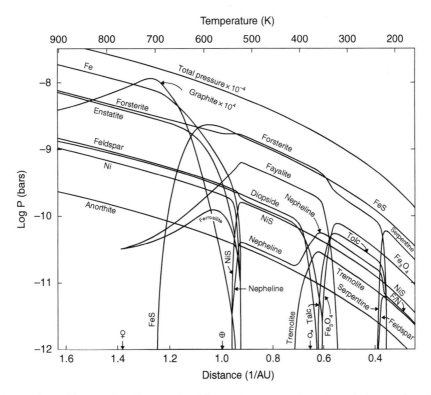

Figure 2.2 Chemical equilibrium abundances of stable condensates in the solar nebula as a function of inverse radial distance (AU^{-1}) along an adiabatic P – T profile. The abbreviation F/N is used where the abundances of feldspar and nepheline are subequal. The total pressure curve gives 10^{-4} times the total pressure at any temperature, e.g., 10^{-4} bar at 600 K. Reproduced from *Barshay* [1981] with the permission of Dr. Barshay.

The major conclusion from Barshay's chemical equilibrium calculations is that the Earth-forming material at 1 AU should have chondritic relative abundances of rock-forming "metals" (e.g., Mg, Si, Fe, S, Ni, Al, Ca, Na, K, Ti) because compounds of these elements are fully condensed at higher temperatures (smaller radial distances) from the proto-Sun (see the discussion of accretion functions below). His calculations did not include other minor and trace elements (e.g., Cr, Mn, P, Co, refractory lithophiles, refractory siderophiles) that are fully condensed at 600 K in the solar nebula but based on other calculations [e.g., *Larimer*, 1967; *Grossman and Larimer*, 1974; *Lewis*, 1972; *Lodders*, 2003], we know which elements are fully condensed by 600 K, 10^{-4} bar, 1 AU. Barshay's main result thus agrees with arguments of other geoscientists that Earth is chondritic to *first approximation* in major element chemistry [e.g., *Wänke*, 1981; *Kargel and Lewis*, 1993; *Palme and O'Neill*, 2014]. His results predict a CI chondritic Mg/Si ratio of 0.90 for the Earth (vs. 0.92 in our model in Table 2.2), which is not the case unless significant amounts of Si are dissolved in the core.

It is likely that the original composition of solids condensed in the solar nebula was altered, to a greater or lesser extent, as accretion proceeded and grains eventually grew into planetesimal-size meteorite parent bodies. Aqueous alteration, dry thermal metamorphism, and shock metamorphism are evident in various types of chondritic meteorites. For example, thermal metamorphism on meteorite parent bodies altered the chondrites in our meteorite collections. Originally, chondritic material was plausibly more volatile-rich than even the most volatile-rich ordinary chondrite samples in meteorite collections today. The larger volatile abundances in the unequilibrated ordinary chondrites (i.e., the H3, L3, LL3 chondrites), which are less metamorphically altered than the grade 4–6 ordinary chondrites, support this argument [*Schaefer and Fegley*, 2007]. However, it is also known that even the extensive aqueous alteration on the CI chondrite parent body did not alter their elemental abundances even though their mineralogy was changed [e.g., *Anders and Grevesse*, 1989; *Lodders*, 2003].

The question of how collisions may have altered planetesimal compositions is important but not yet answered. On the one hand collisional mixing is evident in our meteorite collections through the presence of "meteorites in meteorites"—xenoliths of one type of meteorite enclosed in another [*Wilkening*, 1977]. On the other hand, the "meteorite in meteorite" samples are the exception rather than the rule. Collisions between large bodies, such as the lunar to Mars size planetary embryos that are believed to have formed the terrestrial planets are very energetic. For example, extensive melting and partial vaporization may have occurred on the post Giant Impact Earth [*Lupu et al.*, 2014; *Fegley and Schaefer*, 2014], but the amount of "volatiles" lost from the post-impact Earth is poorly constrained, and may have been small because of Earth's large escape velocity [e.g., see *Genda and Abe*, 2003a, 2005; *Pahlevan*, 2010; *Lupu et al.*, 2014]. Even hydrodynamic escape of vapor from the proto-lunar disk is problematic because of the strong gravity field [*Genda and Abe*, 2003b].

Collisional erosion is another process suggested to alter composition of planetary embryos by loss of the outer layers, perhaps early formed crust, on the embryos [e.g., *O'Neill and Palme*, 2008]. These new models are motivated in part by the ^{142}Nd excess in terrestrial samples [e.g., *Boyet and Carlson*, 2005; *Caro and Bourdon*, 2010], but as argued by *Huang et al.* [2013], the observed excesses are within the range of observed ^{142}Nd variations in chondritic meteorites. Another motivation is the high metal to silicate ratio of the planet Mercury. In this case the geochemical analyses by the MESSENGER spacecraft support the idea that much of its silicate portion was lost by collisional erosion (e.g., see *Cameron et al.*, 1988, for a discussion of alternative models for producing Mercury's high metal to silicate ratio). Excluding the planet Mercury, we conclude that at present we simply do not know how collisions of planetary embryos altered their chemical compositions.

Safronov-type accretion models predict that the terrestrial planets accreted material formed over a range of radial distances in the solar nebula. *Barshay* [1981] modeled this process by convolving his nebular chemistry calculations with simple accretion functions (e.g., square wave, triangular, Gaussian) with variable widths defined by the full width at half maximum (FWHM = 0.06–0.60 AU^{-1}). This may have been the first attempt to combine a model of solar nebula chemistry with a model of planetesimal accretion. *Barshay* [1981] used accretion functions of different geometric shape and size as a proxy for planetary accretion from smaller planetesimals over a range of radial distances (Figures 2.3–2.5). For example, if Earth collected material within a Gaussian accretion function centered at 1 AU^{-1} with FWHM of 0.60 AU^{-1}, its feeding zone stretched from 0.625–2.50 AU (i.e., inward of Venus at 0.723 AU to outward of Mars at 1.52 AU). The smaller width accretion functions simulate smaller feeding zones. The narrowest accretion functions studied correspond to a feeding zone of 0.94–1.06 AU for Earth. It is possible that Earth, Venus, and Mars had different size feeding zones and that the size of the feeding zones of each planet varied with time.

Barshay [1981] compared the zero pressure density, mean atomic weight, and volatile element mass fractions (for water, carbon, and sulfur) for the Earth and other terrestrial planets for different accretion functions of variable width. The zero pressure density is the uncompressed density of the mineralogy for each planet, and

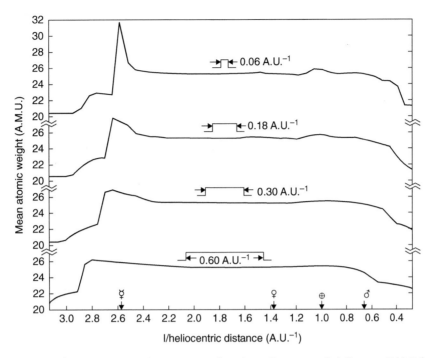

Figure 2.3 The computed mean atomic mass (amu) as a function of inverse radial distance (AU⁻¹) for square wave accretion functions of variable width (see text). The shape of the square wave accretion function used to generate each curve is drawn to scale, and its width is given. The present orbits of the terrestrial planets Mercury, Venus, Earth, and Mars are marked with their astrological symbols along the horizontal scale. Reproduced from *Barshay* [1981] with Dr. Barshay's permission.

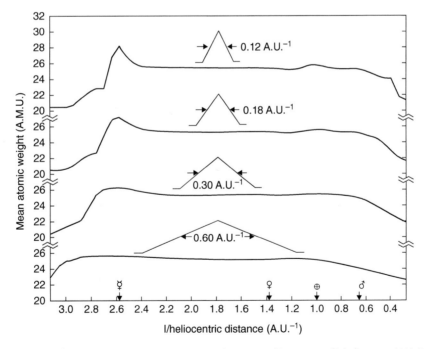

Figure 2.4 The computed mean atomic mass (amu) as a function of inverse radial distance (AU⁻¹) for triangular accretion functions of variable width (see text). The shape of the triangular accretion function used to generate each curve is drawn to scale, and its full width at half maximum (FWHM) is given. The present orbits of the terrestrial planets Mercury, Venus, Earth, and Mars are marked with their astrological symbols along the horizontal scale. Reproduced from *Barshay* [1981] with Dr. Barshay's permission.

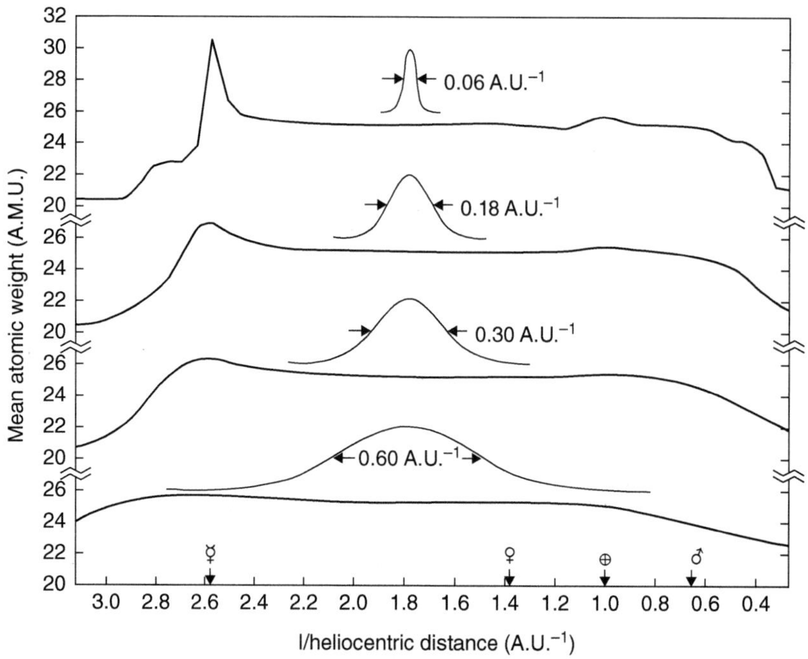

Figure 2.5 The computed mean atomic mass (amu) as a function of inverse radial distance (AU^{-1}) for Gaussian accretion functions of variable width (see text). The shape of the triangular accretion function used to generate each curve is drawn to scale, and its full width at half maximum (FWHM) is given. The wings of the Gaussian accretion functions are drawn out as far as the distance at which ~0.1% of the available mass is accreted into the planet. The present orbits of the terrestrial planets Mercury, Venus, Earth, and Mars are marked with their astrological symbols along the horizontal scale. Reproduced from *Barshay* [1981] with Dr. Barshay's permission.

the mean atomic weight in atomic mass units (amu) is the atom fraction weighted atomic weight of all metal- and rock–forming elements, e.g.,

$$\bar{\mu} = F_O\mu_O + F_{Si}\mu_{Si} + F_{Mg}\mu_{Mg} + F_{Fe}\mu_{Fe} + F_S\mu_S + F_{Al}\mu_{Al} + F_{Ca}\mu_{Ca} + \cdots$$

The general equation for computing the mean atomic weight is thus

$$\bar{\mu} = \sum_i F_i\mu_i$$

The index "i" represents an element, and the summation is carried out for all elements in the calculations. For example, the mean atomic weight of the bulk Earth models in Table 2.2 are 25.2 (MF14), 25.9 [*Kargel and Lewis*, 1993], 26.2 [*McDonough*, 2003], and 26.3 [*Morgan and Anders*, 1980]. The BSE model of *Palme and O'Neill* [2014] has a mean atomic weight of 21.1 amu. Figure 2.5 shows calculated mean atomic weights of ~25.2 to ~25.7 for the widest to narrowest Gaussian accretion functions, respectively. These values fall within the range of mean atomic weights for our bulk Earth model (MF14) and *Kargel and Lewis* [1993]. Higher mean atomic weights,

like those from the *McDonough* [2003] and *Morgan and Anders* [1980] bulk Earth models, are produced in the square wave and triangular accretion function calculations in Figures 2.3 and 2.4.

For comparison the mean atomic weights (in amu) of some of the major chondrite groups are 15.0 (CI), 17.2 (CM), 22.4 (CV), 24.0 (CO), 24.7 (H), 23.3 (L), 22.7 (LL), and 25.9 (EH). The values for the EH and H chondrites are closest to those for the bulk Earth models in Table 2.2 and in Figures 2.3–2.5. The variations in the mean atomic weight of the different chondrite groups are related to their volatile (water, C, N) content (e.g., compare CI vs. CM, CV, CO chondrites) and Fe content (e.g., compare H, L, LL chondrites).

Another important conclusion from *Barshay* [1981] is that the chemical compositions (i.e., mean atomic weight) of the terrestrial planets Venus, Earth, and Mars are similar to one another because their accretion zones overlapped one another in the solar nebula. These results are consistent with prior ideas that the Earth formed from a mixture of nebular materials (e.g., the two component models of *Ringwood* [1978] and *Wänke* [1981]). It would be interesting to update his work using more sophisticated models of planetary accretion.

2.3. BSE AND BULK EARTH COMPOSITION

2.3.1. Similarity to Chondritic Meteorites (Chondrites)

The overall composition of the BSE (in particular see Table 1 and Fig. 2 in *Palme and O'Neill, 2014*) and bulk Earth [e.g., *Larimer*, 1971; *Kargel and Lewis*, 1993] is close to chondritic, i.e., the relative proportions of the major rock-forming elements (Mg, Si, Fe, S, Al, Ca) are close to those found in chondritic meteorites (chondrites). Before discussing this point in more detail, we present some background information about chondritic meteorites.

Chondrites are the most abundant kind of meteorites. They are unmelted stony meteorites that contain Fe-Ni metal alloy, silicate, sulfide, and oxide minerals in variable proportions. The chondrites are the oldest objects in the solar system with ages up to ~4.567 Ga [e.g., *Amelin et al.*, 2002; *Bouvier et al.*, 2007; *Jacobsen et al.*, 2008]. The relative abundance of rock-forming elements in chondrites is close to their relative abundances in the solar photosphere [e.g., *Anders and Grevesse*, 1989; *Lodders*, 2003; *Palme et al.*, 2014; *Suess and Urey*, 1956]. Atmophile elements such as H, C, N, O, and noble gases are depleted relative to solar composition, and the light elements Li and B are enriched relative to solar composition (because they are destroyed by thermonuclear reactions in the Sun), but otherwise there is a close match between the composition of chondrites and that of the Sun. The five CI carbonaceous chondrites (Alais, Ivuna, Orgueil, Revelstoke, and Tonk), all of which are observed falls, are the closest in composition to the solar photosphere [*Anders and Grevesse*, 1989; *Lodders*, 2003; *Palme et al.*, 2014].

The major different classes of chondrites, and the major groups within each class, are (1) carbonaceous chondrites (CI, CM, CO, CV, CK, CH, CB), comprising 3.1% by number of all chondrites, (2) ordinary chondrites (H, L, LL), which are 96.2% of all known chondrites, (3) enstatite chondrites (EH, EL), making up 0.6% of all chondrites, and (4) other chondritic classes (R, K) that are <0.1% of all chondrites [*Krot et al.*, 2014]. The different classes of chondrites are distinguished by different metal/silicate abundances, lithophile element ratios (e.g., Mg/Si, Al/Si, Na/Si), redox state, and oxygen isotopic (16, 17, 18) composition. All chondrites have been heated to varying extents, and chondrite metamorphic grades are distinguished by various chemical and petrographic criteria such as presence or absence of glass, homogeneity of ferromagnesian silicates, abundance of volatile elements and presolar grains [*McSween et al.*, 1988; *Huss et al.*, 2006].

As mentioned above, the chondrites are the most abundant kind of meteorites, and achondrites (melted stony meteorites), stony – iron meteorites, and iron meteorites are less abundant. As of the time of writing (June 2014), there are 44,332 known chondritic meteorites out of 47,520 meteorites of all kinds. Strictly speaking these statistics only give the abundance of meteoritic material in orbits that intersect the Earth and not the abundance of meteoritic material formed in the solar nebula and early solar system or existing in the asteroid belt today [e.g., *Cloutis et al.*, 2014]. We return to this question in the next section.

2.3.2. Chondritic Material

We emphasize that chondrites are not *pristine* samples of nebular material because the compositions of chondrites have been more or less altered by dry thermal metamorphism and/or aqueous alteration on their meteorite parent bodies [*McSween et al.*, 1988; *Huss et al.*, 2006]. As discussed in the rest of this chapter, we are interested in determining the type (or types) of chondritic material that was the building block(s) of the Earth. Following *Schaefer and Fegley* [2007, 2010a] and *Fegley and Schaefer* [2014], we use the term "chondritic material" to refer to the pristine nebular material that formed chondrites. Thus, carbonaceous chondritic material was the nebular precursor to carbonaceous chondrites, ordinary chondritic material was the nebular precursor to ordinary chondrites, and enstatite chondritic material was the nebular precursor to enstatite chondrites. Using this nomenclature, we do not have samples of chondritic material in our meteorite collections. Interplanetary dust particles (IDP) are another analog to the chondritic material that existed in the solar nebula, but their bulk chemical and isotopic compositions are not as well known as that of meteorites. Perhaps in the future returned samples of relatively unaltered interior regions of comets will provide a closer approach to the original composition of some types of chondritic material. At the present time chondritic meteorites are the best samples that we have available, hence we use them in our discussion.

2.3.3. Preferred Bulk Earth Model

Table 2.2 gives our composition (MF14) for the bulk Earth based on BSE and core compositions given in the literature [e.g., *Badro et al.*, 2014; *Lodders and Fegley*, 1998, 2011; *McDonough*, 2003; *Palme and O'Neill*, 2014] and from arguments in this paper. It is calculated using equations analogous to the one for the bulk earth concentration of Mg, i.e., a weighted average of BSE and core concentrations of each element. Our bulk Earth composition uses the best-fit core composition of *Badro et al.* [2014, and solar Fe/Ni and Fe/Co ratios in the core. The bulk Earth oxygen abundance is identical to the arithmetic mean value for H-chondrites (within 1σ uncertainty), but the Mg/Si ratio is much higher than that for H-chondrites (0.82) and is identical to that for the more oxidized CO chondrites.

This may not be unexpected because *Hart and Zindler* [1986] emphasized that "the Earth is not like any chondrite (in major elements) but is composed of its own blend of accretion products, with a bias toward a higher proportion of the high temperature refractory components (or equivalently, a lower proportion of the volatiles and partially refractory components)."

Also, if Earth sampled material from a range of radial distances in the solar nebula, as physical accretion models predict [e.g., see the review by *Chambers*, 2014], it will contain material of variable elemental composition. For example, the amount of oxidized Fe ($FeSiO_3$ in pyroxene, Fe_2SiO_4 in olivine, and magnetite) increases with decreasing temperature in the solar nebula (see Figures 2.1 and 2.2).

Badro et al. [2014] also found a core model containing 5.4% O; remainder Fe-Ni alloy also fit the seismic velocity and density data. In this case we compute a bulk Earth model with ~35.3% O, 14.3% Si, 29.8% Fe, 1.68% Ni, and otherwise identical to that shown in Table 2.2. The oxygen abundance for this model is intermediate between the average values for H (33.86%) and L chondrites (36.93%), but the Mg/Si ratio of 1.05 is significantly higher than that of H (0.82) and L (0.80) chondrites or any of the major chondrite groups.

2.4. CHONDRITIC VS. ACHONDRITIC EARTH

Larimer [1971] considered whether the Earth is chondritic or achondritic (like eucrites) and concluded it is chondritic to first order. His arguments are worth revisiting because dating by three relative chronometers (^{182}Hf–^{182}W, ^{26}Al–^{26}Mg, and ^{53}Mn–^{53}Cr) indicates that the

EHD (eucrite – howardite – diogenite, presumably from the asteroid 4-Vesta) meteorite parent body differentiated during the first 1–10 Ma of solar system history [e.g., *Jacobsen et al.*, 2008; *Schiller et al.*, 2011; *Trinquier et al.*, 2008]. The EHD meteorites are the major kind of achondrites (51% of all achondrites by number). Lead–lead (^{207}Pb–^{206}Pb), ^{26}Al–^{26}Mg, and ^{53}Mn–^{53}Cr dating also indicate that the parent body of the Asuka 881394 basaltic meteorite formed and differentiated within 3 Ma of solar system formation [*Wadhwa et al.*, 2009]. Consequently, Earth may have accreted a mixture of chondritic and achondritic material during its formation.

Table 2.3 compares abundances of volatile lithophile elements in the bulk silicate Earth, average ordinary chondrites, and the silicate portion of the eucrite parent body, here called bulk silicate Vesta BSV. The bulk silicate Vesta models are described in the references cited. Table 2.3 shows that the bulk silicate Earth is depleted in volatile lithophiles relative to ordinary chondrites by factors of 0.04 (for C and N) to 0.66 for F. Zinc, which behaves both as a lithophile and chalcophile, is not depleted relative to ordinary chondrites in the BSE. The large depletions for C and N in the BSE may also reflect their partial sequestration in the Earth's core as both elements occur in meteoritic metal and behave as siderophiles under some conditions. We do not give H contents for BSV because no reliable data are available. *Sarafian et al.* [2013, 2014] report analyses of OH-bearing apatite in eucrites, but the H_2O content of the parental melts cannot be determined uniquely from apatite analyses [*Boyce et al.*, 2014]. The depletions for the alkalis (Na 0.37, K 0.30, Rb 0.29, Cs 0.30) are well known [e.g., *Gast*, 1960; *Hurley*, 1968; *Larimer*, 1971] although the BSE data of *Palme and O'Neill* [2014] correspond to larger alkali abundances (i.e., smaller depletions) than estimated

Table 2.3 Volatile Elements in Bulk Silicate Earth, Bulk Silicate Vesta, and Ordinary Chondrites

Element	BSV (ppm)[a]	BSE (ppm)[b]	BSV/BSE	Chondrites[d]	BSE/chondrites
H	?	120	?	470	0.26
C[c]	10–30	100	0.1–0.3	2570	0.04
N[c]	0.05–1.3	2	0.025–0.65	54	0.04
Zn	0.36	53.5	0.0067	53.2	1.0
Na	757	2590	0.29	6950	0.37
K	66	260	0.25	860	0.30
Rb	0.06	0.605	0.10	2.1	0.29
Cs	0.002	0.018	0.11	0.06	0.30
F	4.8	25	0.19	38	0.66
Cl	4.6	30	0.15	93	0.32
Br	0.03	0.075	0.40	0.5	0.15

[a] BSV values are mainly from *Dreibus and Wanke* [1980] and *Dreibus et al.* [1997].
[b] BSE values are from *Palme and O'Neill* [2014].
[c] Carbon and nitrogen values for BSV are eucrite data from *Grady and Wright* [2003].
[d] Average ordinary chondrite data are from *Schaefer and Fegley* [2007, 2010].

earlier. The explanations proposed for the observed alkali depletions in the BSE include (1) nebular fractionation of proto-Earth material, (2) evaporative loss during accretion, or (3) sequestration into Earth's core [e.g., see *Palme and O'Neill*, 2014; *Kargel and Lewis*, 1993; *Lodders*, 2000]. The first model implies that all elements with similar volatility will be equally fractionated. This is not the case because Zn is much less depleted in the BSE than Rb or Cs, which have similar 50% condensation temperatures [*Lodders*, 2003]. The second model requires loss of vapor from Earth's gravity field either by thermal escape or physical blow-off. The third model requires partitioning into Earth's core during its formation, and there is little experimental evidence for this [e.g., *Corgne et al.*, 2007]. Whatever the cause(s) for alkali depletion in the BSE may be, Table 2.3 shows that volatile lithophile element depletions in bulk silicate Vesta are significantly larger than in the BSE. The volatile lithophile depletions on Vesta are presumably either inherited from the material accreted by Vesta or due to heating of Vesta. In any case, achondritic material like the EHD meteorites cannot supply the terrestrial abundances of carbon, nitrogen, halogens (Cl, Br), Zn, or alkalis (Na, K, Rb, Cs). Other types of achondrites are also depleted in volatile lithophiles relative to the BSE (e.g., angrites, mesosiderites, pallasites, and ureilites). The enstatite achondrites (aubrites) contain more volatile lithophiles than other achondrites and are an exception.

It is unlikely for at least two reasons that the Earth accreted *only* achondritic material during its formation. First, if the Earth did so, then its oxygen isotopic composition would be identical to that of achondritic material (e.g., the eucrites). This is not the case [e.g., see Fig. 1 of *Lodders and Fegley*, 1997]. Second, chondritic material is significantly more abundant in our meteorite collections and in the asteroid belt than achondritic material [e.g., *Cloutis et al.*, 2014]. It would be difficult to believe that the Earth accreted *no* chondritic material. Instead it is more plausible that Earth accreted a mixture of achondritic and chondritic material. Third, the Earth has so many isotopic similarities to enstatite chondritic material that it seems likely that enstatite chondrites and the Earth are somehow related to one another. This could not happen if the Earth formed solely from achondritic material. For example, oxygen isotope mixing models [*Lodders*, 1991, 2000; *Lodders and Fegley*, 1997] predict that E-chondritic material is an important building block of the Earth, Mars, and 4-Vesta.

It is possible that accretion of volatile–poor achondritic material could be balanced by accretion of more volatile–rich chondritic material. In principle the relative mass fractions of achondritic and chondritic components can be constrained by an oxygen isotopic mixing model [e.g., *Lodders*, 1991, 2000; *Lodders and Fegley*, 1997], but this is beyond the scope of this review chapter because the abundances of all natural elements, not only volatile lithophiles, need to be considered simultaneously in the modeling.

2.5. ISOTOPIC ARGUMENTS

Until the early 1970s, it was assumed that the hot solar nebula was chemically and isotopically homogeneous; however, the discoveries of isotopic anomalies in refractory inclusions of meteorites, which cannot be explained by radioactive decay, cosmogenic effects, or mass fractionations, have lead to revisions of this concept [*Clayton et al.*, 1973]. Planetary processing such as core formation or magmatic differentiation can only produce mass-dependent isotopic fractionations and therefore do not modify isotopic anomalies, so terrestrial surface samples can be considered isotopically representative of the bulk Earth for comparison with meteoritic materials.

Following the discovery of oxygen isotopic variations among meteorites and planetary materials, isotopic variability has been found for other elements, and combining these different isotopic systems is one of the best tracers to determine the Earth's building blocks.

Isotopic variability between solar system materials can be accounted for by few mechanisms:

• Stable isotope fractionations, which originate from vibrational energy partitioning between coexisting phases at equilibrium or under kinetic conditions [e.g., *Bigeleisen and Mayer*, 1947]. At first order, this process is proportional to the mass-difference between the isotopes; however, non-mass-dependent effects have been observed in some systems [e.g., *Thiemens*, 1999; *Fujii et al.*, 2006ab; *Moynier et al.*, 2013].

• Nuclear reactions such as the decay of long-lived (e.g., ^{87}Rb-^{87}Sr) or short-lived (e.g., ^{26}Al-^{26}Mg) radioactive isotopes or spallation by cosmic rays or solar wind.

• Inheritance of isotopic heterogeneities, which originated from an incomplete mixing of nucleosynthetic products. These effects are observed after correction of isotopic ratios for stable isotope fractionation, radioactive decay, and spallation effects.

To identify which group(s) of meteorites (if any) represents the composition of the Earth, nucleosynthetic anomalies, radiogenic isotopes produced by short-lived radioactivity (e.g., ^{142}Nd) and stable isotope fractionations (e.g., Si isotopes) can be used. In the following section we will present these different arguments.

2.5.1. Applications of Isotopic Anomalies

Since isotopic anomalies cannot be modified by planetary differentiation they are usually considered to be very robust tracers of the Earth's building blocks.

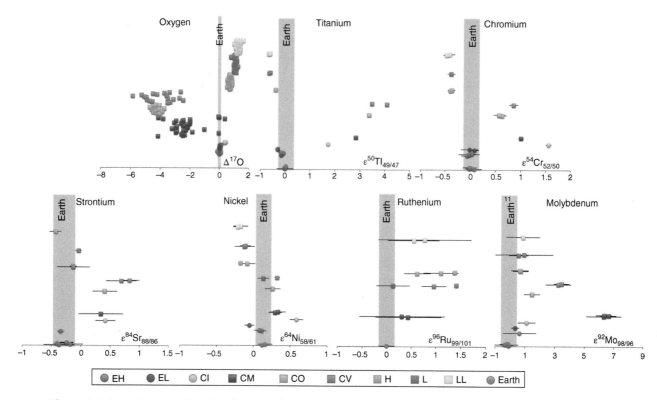

Figure 2.6 Isotopic anomalies for elements that show difference between different meteorite groups. $\Delta^{17}O = \delta^{17}O - 0.52\delta^{18}O$. The $\varepsilon^a X_{b/c}$ corresponds to the per ten thousand deviation of the ratio $^a X/^c X$ internally normalized to the $^c X/^c X$ ratio (see main text for details). Data are from *Clayton and Mayeda* [1991, 1996, 1999], *Burkhardt et al.* [2011], *Chen et al.* [2010], *Zhang et al.* [2012], *Trinquier et al.* [2009], *Moynier et al.* [2012], *Steele et al.* [2012].

Among the different meteorite groups, the Enstatite (E) chondrites are the closest to the terrestrial mantle in term of isotopic composition (see Figure 2.6). Isotopic anomalies are either reported as the deviation from the mass-fractionation line in a three-isotope diagram (the so-called Δ notation [e.g., $\Delta^{17}O$]) or using the ε–notation, which represents the per ten thousand deviation of an isotopic ratio normalized to another isotopic ratio of the same element. Therefore, isotopic anomalies can only be investigated for elements with at least three isotopes.

Elements for which isotopic anomalies have been found in one or more groups of meteorites are oxygen [e.g., *Clayton*, 1993], titanium [*Trinquier et al.*, 2009; *Zhang et al.*, 2012], calcium [*Simon et al.*, 2009; *Moynier et al.*, 2010a], chromium [e.g., *Trinquier et al.*, 2007; *Qin et al.*, 2010], nickel [e.g., *Regelous et al.*, 2008; *Steele et al.*, 2012], strontium [*Moynier et al.*, 2012], zirconium [*Akram et al.*, 2013], molybdenum [e.g., *Dauphas et al.*, 2002; *Yin et al.*, 2002; *Burkhardt et al.*, 2011], ruthenium [*Chen et al.*, 2010], barium [*Carlson et al.*, 2007, *Andreasen and Sharma*, 2007, *Ranen et al.*, 2007], neodymium [*Carlson et al.*, 2007], and samarium [*Andreasen and Sharma*, 2007]. The elements for which all solar system materials have similar isotopic composition within error are zinc [*Moynier et al.*, 2009], iron [*Wang et al.*, 2011; *Tang and*

Dauphas, 2012], tellurium [*Fehr et al.*, 2006], osmium [*Yokoyama et al.*, 2007], silicon [*Pringle et al.*, 2013], and hafnium [*Sprung et al.*, 2010]. The absence of isotopic anomalies for some elements but presence of anomalies for others is not completely understood; to date the best explanation suggests a decoupling between the phases carrying the elements with anomalies and those without, which are homogenized to different degrees.

Most elements show a similar isotopic composition between the Earth (estimated by measurements of mantle rocks) and enstatite chondrites (EH and EL) (see Figure 2.6). This isotopic coincidence has been used to propose a strong genetic link between the Earth and enstatite chondrites, and in some cases, to use enstatite chondrites as the main building blocks for the Earth [*Javoy et al.*, 1995; *Javoy et al.*, 2010; *Jacobsen et al.*, 2012]. The main problem with making the Earth from enstatite chondrites comes when considering the chemical composition (see above); among other dissimilarities, they have insufficient FeO, low Mg/Si ratio, and overabundant volatile elements. In order to reconcile the chemistry of enstatite chondrites with what we know of the terrestrial mantle, *Javoy et al.* [2010] proposed that enstatite chondrites were devolatilized prior to accreting to form the Earth (to account for the depletion of the

Earth in volatile elements) and that the Earth is chemically heterogeneous, and the remainder of the chemical differences are stored in the lower mantle. Alternatively, *Jacobsen et al.* [2012] proposed that the material that formed the Earth and the E-chondrites had a common nebular precursor, but a secondary event modified the chemical composition of the enstatite chondrites without altering the isotopic composition.

It must be noted that three recent studies have shown that the terrestrial isotopic composition of Mo [*Burkhardt et al.*, 2011], Ti [*Zhang et al.*, 2012], and O [*Herwartz et al.*, 2014] do not exactly match E-chondrites and the Earth. Although *Zhang et al.* [2012] do not discuss this possible difference in Ti isotopes between E-chondrites and the Earth, *Burkhardt et al.* [2011] propose that no known type of chondrites are representative of the bulk Earth for Mo isotopes. However, when the analytical errors are taken into account (see Figure 2.6), EHs are similar to Earth for both Ti and Mo isotopes, while ELs are slightly different. In addition, only one EH and one EL have been analyzed for Mo isotopic composition, and more data are needed to resolve this issue.

On the other hand, the recent high precision oxygen isotopes of *Herwartz et al.* [2014] show a clear $\Delta^{17}O$ difference between the Earth, the Moon (12 ppm heavier than Earth), and the E-chondrites (60 ppm heavier than Earth). This represents a major observation that suggests that the Earth cannot be formed from E-chondrites alone as developed above based on major element composition and as we will developed below based on isotope ratio.

Another line of interpretation of the isotopic anomalies is that the Earth represents a mixing between different types of chondrites [*Lodders*, 2000; *Warren et al.*, 2011]. *Lodders* [2000] used O isotopic compositions to calculate the best mixing model in order to account for the composition of the Earth and obtained 70% EH, 21% H, 5% CV, and 4% CI. Following the same logic, *Fitoussi and Bourdon* [2012] expanded this work by combining multiple isotopic systems (O, Cr, and Ni isotopes). They found that a good mixing relation would involve 63% LL, 16% CI, and 23% CO.

Mixing different types of chondrites can clearly reproduce the isotopic composition of the Earth. The very close proximity between E-chondrites and the Earth suggest that these meteorites may have close connection to the Earth in terms of their isotopic composition but does not simply prove that the Earth has been made of E-chondrite materials.

2.5.2. Radiogenic Isotopes

The discovery of a 20-ppm excess of the $^{142}Nd/^{144}Nd$ in terrestrial rocks compared to any chondrite group had fundamental consequences on our understanding of the Earth's mantle [*Boyet and Carlson*, 2005]. ^{142}Nd is produced by the radioactive decay of ^{146}Sm (half life of 103 Ma). This is a short-lived radioactive system that can only record Sm/Nd fractionation that occurred during the first 500Ma of Earth's history. This terrestrial excess in ^{142}Nd has been interpreted as the result of the decay of ^{146}Sm in a suprachondritic Sm/Nd reservoir, which formed by partial melting of the Earth's primitive mantle and is referred to as the early-depleted reservoir (see Chapter 8). Its counterpart, the early enriched reservoir, with a low Sm/Nd ratio and a depletion in ^{142}Nd (compared to chondrites), would have been either incorporated within the Earth's mantle and not recorded in geological samples (a hidden reservoir, *Boyet and Carlson* [2005]) or lost to space following an asteroid impact (impact erosion, e.g., *O'Neill and Palme* [2008]; *Caro and Bourdon* [2010]), both scenario would have to occur within <75Myrs after Earth formation [*Boyet and Carlson*, 2005; *Moynier et al.*, 2010b]. A hidden reservoir would mean that the Earth's mantle has a chondritic composition for the refractory lithophile elements, whereas an impact erosion loss of an early-enriched reservoir would imply that the bulk silicate Earth does not have a chondritic abundance of lithophile elements.

2.5.3. Stable Isotope Fractionation

Stable mass-dependent isotopic variations between meteorites and planetary bodies are attributable to processes that fractionated isotopes during nebular events [e.g., *Luck et al.*, 2003; *Moynier et al.*, 2006] or during planetary events such as core formation [e.g., *Georg et al.*, 2007; *Moynier et al.*, 2011; *Savage et al.*, 2015]. Among the major elements, only O, Si, and Ca show mass-dependent isotopic fractionation between different chondrite groups and can potentially be used as tracers of the Earth's building blocks [e.g., *Clayton et al.*, 1993; *Fitoussi et al.*, 2011; *Valdes et al.*, 2014]. Other multi-isotopic major elements such as Mg [e.g., *Teng et al.*, 2007] and Fe [e.g., *Poitrasson et al.*, 2004; *Craddock and Dauphas*, 2011; *Wang et al.*, 2014] do not show any mass-dependent isotopic variations between chondrite groups and therefore are not useful to investigate the Earth's building blocks and will not be discussed here.

A small heavy Si isotopic enrichment in Bulk Silicate Earth (BSE) compared to chondrites has been interpreted as evidence for the incorporation of ~4–7 wt.% silicon as a light element into the Earth's core [*Georg et al.*, 2007; *Fitoussi et al.*, 2009; *Savage and Moynier*, 2013; *Armytage et al.*, 2011]. *Fitoussi and Bourdon* [2012] used the observation that enstatite chondrites are significantly lighter than Earth in Si isotope compositions to argue that the BSE cannot consist of more than 15% of enstatite chondrite precursor material. Their logic was that since the

BSE has a ^{30}Si/^{28}Si ratio ~0.30 per mil heavier than the E-chondrites, unrealistically large amounts of Si (>20 wt.%) would need to have entered the Earth's core if the bulk Earth had E-chondritic Si isotope compositions; therefore, the density constraints placed on the amount of light elements in the Earth's core coupled with the fact that E-chondrites are Si-rich significantly limits the amount of E-chondritic material as building blocks of Earth. On the other hand, *Savage and Moynier* [2013] have shown that the light Si isotopic composition of the E-chondrites is carried by the metal phases, and enstatite minerals have a similar Si isotopic composition to the carbonaceous chondrites. They argued that E-chondrites could not be the building blocks of the Earth, but materials enriched in refractory siderophile elements that condensed in the same region as the E-chondrites could be a major component of the Earth. In addition, Si isotope composition of angrites is terrestrial-like, which cannot be explained by core formation scenario (because of the high fO2 of the angrites), and most probably reflects isotope fractionation during impact-induced evaporation [*Pringle et al.*, 2014]. These results imply the bulk Earth Si isotopic composition may have been fractionated not only by core formation but also by evaporation. This would lower the Si content required in Earth's core based on the Si isotopic offset between Earth's mantle and bulk Earth [*Pringle et al.*, 2014].

Calcium isotopes are potentially very useful tracers of the Earth's building blocks, but until recently there were no high precision Ca isotope data available. (*Simon and DePaolo* [2011] showed that different chondrite groups have unique stable Ca isotopic compositions. (See Figure 2.7.) The carbonaceous chondrites are isotopically lighter and the E-chondrites are heavier than BSE. On the other hand, the ordinary chondrites were isotopically similar to BSE. This led *Simon and DePaolo* to exclude E-chondrites and C-chondrites as representative building blocks for the Earth. However, a more recent study has shown that although most C-chondrite groups are isotopically lighter than the Earth, the CO chondrites are similar to Earth in Ca isotopes [*Valdes et al.*, 2014, see Figure 2.7]. In addition, this new study has shown that the E-chondrites are similar to the Earth, and therefore it is not possible to exclude the E- chondrites as building blocks of Earth based on Ca isotopes [*Valdes et al.*, 2014; *Huang et al.*, 2012].

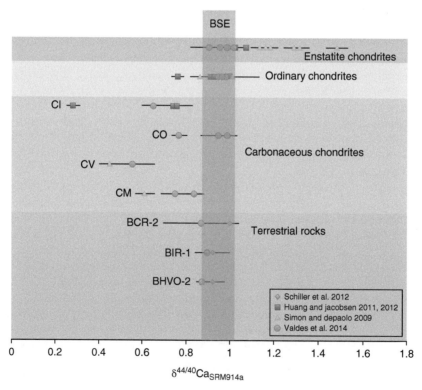

Figure 2.7 Ca isotope composition of various group of meteorites measured by different groups. The data are reported as the permil deviation of the ^{44}Ca/^{40}Ca ratio compared to the SRM914a standard. Contrary to previously proposed, E chondrites have very similar Ca isotope composition with terrestrial rocks (such are ordinary chondrites, and CO chondrites). Data are from *Valdes et al.* [2014], *Simon and DePaolo* [2010], *Schiller et al.* [2012], and *Huang and Jacobsen* [2011, 2012].

In conclusion, stable isotopes are not as powerful tracers for the Earth's building block as nucleosynthetic anomalies since their abundances can be modified by planetary processes. However, it seems clear that based on Si isotopes, it is improbable to form the Earth from pure E-chondrites since this would require amounts of Si in the core in excess of 20 wt.%. It should be noted that a recent theoretical study [Huang et al., 2014] has shown that Si isotopes might be fractionated within the mantle. Due to the large difference in pressure within the Earth's mantle, high-pressure minerals containing Si (e.g., perovskite-structured silicate) have a different structure than those at lower pressure (e.g., olivine or ringwoodite). This phenomenon is known as a phase transition and results in a change in atomic bonding or coordination. The calculations of *Huang et al.* [2014] indicate that Si isotopes may be fractionated between silicate minerals with different coordinations of Si. If these results are correct, they could have important implications for the composition of the core and the conditions of formation of the Earth. Specifically, even a small fractionation of Si isotopes in the mantle could drastically decrease the need for large quantities of Si in the Earth's core to account for the difference between the measurable upper mantle and chondrites in Si isotope composition. This would change the conclusions regarding the building blocks of the Earth based on the Si isotopes. Since these isotopic fractionation results are only theoretical, this effect must be tested experimentally before strong conclusions can be made. In addition, *Pringle et al.* [2014] have found that angrite meteorites, which are highly oxidized achondrites, have the same Si isotopic composition as the Earth and therefore are different from chondrites. Since these meteorites are oxidized it is very unlikely that the core of the angrite parent body is very rich in Si. These results suggest that other mechanisms of isotopic fractionation (e.g., volatilization) may fractionate Si isotopes in the early solar system, and that Si isotopes might not be such a strong argument to exclude enstatite chondrites from the Earth's building blocks.

2.6. CONCLUSIONS

In this review chapter we propose a new estimate for the bulk chemical composition of the Earth by combining seismological observations, chemical analyses of the atmosphere, biosphere, hydrosphere, crust, and mantle, experimental results of elemental partitioning, ratios of elements with very similar geochemical behavior, and recent theoretical estimates of the core composition by molecular dynamics simulations of density and sound speed (Table 2.2). High-precision isotopic studies are fundamental in establishing compositional relationships between meteorites and the Earth. We show that non-mass-dependent isotopic variations, which are not modified by planetary processes, are the best tracers to search for the genetic origin of the Earth since crustal rocks can be directly compared to meteorites. Enstatite chondrites are the meteorites isotopically closest to Earth. However, they are chemically very different from the estimate of the bulk Earth composition (see Table 2.2), which suggests that either (1) the Earth is not made of any known meteorites or from the mixing of different groups of meteorites in proportions that would fit the isotopic composition of the Earth, or (2) the Earth and enstatite chondrites share a common ancestor but their chemical evolution diverged early in the Solar System history.

ACKNOWLEDGMENTS

B. F. was supported by the NASA EPSCOR Program and by NASA Cooperative Agreement NNX09AG69A with the NASA Ames Research Center. FM thanks the European Research Council under the European Community's H2020 framework program/ERC grant agreement #637503 (Pristine) and the Agence Nationale de la Recherche for a chaire d'Excellence Sorbonne Paris Cité (IDEX13C445) and for the UnivEarthS Labex program (ANR-10-LABX-0023 and ANR-11-IDEX-0005-02). We thank K. Lodders, M. Moreira, P. Savage, J. Siebert, M. Chaussidon, and E. Pringle for discussions and the referees (Rick Carlson and anonymous) for their comments, which led to some of our stronger and clearer statements.

REFERENCES

Akram, W., M. Schonbachler, P. Sprung, and N. Vogel (2013), Zircon-hafnium isotope evidence from meteorites for the decoupled synthesis of light and heavy neutron-rich nuclides. *Astrophys. J.*, 777, 169.

Allègre, C. J., J. P. Poirier, E. Humler, and A. W. Hofmann (1995), The chemical composition of the Earth. *Earth Planet. Sci. Lett.*, 134, 515–526.

Allègre, C. J., A. Hofmann, and K. O'Nions (1996), The argon constraints on the mantle structure. *Geophys. Res. Lett.*, 23, 3555–3557.

Albarede, F. (1998), Time-dependent models of U-Th-He and K-Ar evolution and the layering of mantle convection. *Chem. Geol.*, 145, 413–429.

Amelin, Y., A. N. Krot, I. A. Hutcheon, and A. A. Ulianov (2002), Lead isotopic ages of chondriles and Calcium Aluminium Rich inclusions. *Science*, 297, 1678–1683.

Anders, E. and Grevesse, N. (1989), Abundance of the elements: meteoric and solar. *Geochimica et Cosmochimica Acta.*, 53: 197–214.

Anderson, D. L. (2002), The case for irreversible chemical stratification of the mantle. *Intl. Geol. Rev.*, 44, 97–116.

Andreasen, R. and M. Sharma (2007), Mixing and homogenization in the early solar system: clues from Sr, Ba, Sm, and Nd isotopes in meteorites. *Astrophys. J.*, 665, 874–883.

Armytage, R. M. G., A. G. Jephcoat, M. A. Bouhifd, and D. Porcelli (2013), Metal–silicate partitioning of iodine at high pressures and temperatures: Implications for the Earth's core and [129]*Xe budgets. *Earth Planet. Sci. Lett., 373,* 140–149.

Armytage, R. M. G., R. B. Georg, P. S. Savage, H. N. Williams, and A. N. Halliday (2011), Silicon isotopes in meteorites and planetary core formation. *Geochim. Cosmochim. Acta., 75,* 3662–3675.

Badro, J., A. S. Côté, and J. P. Brodholt (2014), A sesimologically consistent compositional model of Earth's core. Proc Natl. Acad Sci USA, doi/10.1073/pnas.1316708111.

Ballentine, C. J. and G. Holland (2008), What CO_2 well gases tell us about the origin of noble gases in the mantle and their relationship to the atmosphere. Phil. Trans. Roy. Soc. London, *366A*: 4183–4203.

Barshay, S. S. (1981), Combined Condensation–Accretion Models of the Terrestrial Planets. PhD Thesis, MIT, Cambridge, MA.

Bigeleisen, J. and M. G. Mayer (1947), Calculation of equilibrium constants for or isotopic exchange reactions. *J. Chem. Phys., 15.* 261–267.

Bina, C. R. and G. Helffrich (2014), Geophysical constraints on mantle composition. Chapter 3.2 in Treatise on Geochemistry, 2nd ed., Elsevier.

Birch, F. (1952), Elasticity and constitution of the Earth's interior. *J. Geophys. Res., 57,* 227–286.

Birch, F. (1964), Density and composition of mantle and core. *J. Geophys. Res., 69,* 4377–4388.

Bouhifd, M. A., A. P. Jephcoat, V. S. Heber, and S. P. Kelley (2013), Helium in Earth's early core. *Nature Geosci., 6,* 982–986.

Bouhifd, M. A., D. Andrault, N. Bolfan-Casanova, T. Hammouda, and J. L. Devidal (2013), Metal–silicate partitioning of Pb and U: Effects of metal composition and oxygen fugacity. *Geochim. Cosmochim. Acta., 114,* 13–28.

Bouvier, A., J. Blichert-Toft, F. Moynier, J. Vervoort, and F. Albarède (2007), New Pb-Pb ages relevant to the accretion and cooling history of chondrites. *Geochimica et Cosmochimica Acta., 71,* 1583–1604.

Boyce, J. W., S. M. Tomlinson, F. M. McCubbin, J.P. Greenwood, and A. H. Treiman (2014), The lunar apatite paradox. *Science., 344,* 400–402.

Boyet, M. and R. Carlson (2005), [142]Nd evidence for early (>4.53Ga) global differentiation of the silicate Earth. *Science, 309,* 575.

Brock, D. S. and G. J. Schrobilgen (2011), Synthesis of the missing oxide of xenon, XeO_2, and its implications for Earth's missing xenon. *J. Am. Chem. Soc., 133,* 6265–6269.

Burkhardt, C., T. Kleine, F. Oberli, A. Pack, B. Bourdon, and R. Wieler (2011), Molybdenum isotope anomalies in meteorites: Constraints on solar nebula evolution and origin of the Earth, *Earth Planet. Sci. Lett., 312,* 390–400.

Cameron, A. G. W. (1995), The first ten million years in the solar nebula. *Meteoritics, 30,* 133–161.

Cameron, A. G. W., B. Fegley, Jr., W. Benz, and W. L. Slattery (1988), The Strange Density of Mercury: Theoretical Considerations, in *Mercury*, eds. M. S. Matthews, C. Chapman, and F. Vilas, pp. 692–708, Univ. of Arizona Press, Tucson, AZ.

Carlson, R. W., M. Boyet, and M. Horan (2007), Chondrite barium, neodymium and samarium isotopic heterogeneity and early Earth differentiation. *Science, 316,* 1175.

Carlson, R. W., E. Garnero, T. M. Harrison, J. Li, M. Manga, W. F. McDonough, S. Mukhopadhyay, B. Romanowicz, D. Rubie, Q. Williams, and S. Zhong (2014), How did early Earth become our modern world? Annu. *Rev. Earth Planet. Sci.,* in press.

Caro, G. and B. Bourdon (2010), Non-chondritic Sm/Nd ratio in the terrestrial planets: Consequences for the geochemical evolution of the mantle-crust system. *Geochim. Cosmochim. Acta., 74,* 3333–3349.

Chambers, J. E. (2014), Planet formation. Chapter 2.4 in *Treatise on Geochemistry*, 2nd ed., Elsevier.

Chen, J., D. Papanastassiou, and G. J. Wasserburg (2010), Ruthenium endemic isotope effects in chondrites and differentiated meteorites. *Geochim. Cosmochim. Acta., 74,* 3851–3862.

Clayton, R. N. L., L. Grossman, and T. K. Mayeda (1973), A component of primitive nuclear composition in carbonaceous chondrites, *Science, 182,* 485–488.

Clayton R. N. (1993), Oxygen isotopes in meteorites. *Ann. Rev. Earth. Planet. Sci., 21,* 115–149.

Clayton, R. N. and T. K. Mayeda (1996), Oxygen isotope studies in achondrites. *Geochim. Cosmochim. Acta., 60,* 1999–2017.

Clayton, R. N. and T. K. Mayeda (1999), Oxygen isotope studies of Carbonaceous chondrites. *Geochim. Cosmochim. Acta., 63,* 2089–2104.

Clayton, R. N., T. K. Mayeda, J. N. Goswami, and E. J. Olsen (1991), Oxygen isotope studies of ordinary chondrites. *Geochim. Cosmochim. Acta., 55,* 2317–2337.

Cloutis, E. A., R. P. Binzel, and Gaffey, M. J. (2014), Establishing asteroid – meteorite links. *Elements, 10,* 25–30.

Corgne, A., S. Keshav, Y. Fei, and W. F. McDonough (2007), How much potassium is in Earth's core? New insights from partitioning experiments. *Earth Planet. Sci. Lett., 256,* 567–576.

Craddock, P. and N. Dauphas (2011), Iron isotopic composition of geological reference materials and chondrites. Geostandard and geoanalitycal research., 35, 101–123.

Dauphas, N., B. Marty, and L. Reisberg (2002), Molybdenum evidence for inherited planetary scale isotope heterogeneity of the protosolar nebula. *Astrophys. J., 565,* 640–644.

Dauphas, N. and A. Morbidelli (2014), Geochemical and planetary dynamical views on the origin of Earth's atmosphere and oceans. Chapter 6.1 in *Treatise on Geochemistry*, 2nd ed., Elsevier.

DeCampli, W. M. and A. G. W. Cameron (1979), Structure and evolution of isolated giant gaseous protoplanets. *Icarus, 38,* 367–391.

Dreibus, G. and H. Wänke (1980), The bulk composition of the eucrite parent asteroid and its bearing on planetary evolution. *Z. Naturf., 34a,* 204–216.

Dreibus, G., J. Brückner, and H. Wänke (1997), On the core mass of asteroid *Vesta. Meteoritics Planet. Sci., 32,* A36.

Fegley, B., Jr. and L. K. Schaefer (2014), Chemistry of Earth's earliest atmosphere. Chapter 6.3 In *Treatise on Geochemistry*, 2nd ed., Elsevier.

Fehr, M., M. Rekhamper, A. N. Halliday, U. Wiechter, B. Hattendorf, D. Gunther, S. Ono, J. Eigenbrode, and D. Rumble III (2005), Tellurium isotopic composition of the early solar system—A search for effect resulting from stellar nucleosynthesis, ^{126}Sn decay, and mass-independent fractionation. *Geochim. Cosmochim. Acta.*, *69*, 5099–5112.

Fitoussi, C. and B. Bourdon (2012), Silicon isotope evidence against an enstatite chondrites Earth. *Science.*, *335*, 1477–1480.

Fujii, T., F. Moynier, and F. Albarede (2006a), Nuclear field vs nucleosynthetic effects as cause of isotopic anomalies in the early Solar System. *Earth and Planetary Science Letters*, *247*, 1–9.

Fujii, T., F. Moynier, and F. Albarède (2006b), Mass-Independent Isotope Fractionation of Molybdenum and Ruthenium and the Origin of Isotopic Anomalies in Murchison. *The Astrophysical Journal*, *647*, 1506–1516.

Gast, P. W. (1960), Limitations on the composition of the upper mantle. *J. Geophys. Res.*, *65*, 1287–1297.

Genda, H. and Y. Abe (2003a), Survival of a proto-atmosphere through the stage of giant impacts: The mechanical aspects. *Icarus, 164*: 149–162.

Genda, H. and Y. Abe (2003b), Modification of a proto-lunar disk by hydrodynamic escape of silicate vapor. *Earth Planets Space*, *55*, 53–57.

Genda, H, and Y. Abe (2005), Enhanced atmospheric loss on protoplanets at the giant impact phase in the presence of oceans. *Nature, 433*: 842–844.

Grossman, L. and J. W. Larimer (1974), Early chemical history of the solar system. *Rev. Geophys. Space Phys.*, *12*, 71–101.

Hart, S. R. and A. Zindler (1986), In search of a bulk-earth composition. *Chemical Geology*, *57*: 247–267.

Helffrich, G. R. and B. J. Wood (2001), The Earth's mantle. *Nature*, *412*, 501–507.

Henning, T. and D. Semenov (2013), Chemistry in protoplanetary disks. *Chem. Rev.*, *113*, 9016–9042.

Herwatz, D., A. Pack, B. Friedrich, and A. Bischoff (2014), Indentification of the giant impactor Theia in lunar rocks. *Science, 344*, 1146–1150.

Hirschmann, M. M. and R. Dasgupta (2009), The H/C ratios of Earth's near-surface and deep reservoirs, and consequences for deep Earth volatile cycles. *Chem. Geol.*, *262*, 4–16.

Huang, S. and S. B. Jacobsen (2012), Calcium isotopic variations in chondrites: implications for planetary isotope compositions. *Lunar Planet. Sci. Conf.*, *43*, #1334 (abstr.).

Huang, S., J. Farkas, and S. Jacobsen (2011), Stable calcium isotopic composition of Hawaiian shield lavas: Evidence for recycling of ancient marine carbonates into the mantle. *Geochim. Cosmochim. Acta.*, *75*, 4987–4997.

Huang, F., Z. Wu, S. Huang, and F. Wu (in press), First Calculations of equilibrium silicon isotope fractionation among mantle minerals, *Geochim. Cosmochim. Acta.*

Huang, S., S. B. Jacobsen, and S. Mukhopadhyay (2013), ^{147}Sm–^{143}Nd systematics of Earth are inconsistent with a superchondritic Sm/Nd ratio. *PNAS, 110*(13), 4929–4934.

Hurley, P. M. (1968), Absolute abundance and distribution of Rb, K, and Sr in the Earth. *Geochim. Cosmochim. Acta.*, *32*, 273–283.

Huss, G. R., A. E. Rubin, and J. N. Grossman (2006), Thermal metamorphism in chondrites. In *Meteorites and the Early Solar System II* (Eds. D. S. Lauretta, H. Y. McSween, and R. P. Binzel), pp. 567–586, University of Arizona Press, Tucson.

Jacobsen S. B., M. C. Ranen, M. I. Petaev, J. L. Remo, R. J. O'Connell, and D. D. Sasselov (2008), Isotopes as clues to the origin and earliest differentiation history of the Earth. Phil. Trans. Royal Soc. of London, *366A*, 4129–4162.

Jacobsen, S. B., M. I. Petaev, S. Huang, and D. D. Sasselov (2013), An isotopically homogeneous region of the inner terrestrial planet region (Mercury to Earth): Evidence from E chondrites and implications for giant-Moon impact forming scenario. Lunar and Planetary Sci. Conf. 44, #2344.

Jagoutz, E., H. Palme, H. Baddenhausen, K. Blum, M. Cendales, G. Dreibus, B. Spettel, V. Lorenz, and H. Wänke (1979), The abundances of major, minor, and trace elements in the earth's mantle as derived from primitive ultramafic nodules. Proc. Lunar Planet. Sci. Conf. 10th, 2031–2050.

Javoy, M. (1995), The integral enstatite chondrites model of the Earth. *Geophys. Res. Lett.*, *22*, 2219–2222.

Javoy, M., E. Kaminski, F. Guyot, D. Andrault, C. Sanloup, M. Moreira, S. Labrosse, A. Jambon, P. Agrinier, A. D'Availle, and C. Jaupart (2010), The chemical composition of the Earth: enstatite chondrite models. *Earth Planet. Sci. Lett.*, *293*, 259–268.

Javoy, M., E. Balan, M. Meheut, M. Blanchard, and M. Lazzeri (2012), First principles investigation of equilibrium isotopic fractionation of O and Si isotopes between refractory solids and gases in the solar nebula. *Earth Planet. Sci. Lett., 319–320*, 118–127.

Kargel, J. S. and Lewis, J. S. (1993), The composition and early evolution of the Earth. *Icarus, 105*, 1–25.

Krot, A. N., K. Keil, E. R. D. Scott, C. A. Goodrich, and M K. Weisberg (2014), Classification of meteorites and their genetic relationships. Chapter 1.1 in *Treatise on Geochemistry* 2nd ed. Elsevier.

Larimer, J. W. (1967), Chemical fractionations in meteorites, 1. Condensation of the elements. *Geochim. Cosmochim. Acta.*, *31*, 1215–1238.

Larimer, J. W. (1971) Composition of the Earth: Chondritic or achondritic. *Geochim. Cosmochim. Acta.*, *35*, 769–786.

Lewis, J. S. (1972), Metal/silicate fractionation in the solar system. *Earth Planet. Sci. Lett.*, *15*, 286–290.

Lewis, J. S. (1974), The temperature gradient in the solar nebula. *Science, 186*, 440–443.

Lewis, J. S. and R. G. Prinn (1984), Planets and their atmospheres: Origin and evolution. Academic Press, NY.

Lodders, K. (1991), *Spurenelementverteilung zwischen Sulfid und Sili- katschmelze und kosmochemische Anwendungen.* Ph.D. thesis, Univ. Mainz, Germany.

Lodders, K. (2000), An oxygen isotope mixing model for the accretion and composition of rocky planets. *Space Sci. Rev.*, *92*, 341–354.

Lodders, K. (2003), Solar system abundances and condensation temperatures of the elements. *Astrophys. J.*, *591*, 1220–1247.

Lodders, K. and B. Fegley, Jr. (1997), An oxygen isotope model for the composition of Mars. *Icarus, 126*, 373–394.

Lodders, K. and B. Fegley, Jr. (1998), *The Planetary Scientist's Companion.* New York: Oxford University Press.

Lodders, K. and B. Fegley, Jr. (2011), *Chemistry of the Solar System.* Cambridge: RSC Publishing.

Luck, J. M., D. Ben Othman, J. A. Barrat, and F. Albarede (2003), Coupled ^{63}Cu and ^{16}O excesses in chondrites. *Geochim. Cosmochim. Acta.*, *67*, 143–151

Lupu, R. E., K. Zahnle, M. S. Marley, L. Schaefer, B. Fegley, C. Morley, K. Cahoy, R. Freedman, and J. J. Fortney (2014), The atmospheres of Earthlike planets after giant impact events. *Astrophys. J.*, *784*:27, doi:10.1088/0004-637X/784/1/27.

Lyons, J. R. and E. D. Young (2005), CO self-shielding as the origin of oxygen isotopic anomalies in the early solar nebula. *Nature*, *435*, 317–320.

Marty, B. (2012), The origins and concentrations of water, carbon, nitrogen and noble gas on Earth. *Earth Planet. Sci. Lett.*, *313–314*, 56–66.

McDonough, W. F. (2003), Compositional models for the Earth's core. In: Carlson R. W. (Ed.) *Treatise on Geochemistry*, *Vol. 2*, The Mantle and Core, pp. 547–568. Oxford: Elsevier-Pergamon.

McDonough, W. F. and S. S. Sun (1995), The composition of the Earth. Chem. Geol., *120*, 223–253.

McSween Jr., H. Y., D. W. G. Sears, and R. T. Dodd (1988), Thermal metamorphism. In: Kerridge, J. F., M. S. Matthews (Eds.), Meteorites and the Early Solar System. University of Arizona Press, Tucson, pp. 102–113.

Morgan, J. W. and E. Anders (1980), Chemical composition of Earth, Venus, and Mercury. Proc Natl. Acad Sci USA, *77*, 6973–6977.

Moynier, F., N. Dauphas, and Podosek, F.A. (2009), A search for ^{70}Zn anomalies in meteorites. *Astrophys. J.*, *700*, L92–L95.

Moynier, F., Q. Z. Yin, K. Irisawa, M. Boyet, B. Jacobsen, and M. Rosing (2010a), A coupled 182W-142Nd constraint for early Earth differentiation. Proceeding of the National Academy of Sciences of the United States of America, *107*, 10810–10814

Moynier, F., J. I. Simon, F. A. Podosek, B. S. Meyer, J. Brannon, D. DePaolo (2010b), Ca isotope effect in Orgueil leachates and the implications for the carrier phases of ^{54}Cr anomalies. *Astrophys. J. Lett.*, *719*, L7–L13.

Moynier, F., Q. Z. Yin, and E. Schauble (2011), Isotopic evidence for Cr partitioning into the Earth core. *Science*, *331*, 1417–1420.

Moynier, F., J. Day, W. Okui, T. Yokoyama, A. Bouvier, R. J. Walker, and F. A. Podosek (2012), Planetary scale Sr isotopic heterogeneity and the age of volatile depletion of the Early solar system materials. *Astrophys. J.*, *758*, 45–51.

Moynier, F., T. Fujii, G. A. Brennecka, and S. Nielsen (2013), Nuclear field shift in natural environments. *Comptes Rendus Geoscience, 345*, 150–159.

Nishio-Hamane, D., T. Yagi, N. Sata, T. Fujita, and T. Okada (2010), No reactions observed in Xe–Fe system even at Earth core pressures. *Geophys. Res. Lett.*, *37*: L04302.

O'Neill, H. St. C. and H. Palme (2008), Collisional erosion and the non-chondritic composition of the terrestrial planets. Phil. Trans. R. Soc., *366*, 4205–4233.

Ozima, M. and F. A. Podosek (2001), Noble Gas Geochemistry, 2nd ed. Cambridge University Press.

Pahlevan, K. (2010), Chemical and isotopic consequences of lunar formation via giant impact. Ph.D. thesis, Caltech, Pasadena, CA.

Palme, H. and K. G. Nickel (1985), Ca/Al ratio and the composition of the Earth's upper mantle. *Geochim. Cosmochim. Acta.*, *49*, 2123–2132.

Palme, H. and H. St. C. O'Neill (2014), Cosmochemical estimates of mantle composition. Chapter 3.1 in *Treatise on Geochemistry*, 2nd ed., Elsevier.

Pepin, R. (2006), Atmospheres on the terrestrial planets: Clues to origin and evolution. *Earth Planet. Sci. Lett.*, *252*, 1–14.

Perryman, M. (2011), The Exoplanet Handbook. Cambridge University Press.

Poitrasson, F., A. N. Halliday, D. C. Lee, S. Levasseur, and N. Teush, (2004), Iron isotope differences between Earth, Moon, Mars, and Vesta of possible records of contrasted accretion mechanisms. *Earth Planet. Sci. Lett.*, *223*, 253–266.

Pringle, E., P. Savage, M. G. Jackson, J. A. Barrat, and F. Moynier (2013), Si isotope homogeneity of the Solar Nebula. *Astrophys. J.*, *779*, 123.

Pringle, E.A., P. Savage, J. Badro, J. A. Barrat, and F. Moynier (2014), Silicon isotopes in angrites and volatile loss in planetesimals. Proceeding of the National Academy of Sciences of the United States of America, 111, 17029–17032, doi/10.1073/pnas.1418889111.

Prinn, R. G. and B. Fegley, Jr. (1989), Solar nebula chemistry: Origin of planetary, satellite, and cometary volatiles. In *Origin and Evolution of Planetary and Satellite Atmospheres*, Eds. S. Atreya, J. Pollack and M.S. Matthews, pp. 78–136, Univ. of Arizona Press, Tucson, AZ.

Pujol, M., B. Marty, and R. Burgess (2011), Chondritic-like xenon trapped in Archean rocks: A possible signature of the ancient atmosphere. *Earth Planet. Lett.*, *208*, 398–406.

Qin, L., C. M. O. D. Alexander, R. W. Carlson, M. F. Horan, and T. Yokoyama (2010), Contributors to chromium isotope variations in meteorites. *Geochim. Cosmochim. Acta.*, *74*, 1122–1145.

Ranen, M. and S. Jacobsen (2007), Barium isotopes in chondritic meteorites: implications for planetary reservoirs models. *Science*, *314*, 809–812.

Regelous, M., T. Elliott, and C. D. Coath (2008), Nickel isotope heterogeneity in the early solar system, *Earth Planet. Lett.*, *372*, 330–338.

Ringwood, A. E. (1979) Origin of the Earth and Moon. Springer-Verlag, Berlin.

Saal, A. E., E. H. Hauri, C. H. Langmuir, and M. R. Perfit (2002), Vapour undersaturation in primitive mid-ocean-ridge basalt and the volatile content of Earth's upper mantle. *Nature*, *419*: 451–455.

Sanloup, C., B. C. Schmidt, E. M. Chamorro Perez, et al. (2005), Retention of xenon in quartz and Earth's missing xenon. Science, *310*: 1174–1177.

Sanloup, C., B. C. Schmidt, G. Gudfinnsson, A. Dewaele, and M. Mezouar (2011), Xenon and argon: A contrasting behavior in olivine at depth. *Geochim. Cosmochim. Acta.*, *75*, 6271–6284.

Sarafian, A.R., M. F. Roden, and A. Patino-Douce (2013), The volatile content of Vesta: clues from apatite in eucrites. *Met. and Planet. Sci.*, *48*, 2135–2154.

Sarafian, A.R., H. R. Marschall, S. G. Nielsen, F. M. McCubbin, and B. Monteleone (2014), An Earth-like hydrogen isotopic composition of Vesta as revealed by apatites. 45th Lunar Planet. Sci. abstract #2106.

Savage, P., and F. Moynier, (2013), Silicon isotopic variations in enstatite meteorites: Clues to their origin and Earth forming materials. *Earth Planet. Sci. Lett.*, *361*: 487–496.

Savage, P., F. Moynier, H. Chen, G. Shofner, J. Siebert, J. Badro, I. Puchtel, (2015), Copper isotope evidence for large-scale sulphide fractionation during Earth's differentiation. *Geochemical Perspective Letters*, *1*, 53–64.

Schaefer, L. and B. Fegley, Jr. (2007), Outgassing of ordinary chondritic material and some of its implications for the chemistry of asteroids, planets, and satellites. *Icarus*, *186*, 462–483.

Schaefer, L. and B. Fegley, Jr. (2010a), Volatile element chemistry during metamorphism of ordinary chondritic material and some of its implications for the composition of asteroids. *Icarus*, *205*, 483–496.

Schaefer, L. and B. Fegley, Jr. (2010b), Cosmochemistry. pp. 347–377, In *Principles and Perspectives in Cosmochemistry: Lecture Notes of the Kodai School on "Synthesis of Elements in Stars"* (Eds. A. Goswami and B. E. Reddy), Springer.

Schiller M., J. Baker, J. Creech, et al. (2011), Rapid timescales for magma ocean crystallization on the howardite-eucrite-diogenite parent body. *Astrophys. J.*, *740*, L22.

Schiller, M., C. Paton, and M. Bizzarro (2012), Calcium isotope measurements by combined HR-ICP-MS and TIMS. JAAS, doi: 10.1039/c1ja10272a.

Simon, J., D. J. DePaolo, and F. Moynier (2009), Calcium isotope composition of meteorites, Earth, and Mars, *Astrophys. J.*, *702*, 707–715.

Simon, J.I. and D. J. DePaolo (2010), Stable calcium isotopic composition of meteorites and rocky planets. *Earth Planet. Sci. Lett.*, *289*, 457–466.

Slattery, W. L. (1978), Protoplanetary core formation by rain-out of iron drops. *Moon Planets*, *19*, 443–457.

Slattery, W. L., W. M. DeCampli, and A. G. W. Cameron (1980) Protoplanetary core formation by rain-out of minerals. *Moon Planets*, *23*, 381–390.

Sprung, P., E. E. Scherer, D. Upadhyay, I. Leya, and K. Mezger (2010), Non-nucleosynthetic heterogeneity in non-radiogenic stable Hf isotopes: Implications for early solar system chronology. *Earth Planet. Sci. Lett.*, *295*, 1–11.

Steele, R. C. J., C. D. Coath, M. Regelous, S. Russell, and T. Elliott (2012), Neutron-poor nickel isotope anomalies in meteorites. *Astrophys. J.*, *758*, 59.

Suess, H. E. and H. C. Urey (1956), Abundances of the elements. *Rev. Mod. Phys.*, *28*, 53–74.

Tang, H. and N. Dauphas (2012). Abundance, distribution, and origin of ^{60}Fe in the solar protoplanetary disk. *Earth Planet. Sci. Lett.*, *359–360*, 248–263.

Teng, F.-Z., M. Wadhwa, and R. T. Helz (2007), Investigation of magnesium isotope fractionation during basalt differentiation: implications for a chondritic composition of the terrestrial mantle. *Earth Planet. Sci. Lett.*, *261*, 84–92.

Thiemens, M. H. (1999), Mass-independent isotope effects in planetary atmospheres and the early Solar System. *Science*, *283*, 341–345.

Torgersen, T. (1989), Terrestrial helium degassing fluxes and the atmospheric helium budget: Implications with respect to the degassing processes of continental crust. *Chem. Geol.*, *79*, 1–14.

Trinquier, A., J.-L. Birck, C. J. Allegre, C. J. Gopel, and D. Ulfbeck (2008), ^{53}Mn ^{53}Cr systematics of the early solar system revisited. *Geochim. Cosmochim. Acta*, *72*, 5146–5163.

Trinquier, A., T. Elliott, D. Ulfbeck, C. Coath, A. N. Krot, and M. Bizzarro (2009), Origin of nucleosynthetic isotope heterogeneity in the solar protoplanetary disk. *Science*, *324*, 374–376.

Valdes, M., M. Moreira, J. Foriel, and F. Moynier (2014), The nature of the Earth's building blocks as revealed by calcium isotopes. *Earth Planet. Sci. Lett*, *394*, 135–145.

Wadhwa, M., Y. Amelin, O. Bogdanovski, A. Shukolyukov, G. W. Lugmair, and P. Janney (2009), Ancient relative and absolute ages for a basaltic meteorite: Implications for timescales of planetesimals accretion and differentiation. *Geochim. Cosmochim. Acta.*, *73*, 5189–5201.

Wang, K., F. Moynier, F. Podosek, and J. Foriel (2011), ^{58}Fe and ^{54}Cr in early Solar System materials. *Astrophys. J. Lett.*, *738*, L58.

Wang, K., F. Moynier, J. A. Barrat, B. Zanda, R. C. Paniello, P. S. Savage (2014), Homogeneous distribution of Fe isotopes in the early solar nebula. *Meteoritics and Planetary Sciences*, *48*, 354–364.

Wänke, H. (1981), Constitution of terrestrial planets. Phil. Trans. Roy. Soc. London, *A303*, 287–302.

Wänke, H., G. Dreibus, and E. Jagoutz (1984), Mantle chemistry and accretion history of the Earth. In: A. Kröner (Ed.) Archean Geochemistry. Berlin: Springer Verlag.

Warren, P. (2011), Stable isotopic anomalies and the accretionary assemblage of the Earth and Mars: A subordinate role of Carbonaceous chondrites, *Earth Planet. Sci let.*, *311*, 93–100.

Wilkening, L. L. (1977), Meteorites in meteorites—Evidence for mixing among the asteroids. In: *Comets, Asteroids, Meteorites* (Ed. A. H. Delsemme), pp. 389–396, Univ. Toledo Press, Toledo, OH.

Yin, Q., S. B. Jacobsen, K. Yamashita, J. Blichert-Toft, P. Télouk, F. Albarède, and F. Yokoyama (2007), A short timescale for terrestrial planet formation from Hf-W chronometry of meteorites. *Nature*, *418*, 949–952.

Zhang, Y. (2014), Degassing history of the Earth. Chapter 6.2 in *Treatise on Geochemistry*, 2nd ed., Elsevier.

Zhang, J., N. Dauphas, A. M. Davis, I. Leya, and A. Fedkin (2012), The proto-Earth as a significant source of lunar material. *Nature Geos.*, *5*, 251–255.

3

Earth and Terrestrial Planet Formation

Seth A. Jacobson[1,2] and Kevin J. Walsh[3]

ABSTRACT

The growth and composition of Earth is a direct consequence of planet formation throughout the Solar System. We discuss the known history of the Solar System, the proposed stages of growth, and how the early stages of planet formation may be dominated by pebble growth processes. Pebbles are small bodies whose strong interactions with the nebular gas lead to remarkable new accretion mechanisms for the formation of planetesimals and the growth of planetary embryos.

Many of the popular models for the later stages of planet formation are presented. The classical models with the giant planets on fixed orbits are not consistent with the known history of the Solar System, fail to create a high Earth-to-Mars mass ratio, and, in many cases, are also internally inconsistent. The successful Grand Tack model creates a small Mars, a wet Earth, a realistic asteroid belt, and the mass-orbit structure of the terrestrial planets.

In the Grand Tack scenario, growth curves for Earth most closely match a Weibull model. The feeding zones, which determine the compositions of Earth and Venus, follow a particular pattern determined by Jupiter, while the feeding zones of Mars and Theia, the last giant impactor on Earth, appear to randomly sample the terrestrial disk. The late accreted mass samples the disk nearly evenly.

3.1. INTRODUCTION

The formation and history of Earth cannot be studied in isolation. Its existence, composition and dynamics are a direct consequence of events that occurred throughout the Solar System. Planet formation is a combination of local and global processes at every stage of growth. Therefore, the early Earth must be placed in the context of the history of the Solar System.

3.1.1. History of the Solar System

There are very few firmly established temporal events during planet formation, however, one that seems irrefutable is that the giant planets must have accumulated all of their gas envelopes, nearly all of their mass, by the time the nebular gas is removed from the disk. Observations of nebular disks about other stars indicate that disk survival times are typically only millions of years [*Haisch et al.*, 2001; *Briceño et al.*, 2001; *Mamajek*, 2009], although there is some evidence that disks can last for tens of millions of years [*Pfalzner et al.*, 2014]. Although we do not have a direct observation of our own disk survival time, we do know that chondrule formation, a likely nebular process [*Alexander et al.*, 2008], ended after a few million years [*Kita et al.*, 2005; *Villeneuve et al.*, 2009]. Furthermore, there is evidence from modeling of the

[1]*Bayerisches Geoinstitut, Universität Bayreuth, Bayreuth, Germany*

[2]*Laboratoire Lagrange, Observatoire de la Côte d'Azur, Nice, France*

[3]*Planetary Science Directorate, Southwest Research Institute, Boulder, CO, USA*

The Early Earth: Accretion and Differentiation, Geophysical Monograph 212, First Edition.
Edited by James Badro and Michael Walter.
© 2015 American Geophysical Union. Published 2015 by John Wiley & Sons, Inc.

formation of Iapetus including ^{26}Al decay that it must have formed ~3.4 to 5.4 Myr after CAIs [*Castillo-Rogez et al.*, 2007, 2009]. Since the formation of Iapetus requires Saturn to be near completion, this also constrains the age of Saturn.

The last giant impact on Earth, the Moon-forming impact, has a date established to be approximately 70–110 Myr after CAIs from three lines of evidence: radiometric [*Touboul et al.*, 2007; *Allègre et al.*, 2008; *Halliday*, 2008], dynamical-elemental [*Jacobson et al.*, 2014], and dynamical-isotopical [*Bottke et al.*, 2014]. There is some disagreement; others have suggested earlier dates using different models to interpret radiometric data [*Yin et al.*, 2002; *Jacobsen*, 2005; *Taylor et al.*, 2009], but even these 20–60 Myr estimates are significantly after the few million year lifetime of the nebular gas disk. Interestingly, the age of Mars is considerably different than that of Earth. Radiometric evidence from the Hf-W system indicates that it likely formed within 10 Myr [*Nimmo and Kleine*, 2007] and perhaps it was within a few percent of its final mass by the time the disk dissipated [*Dauphas and Pourmand*, 2011], so Mars may have nearly completed its formation in the presence of the nebular gas. The formation dates of Mercury and Venus are unknown.

The other temporal event with significant evidence in the history of the Solar System is the late heavy bombardment. This intense period of impacts ~400 Myr after CAIs left a record on the Moon [*Tera et al.*, 1974; *Ryder*, 1990; *Cohen et al.*, 2000; *Ryder*, 2002], on Earth [*Marchi et al.*, 2014], and on Vesta through analysis of the HED meteorites [*Marchi et al.*, 2013]. From the crater record on the Moon, it appears that ~$10^{-4} M_{\oplus}$ of material was delivered to Earth during this bombardment [*Morbidelli et al.*, 2012a; *Marchi et al.*, 2014]. For a contrasting interpretation of the lunar cratering record and asteroid melt chronology, see *Chapman et al.* [2007], although this review occurred before the terrestrial and HED evidence was presented.

3.1.1.1. The Nice Model

There is significant evidence in the orbital structure of the Kuiper Belt that the giant planets migrated due to interactions with this disk of outer planetesimals [*Fernandez and Ip*, 1984; *Malhotra*, 1995; *Gomes*, 2003; *Levison and Morbidelli*, 2003]. This interaction caused the outward migration of Saturn, Uranus, and Neptune and the inward migration of Jupiter because the planetesimals are first inwardly scattered by the outer three giant planets and then scattered out of the Solar System by Jupiter. When this migration occurred is not directly known, but it has been associated with the origin of the late heavy bombardment through a dynamical instability [*Strom et al.*, 2005]; the 'Nice' model stipulates that an early compact configuration of the giant planets undergoes

an instability [*Gomes et al.*, 2005]. This giant planet instability explains a number of features of the Solar System including the Trojans of Jupiter [*Morbidelli et al.*, 2005] and Neptune [*Tsiganis et al.*, 2005], the number and inclination distribution of the irregular satellites of the giant planets [*Nesvorný et al.*, 2007; *Bottke et al.*, 2010], and dynamical features of the Kuiper Belt [*Levison et al.*, 2008; *Morbidelli et al.*, 2014]. Furthermore, close encounters between an ice giant and Jupiter reproduce the structure of the main asteroid belt [*Morbidelli et al.*, 2010]. More recently, the model has been updated, Nice 2.0, to better reflect the likely orbital spacing of the giant planets in mean motion resonances after removal of the gas [*Levison et al.*, 2011].

Although a complete rejection of the Nice model would require a number of less likely or so-far incomplete hypotheses to take its place, the connection between the late heavy bombardment and the giant planet instability of the Nice model is not required. For instance, a Nice model-like instability may occur immediately after nebula gas removal, thereby creating the Solar System architecture and small body populations that are observed today. In this alternative scenario, the late heavy bombardment would either need to be triggered by the instability of a small fifth planet at ~2 AU [*Chambers*, 2007], to not occur due to a different interpretation of the absolute age evidence on the Moon [*Chapman et al.*, 2007], or to be triggered by the catastrophic break-up of a Vesta-sized Mars-crossing asteroid due to a different interpretation of the cratering evidence on the Moon [*Ćuk et al.*, 2010; *Ćuk*, 2012]. The latter two alternatives are unlikely because of new evidence that the late heavy bombardment is recorded in the HED meteorites [*Marchi et al.*, 2013].

It is important to note that the Nice model and the Grand Tack model are not the same and are independent of one another. The inward-then-outward migration of Jupiter and Saturn called the 'Grand Tack' [*Walsh et al.*, 2011], which is described in Section 3.4, is proposed to take place in the presence of the nebular gas, much earlier in the history of the Solar System than the Nice model. The Grand Tack scenario is consistent with alternatives to the Nice model such as the conjecture that the giant planet instability occurs shortly after gas removal and the evidence for the late heavy bombardment has some other explanation such as the Planet V hypothesis [*Chambers*, 2007].

3.1.2. Stages of Planet Formation

For convenience, we divide the history of planet formation into four stages: dust sedimentation and growth, planetesimal growth, planetary embryo growth, and planet growth. Although it may be unlikely that these stages are coincident locally, it's very likely that they are occurring simultaneously at different radii across the disk since the orbital period, solid surface density, and gas density, viscosity, and temperature change dramatically with semi-major axis as well as time. Particularly, there is a divide in the Solar System between the giant planets, Jupiter and Saturn, which grew large enough to accrete most of their mass from the solar nebula, doing so during the first few million years of Solar System history before the nebular gas dissipated, and the terrestrial planets, which did not grow large enough to accrete a significant amount of solar nebula gas. Of course, the sub-giants Uranus and Neptune grew at a rate somewhere in between, but exterior to Jupiter and Saturn.

It is computationally difficult (and impossible as of now) for a single numerical simulation to cover all stages of growth, so typically each stage is handled separately. Earlier processes such as dust growth and planetesimal formation are often studied in local patches of the disk or in small annuli. This is thought to be acceptable since perturbations from elsewhere in the disk are homogenous across the studied zone. Even global simulations of the later stages are computationally expensive.

Each growth stage is marked by a different dominant accretion mechanism for the largest bodies, those we typically care the most about, but smaller bodies may continue to accrete according to earlier paradigms. We review each growth stage, in turn, focusing on new developments, but we begin by defining some new terminology that has appeared in the planet formation literature.

The motion of solid objects in a gas disk is dictated by the Stokes number. It is a dimensionless number with a long history in fluid mechanics that quantifies the momentum coupling between an object of approximate radius R and the surrounding fluid. In the context of a gas disk, it is defined: $St = (\rho_s / \rho_0) R\Omega / c_0$, where ρ_s/ρ_0 is the ratio of the solid to mid-plane gas density, Ω is the Keplerian orbital frequency, and c_0 is the mid-plane sound speed. This formula is correct for the Epstein drag regime, which assumes that the gas mean free path is longer than the object (for a review of other drag regimes and the appropriate formulae for determining the Stokes number, see *Youdin*, 2010). Accretion depends both on the Stokes number of the target and projectile (for recent examples see *Chambers*, 2014; *Guillot et al.*, 2014). When $St \ll 1$, the object is completely entrained in the gas and when $St \gg 1$, the object's motion is independent of the presence of the gas. Thus, different regimes are identified: objects with Stokes numbers much less than 0.01 are 'dust,' those with Stokes numbers between 0.01 and 1 are 'pebbles,' and those with Stokes numbers just greater than 1 are 'boulders.' Planetesimals and embryos have Stokes numbers much greater than 1. From the definition of the Stokes number, an object's membership in each regime is dependent on the disk properties. The Stokes number is the fundamental quantity determining the significance of passive concentration processes, streaming instabilities, and pebble accretion; all discussed in detail in Section 3.1.3.

3.1.2.1. Dust Sedimentation and Growth

Planet formation begins in the nebular disk that forms simultaneously with the protostar from a collapsing molecular cloud as a result of the conservation of angular momentum. Pre-existing dust grains as well as condensates from the gas sediment toward the mid-plane of the protoplanetary disk [*Weidenschilling*, 1980]. Turbulent diffusion of solid material due to entrainment with the gas eventually balances gravity, and so a vertical equilibrium structure is assembled in the disk [*Weidenschilling and Cuzzi*, 1993; *Cuzzi et al.*, 1993; *Dubrulle et al.*, 1995; *Carballido et al.*, 2006]. In this structure, dust grains can grow via binary collisions accreting through adhesive forces, 'fluffy' aggregation and compaction, and by mass transfer during fragmenting collisions (for reviews, see *Dominik et al.*, 2007; *Johansen et al.*, 2014). From these growth mechanisms, a population of pebbles and boulders emerges from the coagulating dust.

3.1.2.2. Planetesimal Growth

All objects experience some orbital drag because the nebular gas is partially pressure supported; a parcel of gas orbits the Sun at sub-Keplerian speeds, but boulders orbit at Keplerian speeds. Thus, boulders experience a headwind, which generates energy loss, and their orbits spiral into the Sun [*Whipple*, 1972]. This effect is often referred to as the 'radial drift' or 'meter' barrier, when considering specific disk conditions. Furthermore, as objects grow aerodynamic drag decreases, impact velocities increase, and binary collisions are no longer an efficient growth mechanism because colliding boulders bounce or even fragment (for review, see *Blum and Wurm*, 2008). This is often referred to as the 'bouncing' barrier [*Zsom et al.*, 2010].

These boulder barriers restrict growth to $St \sim 1$ or about ~ 0.1 to 1 m at an AU for a standard model disk [*Weidenschilling*, 1977; *Zsom et al.*, 2010]. It is possible that fluffy aggregates of dust grains can grow to km or even larger sizes if ice enhances stickiness and collision velocities remain below ~ 50 m s^{-1} [*Wada et al.*, 2009, 2013; *Okuzumi et al.*, 2012]; whether these planetesimals would be too water rich to contribute significantly to terrestrial planet formation is an open question. In Section 3.1.3, we discuss new ideas regarding the interactions between gas and 'pebbles' that would create ~ 100 km planetesimals directly from <1 m 'pebbles' skipping over these barriers [*Johansen et al.*, 2007; *Cuzzi et al.*, 2008].

3.1.2.3. Planetary Embryo Growth

Objects that grow beyond the boulder barriers are planetesimals, and they grow at different rates depending on their size. Smaller planetesimals in the 'orderly' growth regime accrete according to their physical cross-sections alone and so growth proceeds slowly [Safronov, 1972; Nakagawa et al., 1983]. It's possible that most large planetesimals including those that continued to grow into planetary embryos and planets bypassed much of this growth regime and grew suddenly through gravoturbulent growth mechanisms from pebble and boulder sizes to hundreds or thousands of kilometers across (a description of this growth and its consequences is in Section 3.1.3).

If planetesimals have grown large enough that the escape velocities from their surfaces match or exceed their relative velocities, then gravitational focusing is important [Safronov and Zvjagina, 1969; Greenberg et al., 1978]. The enhancement of the collision cross-section due to gravitational focusing is proportional to the square of the ratio of escape to relative velocities (for a derivation and further review, see Armitage, 2014). Since dynamical friction reduces the relative velocities of the largest bodies compared to the rest and those bodies already have the highest escape velocities, planetesimal growth transforms into a 'runaway' growth process; the largest bodies have the largest relative binary collision cross-sections and hence grow the fastest and become planetary embryos [Wetherill and Stewart, 1989, 1993; Ida and Makino, 1992; Kokubo and Ida, 1996]. Although gas drag is less efficient for larger bodies, dynamical friction efficiently transfers energy from these larger bodies to smaller bodies that interact with the gas more strongly [Wetherill and Stewart, 1993; Kokubo and Ida, 1996]. Dynamical friction is the damping of the eccentricities and inclinations of larger bodies due to gravitational interactions with a large number of smaller bodies. Smaller bodies are also more likely to fragment rather than grow due to these higher relative velocities and their lower gravitational binding energies, further increasing the number of even smaller bodies and the effectiveness of dynamical friction and gas drag [Wetherill and Stewart, 1989].

Runaway growth produces a bi-modal mass distribution in the disk of planetary embryos and planetesimals. The formation of planetary embryos is self-limiting and slows down runaway growth by viscously stirring the nearby planetesimal population increasing relative velocities [Lissauer, 1987; Ida and Makino, 1993], so growth is slower and similar between neighboring embryos. This 'oligarchic' growth maintains the bi-modal mass distribution of planetary embryos so individually each embryo is about 100 times as massive as the average planetesimal. Orbital repulsion keeps planetary embryos about 10 mutual Hill radii apart from each other [Kokubo and Ida, 1995,

1998]. See Morbidelli et al. [2012b] and Raymond et al. [2013] for more mathematical reviews of the runaway and oligarchic dynamical processes.

3.1.2.4. Planet Growth, the Giant Impact Phase

Oligarchic growth ends when energy transfer via dynamical friction can no longer continue circularizing the orbits of the planetary embryos. Typically, this occurs because the planetesimal population becomes too depleted as planetesimals are accreted onto planets and the Sun or scattered from the Solar System. Many simulations do not include the role of collisional grinding but this also leads to a reduction of the planetesimal population [Levison et al., 2012]. If gas drag was a major source of dissipation, then the sudden removal of the solar nebula can trigger the end of the oligarchic regime [Iwasaki et al., 2002; Zhou et al., 2007].

Without dynamical friction, planetary embryos perturb each other onto crossing orbits leading to either giant impacts or scattering events [Wetherill, 1985]. The planetesimal population steadily declines during this phase until only a few planetesimals are left on semi-stable orbits in the Main Belt. These giant impacts are the last phase of planet formation. Once the remaining planetary embryos have found stable orbits and finished accreting, we call them planets. An extended giant impact phase is consistent with the late formation of Earth ~70–110 Myr [Touboul et al., 2007; Allègre et al., 2008; Halliday, 2008; Jacobson et al., 2014; Bottke et al., 2014]. The early formation of Mars (~1 to 3 Myr [Nimmo and Kleine, 2007; Dauphas and Pourmand, 2011]) may imply that it finished its growth during the oligarchic growth phase, so it is a stranded embryo that never participated in the giant impact phase [Jacobson and Morbidelli, 2014].

In the outer Solar System, the embryos need to grow to ~10 $M_\oplus$ giant planet cores so that gas accretion occurs [Pollack et al., 1996]. The core accretion scenario of dust growth, runaway growth, and oligarchic growth is too slow (for a review of the difficulties in this scenario, see Levison et al., 2010). Recently, an alternative solution has been proposed called pebble accretion [Ormel and Klahr, 2010; Johansen and Lacerda, 2010; Murray-Clay et al., 2011; Lambrechts and Johansen, 2012; Chambers, 2014]. Although this pebble growth process was developed to solve a problem in the outer Solar System, it has strong implications for terrestrial planet formation.

3.1.3. Pebble Growth Processes

In the terrestrial protoplanetary disk, radial drift and the bouncing barrier frustrate growth at boulder sizes preventing planetesimal formation. These barriers may be overcome by what may be broadly referred to as gravoturbulent growth (for a thorough review, see Johansen *et al.*, 2014), a possibly misleading nomenclature since turbulence isn't always necessary. Gravo-turbulent mechanisms bring together such great particle concentrations that the collection of small particles collapses under self-gravity. By gathering enough dust or pebbles, a single body forms, which is significantly larger than either of the barriers, skipping the boulder size regime altogether. The initial asteroid size-frequency distribution of the Main Belt may have been very shallow and dominated by asteroids larger than 100 km [*Morbidelli et al.*, 2009], but this is not conclusive since growth from km-sized asteroids can also match the current size-frequency distribution of the Main Belt [*Weidenschilling*, 2011]. However, there is also evidence from the Kuiper Belt that large equal mass binary systems with similar surface appearances were likely formed from a gravitationally collapsing cloud of particles that had too much angular momentum for a single body [*Nesvorný et al.*, 2010]. These clues suggest that planetesimals appeared as 100 to 1000 km bodies.

The importance of gas-particle interactions have a long history [e.g., *Goldreich and Ward*, 1973; *Weidenschilling and Cuzzi*, 1993], but older ideas focused on creating a very thin, dynamically cold mid-plane layer of planetesimals. This disk could undergo a gravitational instability, but turbulent coupling with the gas and dynamical self-stirring makes this process difficult [*Weidenschilling and Cuzzi*, 1993; *Youdin and Shu*, 2002]. The more recently developed pebble mechanisms that allow growth beyond the boulder barriers fall into two categories. First, gas dynamics drive particles together via 'passive' concentration mechanisms. In regions of the disk with significant turbulence, eddies form at the smallest turbulent scales, and these drive dust into the regions between the eddies [*Cuzzi et al.*, 2008; *Chambers*, 2010]. This mechanism creates smaller planetesimals, which may not be consistent with the evidence above. Larger vortices or radial velocity variations created by gap-clearing giant planets, phase transitions in the disk chemistry, or ionization levels of the disk, so called 'dead zones' create pressure bumps that collect particles (for a list of references see Section 4.1.4 in *Johansen et al.*, 2014). A pressure bump creates a zonal flow that slows and possibly reverses the radial drift due to gas drag of pebbles [*Whipple*, 1972]. As mentioned, the pressure bumps are associated with specific structures in the disk, so if this mechanism is dominant, growth beyond the boulder barriers only occurs at unique locations. How numerous these locations are and where they are in the disk are open questions, but the answers so far appear to be few and typically they rest outside of the terrestrial planet formation region [*Johansen et al.*, 2014].

The second concentration mechanism is not associated with a pre-existing disk structure, instead the 'streaming instability' is a consequence of pebble and gas interactions.

A single pebble causes almost no back-reaction on the gas dynamics, but a small clustering of pebbles causes the local gas to move with the pebbles at Keplerian orbital rates. Exterior pebbles catch up with this cluster via radial drift and increase the size of the back-reaction, further retarding the radial drift of the cluster and increasing the number of pebbles that catch up with the cluster [*Johansen et al.*, 2007]. Eventually the cluster grows large enough to collapse due to self-gravity creating a 100 to 1000 km body directly from pebbles (for a thorough review of the material in the last two paragraphs, see *Johansen et al.*, 2014).

Pebble-pile planetesimals are an open topic of research [*Hopkins*, 2014; *Jansson and Johansen*, 2014], but if this mechanism is important there are exciting repercussions. For instance, a fundamental unit of the Solar System is chondrule. Chondrules are small melt spherules found in the most abundant meteorite classes. If chondrules are the pebbles in these processes that would wonderfully explain their ubiquity, but currently most pebble concentration models require particles to be about an order or two of magnitude larger than a typical chondrule given standard disk models [*Johansen et al.*, 2007]. Another repercussion of pebble concentration models are that the variations between planetesimal compositions may be a consequence of where in the disk it is most likely for the streaming instability to occur at a particular time. Due to radial drift, pebbles are constantly migrating from the exterior of the disk inward. If there is an initial compositional gradient or a temporal gradient associated with dust sedimentation, then this gradient may be frozen in as a function of streaming instability location. Alternatively, pebbles may be the fragments from collisions between larger bodies [*Kobayashi and Tanaka*, 2010; *Chambers*, 2014] or fluffy aggregates created from icy dust grains [*Wada et al.*, 2008; *Okuzumi et al.*, 2012]. It is not clear what the dominant pebble creation process is or, even, their appearances.

Pebbles may also be important for the rapid growth of planetary embryos [*Ormel and Klahr*, 2010; *Johansen and Lacerda*, 2010; *Murray-Clay et al.*, 2011; *Lambrechts and Johansen*, 2012; *Chambers*, 2014]. Pebbles are not entrained along streamlines in the nebular gas but neither are their motions independent of the gas. This has consequences for embryo accretion. As described above, planetesimals are accreted by embryos according to their mutual gravitational cross-sections, however this does not account for gas-solid interactions in the presence of a growing embryo. Pebbles can be accreted from a much larger cross-section than the gravitational focusing cross-section because as they react to the gravitational acceleration from the embryo, they experience gas drag, which reduces their velocity relative to the embryo and this increases the time at which the embryo's gravity acts on them, and so on.

In the outer Solar System, this process may allow the formation of the giant cores early enough to accrete gas envelopes [*Lambrechts and Johansen*, 2014; *Lambrechts et al.*, 2014], although the cores could set-up a different persistent oligarchic growth regime [*Kretke and Levison*, 2014]. In the inner Solar System, it also could play a significant role [*Levison et al.*, 2014]. If pebbles are a significant portion of the accreted mass, then embryos could become layered with pebbles. If these pebbles are chondrules or clusters of chondrules, then these layers may be an explanation for the remnant magnetic fields and other evidence that certain chondrites were at the surface of larger bodies (this also assumes that these growing embryos have internally generated magnetic fields [*Weiss et al.*, 2010; *Carporzen et al.*, 2011; *Elkins-Tanton et al.*, 2011]). Then again, these larger bodies could have formed directly from pebbles via a passive concentration or streaming instability mechanism. Most radically, if the streaming instability is inefficient, then, perhaps, the five largest planetesimals grew via pebble accretion into the four terrestrial planets and Theia, the Moon-forming impactor—these are the only planet-sized bodies in the inner Solar System for which we have direct evidence of their existence.

The consequences for pebble processes are revolutionizing planetesimal formation and the giant planets, but their repercussions for the terrestrial planets have not been fully explored.

3.1.4. New Pebble Model from Dust to Embryo

Assuming that pebble processes are the key to bypassing the boulder barriers, modified planet formation timeline stages emerge. Pre-existing grains or condensates settle into a dust structure centered on the mid-plane, and grow to pebble and boulder sizes from binary collisions. However, some as-yet-unknown fraction of pebbles are concentrated by the streaming instability and collapse directly into large 100 to 1000 km planetesimals. If this process is successful, then runaway growth occurs from this sea of planetesimals and planetary embryos emerge as oligarchs. If the population of planetesimals remains sparse, that is, the streaming instability is rare, then embryo formation relies upon pebble accretion. Since the efficiency of pebble accretion is a sharp function of the size of the protoplanet, then planetesimals can be stranded at small sizes never experiencing rapid pebble accretion before the gas in the disk dissipates [*Lambrechts and Johansen*, 2012]. Interestingly, a bi-modal mass distribution between planetesimals and embryos seems inevitable regardless of the growth paradigm. In either paradigm, though, it's unclear what the distribution of embryo masses are and how much mass is in the embryo population compared to the planetesimal population.

This new model regarding the early stages of planet formation has strong implications for how possible initial compositional gradients are emplaced in the disk, for the timing of different accretion events relative to the decay of radioactive elements, etc. Thus, it requires much future study, but rather than waiting until these early stages are completely understood, studies of terrestrial planet formation have plunged ahead experimenting with different assumed initial distributions of planetesimals and embryos.

3.2. MODELS OF THE GIANT IMPACT PHASE OF TERRESTRIAL PLANET FORMATION

Unlike earlier stages, which are often studied in local patches of the disk or in small annuli, the giant impact stage of terrestrial planet formation is usually modeled as a nearly global simulation. It is 'nearly global' because most N-body simulations simplify the treatment of the giant planets. They are often assumed to be fully formed, because they needed to accrete their gaseous envelopes during the lifetime of the gas disk, and often Uranus and Neptune are neglected since their influence on the terrestrial disk is weak (e.g., this was found to be so for the Grand Tack [*Walsh et al.*, 2011]).

The terrestrial protoplanetary disk at this time is assumed to have a bi-modal mass distribution containing planetary embryos and planetesimals. Such a distribution is produced by both the ordered, runaway, and oligarchic growth scenario and a pebble-dominated scenario, and it is not clear which scenario is correct. The key parameter is the ratio of total mass in the embryo population to that of the total mass in the planetesimal population. It determines the amount of dynamical friction in the disk. Assuming ordered planetesimal growth, N-body simulations of the planetary embryo growth stage predicted a ratio near unity [*Kokubo and Ida*, 1998], but planetesimal grinding may significantly enhance this ratio over time [*Levison et al.*, 2012] or perhaps, the planetesimals were only ever very small and collisional grinding and radial drift leaves behind a population of only Mars-sized embryos [*Kobayashi and Dauphas*, 2013]. For pebble accretion, this ratio is unknown. Modern numerical N-body simulations have experimented with a large number of ratio choices [*Jacobson and Morbidelli*, 2014] or self-consistently modeled all the stages of planet formation [*Walsh and Levison*, 2014], and they are concluding that a high total embryo mass relative to the total planetesimal mass best reproduces the late accretion record on Earth [*Jacobson et al.*, 2014].

It is the orbits of Jupiter and Saturn that primarily distinguish the different models of the giant impact phase of terrestrial planet formation. The dynamical excitation of the giant planets is directly reflected in the dynamics of the terrestrial disk via secular resonances (particularly, the ν_5 and ν_6) and scattered embryos [*Raymond et al.*, 2009]. We discuss three classical models, in each the terrestrial disk extends from an inner edge to an outer edge carved by Jupiter. In the classical scenarios, the giant planets do not migrate, but they have significantly different orbits in each model due to different assumptions about past and future evolution of the Solar System. Each assumption has consequences for matching constraints such as the late heavy bombardment. Truncated disk models assume that either the terrestrial disk does not extend out to the giant planets or that the migration of the giant planets truncates the disk. This Grand Tack model is the most successful and is described in great detail.

3.2.1. Comparing Terrestrial Planet Systems

Numerical simulations never exactly reproduce the terrestrial planets, and cross-comparing the simulated planets with the true distribution of masses and orbits one-by-one is a complicated task. Although we can directly compare the orbits and masses of the planets in each system as shown in Figure 3.1, evaluating which model best reproduces the Solar System is a matter of statistics and gross metrics. To help with this process, *Chambers* [2001] started using the following sophisticated quantities.

The angular momentum deficit S_d of the inner terrestrial planets measures their dynamical excitation [*Laskar*, 1997]. It is defined as the difference between the sums of the angular momentum of the corresponding circular and in-plane orbits and the actual orbit normalized by the sum of the angular momentum of the corresponding circular and in-plane orbits [*Chambers*, 2001]:

$$S_d = \frac{\Sigma_j m_j \sqrt{a_j} \left(1 - \sqrt{1 - e_j^2} \cos i_j\right)}{\Sigma_j m_j \sqrt{a_j}}$$

where m_j is the mass, a_j is the semi-major axis, e_j is the eccentricity, and i_j is the inclination of the jth planet. The inner planets do strongly perturb one another exchanging angular momentum, but exchanges with the outer planets are limited so the quantity S_d is conserved within a factor of two [*Laskar*, 1997]. The current value for the Solar System is $S_d = 0.0018$ and is marked by a dashed horizontal line in Figure 3.2. If the Solar System undergoes a giant planet instability as supposed in the Nice model, then the angular momentum deficit after terrestrial planet formation needs to be between 10% (0.00018) and 70% (0.00126) of the deficit after the instability (shown in Figure 3.2 as the gray rectangle [*Brasser et al.*, 2013]).

The other useful but unusual quantity is the concentration statistic S_c. The terrestrial planetary system of

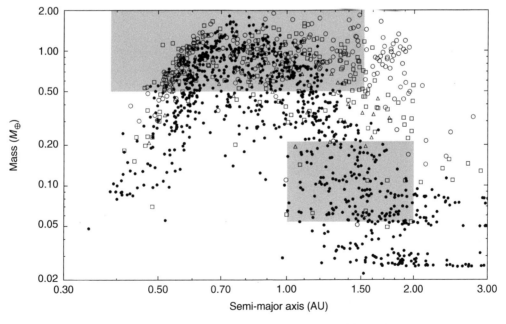

Figure 3.1 The mass and semimajor axis distribution of planets formed from the eccentric [empty squares; *O'Brien et al.*, 2006; *Raymond et al.*, 2009; *Fischer and Ciesla*, 2014], circular [empty circles; *O'Brien et al.*, 2006; *Raymond et al.*, 2009; *Fischer and Ciesla*, 2014], and extra-eccentric [empty triangles; *Raymond et al.*, 2009] Jupiter and Saturn models and the Grand Tack model [solid points; *Walsh et al.*, 2011; *Jacobson and Morbidelli*, 2014; *Jacobson et al.*, 2014; *O'Brien et al.*, 2014]. Planets within the upper gray rectangle are considered Earth-like and those within the lower gray rectangle are considered Mars-like.

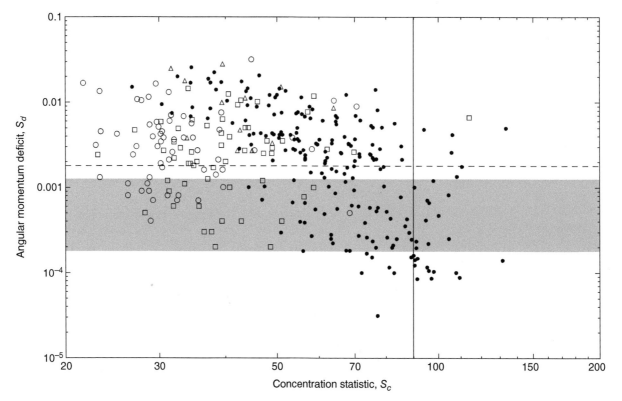

Figure 3.2 The angular momentum deficit S_d and concentration statistic S_c for terrestrial planet systems formed from the eccentric [empty squares; *O'Brien et al.*, 2006; *Raymond et al.*, 2009; *Fischer and Ciesla*, 2014], circular [empty circles; *O'Brien et al.*, 2006; *Raymond et al.*, 2009; *Fischer and Ciesla*, 2014], and extra-eccentric [empty triangles; *Raymond et al.*, 2009] Jupiter and Saturn models and the Grand Tack model [solid points; *Walsh et al.*, 2011; *Jacobson and Morbidelli*, 2014; *Jacobson et al.*, 2014; *O'Brien et al.*, 2014]. The horizontal dashed line is the angular momentum deficit of the current terrestrial planet system $S_d = 0.0018$ and so the target value for the eccentric and extra-eccentric Jupiter and Saturn models. The gray rectangle marks the region with 10 to 70% of the current angular momentum deficit [*Brasser et al.*, 2013] and so is the target for models like the Grand Tack and circular Jupiter and Saturn that include a giant planet instability. The vertical solid line indicates the current terrestrial planet system's concentration statistic $S_c = 89.9$.

the Solar System is interesting because its mass is almost entirely in Earth and Venus with only a little bit in Mercury and Mars. The concentration statistic attempts, albeit degenerately, to capture this mass-orbit distribution with a single number. It is defined [*Chambers*, 2001]:

$$S_c = \max \left[\frac{\Sigma_j m_j}{\Sigma_j m_j \left(\log \frac{a}{a_j} \right)^2} \right]$$

where like before m_j is the mass and a_j is the semi-major axis of the jth planet. S_c is the maximum of the bracketed quantity as a is varied. The terrestrial planets in the Solar System have $S_c = 89.9$, and this value is marked by a solid vertical line in Figure 3.2.

3.3. CLASSICAL MODELS

In the classical models, the terrestrial disk of planetesimals and embryos has an outer boundary set by Jupiter, since most orbits exterior to the inner 3:2 mean-motion resonance with Jupiter are unstable. However, the inner boundary is more mysterious, and there is no conclusive theory regarding why the Solar System doesn't have planets interior to Mercury. Existing theories include high collision velocities that fragment and grind away the planetesimal population before embryos can form [*Chambers*, 2001], gas-driven migration of embryos into the Sun [*Ida and Lin*, 2008], an inner edge to the gas disk, or a silicate evaporation front that doesn't allow dust or pebbles to exist interior to form planetesimals and then embryos. Most N-body simulations place the inner edge at 0.3 to 0.7 AU to avoid forming planets interior to Mercury. Forming a

Mercury-sized planet in the correct orbit with a satisfactory explanation for its high metal content is an open problem.

3.3.1. Eccentric Jupiter and Saturn, Current Orbits

The first giant impact phase models considered disks that contained about half the mass in the planetesimal population and half in the embryo population, which is consistent with the mass ratio found when oligarchic disks go unstable [*Kokubo and Ida*, 1998]. Due to dynamical friction, these bodies start on low eccentricity and low inclination orbits. Importantly, Jupiter and Saturn are assumed to have formed on their current orbits, and the N-body simulations begin after the gas has been removed from the disk.

This model is remarkably successful, and it provided confidence that these models in general are on the right path. Early numerical simulations that only included a limited number of embryos naturally formed about four planets near the correct semi-major axes of the terrestrial planets (see Figure 3.1), had giant impacts similar to the Moon-forming impact, and had a formation timescale of tens of millions of years for Earth-like planets [*Agnor et al.*, 1999; *Chambers*, 2001]. When enough planetesimals are included in the simulations, dynamical friction removes excess energy and angular momentum from the orbits of the growing planets transferring it to the planetesimal population and the simulations reproduce the low eccentricities and low inclinations of the terrestrial planets [*O'Brien et al.*, 2006; *Morishima et al.*, 2008]. Some but not all terrestrial planet systems created by the eccentric Jupiter and Saturn model have angular momentum deficits consistent with the Solar System [*O'Brien et al.*, 2006; *Raymond et al.*, 2009; *Fischer and Ciesla*, 2014]; since the eccentric Jupiter and Saturn model is inconsistent with a Nice model-like giant planet instability, the most consistent angular momentum deficits are those closest to that of the current terrestrial planets (the dashed line in Figure 3.2).

This model is haphazardly successful at delivering water to Earth-like planets from the outer asteroid belt via the ν_6 secular resonance [*Morbidelli et al.*, 2000; *Raymond et al.*, 2004; *O'Brien et al.*, 2006; *Raymond et al.*, 2009]. If most of Earth's water is delivered from the planetesimal carbonaceous chondrite parent bodies, then the D/H ratio of Earth's water is naturally explained (see *Morbidelli et al.*, 2000, for greater detail and a discussion of the less likely alternatives). Inspired by the data from primitive meteorites, a simple model for the water content of planetesimals as a function of radius is that the water mass fraction is 0.001% interior of 2 AU, 0.1% between 2 and 2.5 AU, and 5% exterior to 2.5 AU [*Raymond et al.*, 2004]. This model is used for all of the classical scenario simulations in Figure 3.3, but not for the Grand Tack simulations since that model assumes that the carbonaceous chondrite parent bodies originate from exterior to Jupiter and Saturn.

There are four outstanding issues with the eccentric Jupiter and Saturn model. First, this model systematically creates a planet at the position of Mars with a mass between 2 and 12 times that of Mars. As shown in Figure 3.4, there are a few exceptions to this rule, and this directly leads to low values of the concentration statistic as shown in Figure 3.2. These were discovered by simulating a much larger number of solar systems than had historically been studied before, emphasizing the need for this approach [*Fischer and Ciesla*, 2014]. Of the 62 numerical simulations, 6 (~10%) have Mars-like planets exterior of all Earth-like planets in the system. If Mars is an outlier of the eccentric Jupiter and Saturn model, then this is a challenge to the Copernican principle—that the Solar System is a usual rather than special outcome of planet formation processes. Before such a conclusion should be drawn, a more thorough investigation into what happened differently in these 6 simulations needs to be completed. *Raymond et al.* [2009] demonstrated that imperfect accretion via high velocity impacts cannot be responsible for the difference between the masses of the planets formed at Mars location and the mass of Mars. This was confirmed by *Chambers* [2013], who found that while imperfect accretion makes smaller planets overall including at Mars' location, it tends to decrease the mass ratio between Earth and Mars rather than increasing it relative to the perfect accretion simulations. Also, unlike Mercury, Mars appears inconsistent with this type of impact-driven mass loss.

Second, it does not explain how Jupiter and Saturn obtained their current eccentricities. Interactions with the nebular gas disk should leave the gas giants on quasi-circular orbits [*Kley*, 2000; *Bryden et al.*, 2000; *Snellgrove et al.*, 2001; *Papaloizou*, 2003; *Crida et al.*, 2008]. Also, giant planet migration is likely to move Jupiter and Saturn into a mean motion resonant configuration, and not into the near 5:2 mean motion configuration they currently occupy [*Papaloizou et al.*, 2001; *Kley and Dirksen*, 2006; *Bitsch et al.*, 2013].

Third, this model is inconsistent with the Nice model, and so does not directly account for the late heavy bombardment. As discussed in Section 3.1.1, alternative explanations for the late heavy bombardment exist although none are as complete or successful as the Nice model. Only if an alternative is accepted can the eccentric Jupiter and Saturn model be consistent with the evidence for the late heavy bombardment.

Fourth, the eccentric Jupiter and Saturn model is internally inconsistent since interactions with planetesimals and embryos in the terrestrial disk tend to circularize the giant planets [*Chambers*, 2001].

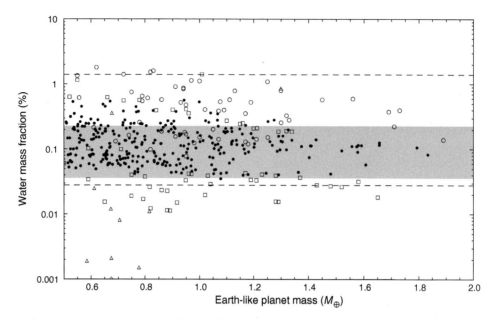

Figure 3.3 The water mass fraction for each Earth-like planet formed from the eccentric [empty squares; *Raymond et al.*, 2009; *Fischer and Ciesla*, 2014], circular [empty circles; *Raymond et al.*, 2009; *Fischer and Ciesla*, 2014], and extra-eccentric [empty triangles; *Raymond et al.*, 2009] Jupiter and Saturn models and the Grand Tack model [solid points; *Walsh et al.*, 2011; *Jacobson and Morbidelli*, 2014; *Jacobson et al.*, 2014; *O'Brien et al.*, 2014] as a function of the mass of the planet. The water mass fraction is the water mass delivered to the planet assuming perfect accretion and either the water model of *Raymond et al.* [2004] for the non-Grand Tack simulations or the model of *O'Brien et al.* [2014] for the Grand Tack simulations. Earth-like planets from the numerical simulations have masses between 0.5 and 2 $M_\oplus$ and orbits between the current orbits of Mercury and Mars as shown in Figure 3.1. The upper dashed line is a liberal estimate of Earth's water content [*Marty*, 2012], the gray region is a more probable estimate [*Lécuyer*, 1998], and the lower dashed line is a minimum [*Lécuyer et al.*, 1998], that is, an estimate of the known water reservoirs. The water content of Earth's mantle reservoir is unknown.

3.3.2. Extra-eccentric Jupiter and Saturn

The extra-eccentric Jupiter and Saturn model was devised by *Chambers and Cassen* [2002] to alleviate the fourth outstanding issue of the previous model. The giant planets are assumed to have formed at their current semi-major axes but are on more excited orbits. This dynamical excitation is transferred through both scattering events and secular resonances, principally the ν_5 and the ν_6, to the terrestrial disk, and this creates angular momentum deficits much higher than those possessed by current terrestrial planets in the Solar System [*Raymond et al.*, 2009], as shown in Figure 3.2.

The rapid depletion of the outer asteroid belt typically leads to a very dry Earth inconsistent with current water fraction estimates as shown in Figure 3.3. If the gas disk disperses in such a way that Mars mass embryos in the outer asteroid belt region, assumed to be water-rich, are swept up in evolving secular resonances with Jupiter, then they may be incorporated into Earth delivering the needed water [*Thommes et al.*, 2008; *Raymond et al.*, 2009]. Alternatively, water may be delivered by mechanisms listed by *Raymond et al.* [2009] such as water adsorption

of small silicate grains [*Muralidharan et al.*, 2008], comet impacts [*Owen and Bar-Nun*, 1995], or oxidation of an H-rich atmosphere [*Ikoma and Genda*, 2006].

Sweeping secular resonances combined with tidal gas drag can effectively deplete the asteroid belt by driving the inward migration of planetary embryos [*Nagasawa et al.*, 2005] and reproducing the small mass of Mars [*Thommes et al.*, 2008], but this requires a rather specifically tuned gas disk, particularly the gas removal rate [*Thommes et al.*, 2008]. Sweeping resonances due to changing precession rates as the gas is dispersed is not necessary, fixed frequency resonant and secular perturbations do clear out the asteroid belt and Mars-region occasionally creating a small Mars, as shown in Figure 3.4; however, there are often planetary embryos remaining in the asteroid belt [*Raymond et al.*, 2009]. Thus, the extra-eccentric Jupiter and Saturn scenario weakly alleviates the first outstanding issue identified above, but many final planets in the Mars region are still too large and since the excitation retards the growth of Earth-like planets, they tend to be too numerous and too small. This leads to unchanged concentration statistics as shown in Figure 3.2.

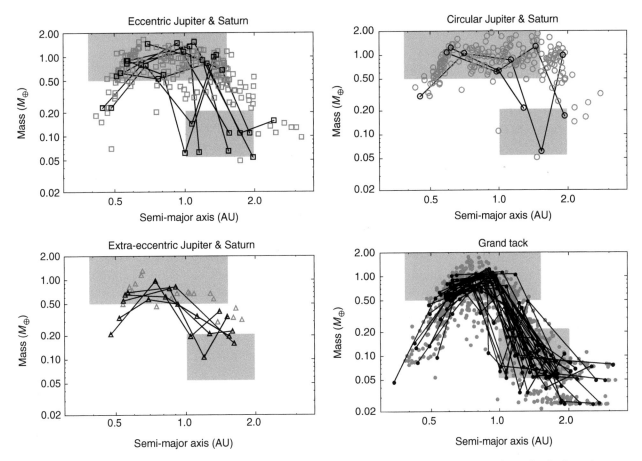

Figure 3.4 Lines connect the final planets of the good terrestrial system analogs. Otherwise, the individual graphs are very similar to Figure 3.1. For the classical models, good terrestrial system analogs have at least one Mars-like planet. For the Grand Tack model, good terrestrial system analogs have at least one Mars-like planet and both a Venus-like planet and Earth-like planet with no planets in between.

The extra-eccentric Jupiter and Saturn scenario accentuates the second problem identified with the current orbit scenario; the assumption of high initial eccentricities for the giant planets is inconsistent with models of their growth in a gas disk [*Morbidelli and Crida*, 2007; *Pierens and Raymond*, 2011; *Bitsch et al.*, 2013]. Similar to the eccentric Jupiter and Saturn scenario, this model is not coherent with the Nice model.

Finally, *Raymond et al.* [2009] identifies a number of other explanations for extra-eccentric giant planets such as an early giant planet instability leaving the giant planets on very eccentric orbits and the corresponding problems with these models.

3.3.3. Circular Jupiter and Saturn, pre-Nice 2.0 Model

Numerical simulations consistent with the Nice model automatically solve all of the latter three outstanding issues of the eccentric Jupiter and Saturn model [*Raymond et al.*, 2009]. Giant planets growing and migrating in a gas

disk circularize and enter resonances, and it is natural for Saturn and Jupiter to enter into a 3:2 mean motion resonance near 5.4 AU and 7.3 AU [*Morbidelli et al.*, 2007]. This configuration of the giant planets is primed for a giant planet instability at ~400 Myr due to interactions with a massive outer disk of which the Kuiper Belt is a remnant [*Levison et al.*, 2011].

This model has many of the same successes as the eccentric scenario: correct number of planets including Earth-like planets near 1 AU as shown in Figure 3.1, giant impacts similar to the Moon-forming impact [*Raymond et al.*, 2009], and growth timescales of tens of millions of years [*Raymond et al.*, 2009; *Fischer and Ciesla*, 2014]. Since the giant planets are on circular orbits, excitation due to secular resonances is reduced [*Raymond et al.*, 2009; *Fischer and Ciesla*, 2014], but since the giant planet instability in the Nice model will dynamically excite those planets, the target angular momentum deficit value is 10 to 70% lower than for the eccentric Jupiter and Saturn model [*Brasser et al.*, 2013]. Such systems are successfully created as shown in Figure 3.2. Another success is the delivery of water to Earth from the asteroid belt. Using the same model as before [*Raymond et al.*, 2004], enough water from the outer asteroid belt is always delivered to Earth as shown in Figure 3.3 [*Raymond et al.*, 2009; *Fischer and Ciesla*, 2014].

Notably though, the small Mars problem is accentuated, as shown in Figure 3.4. Planets at the location of Mars are almost all between 2 and 20 times too massive. Only 1 (~2%) of 62 systems has a Mars-like planet without also containing a third near-Earth mass planet. This reveals itself also as very low concentration statistics, as shown in Figure 3.2.

Although this model is consistent with the Nice model, it is only partially consistent with evolution in a gas disk. The circularization and migration of the giant planets into mean motion resonances in the circular Jupiter and Saturn model is assumed to have no effect on the evolution of the terrestrial disk. The Grand Tack model is the natural consequence of breaking this assumption.

3.4. TRUNCATED DISK MODELS

As mentioned, one of the outstanding problems with the classical models of terrestrial planet formation is the size of the planet formed near 1.5 AU. As shown in Figure 3.1, these models systematically produce planets at the location of Mars that are about 5 to 10 times too massive. *Raymond et al.* [2009] considered a broad range of models including those discussed above and a few others, but none consistently solved this problem and most introduced other issues as well. This problem is an old problem that has been recognized at least as far back as *Wetherill* [1991]. *Walsh and Morbidelli* [2011] considered planet migration and the resultant sweeping secular resonances as a way to deplete the region around 1.5 AU of solid material early enough to frustrate the accretion of Mars-analogs. This failed as the migration of the giant planets would happen too late compared to the accretion of Mars-like planets, leaving them at similar mass as in classical scenarios.

Hansen [2009] proposed a dramatic solution to this problem demanding wholesale changes to the surface density of the solid material in the disk. They found that if the initial disk is truncated at 1 AU then the mass distribution of the terrestrial planets could be recovered including the Earth/Mars mass ratio. The systematic success to the Hansen model stems not just from the disk truncation but the unstable dynamical over-packing of the 400 equal mass embryos, which begin the simulation with overlapping mutual Hill radii [*O'Brien et al.*, 2014]. These initial conditions lead to the rapid scattering of some embryos out of the narrow annulus. The bodies scattered out beyond 1 AU were typically stabilized due

to interactions with other scattered bodies and isolated beyond the edge of the disk. This allowed these bodies to stop their accretion early and remain at a low mass, resulting in a short accretion time relative to the planets that accrete within the annulus. This is similar physics to that observed in previous works that had disk edges around 1 or 1.5 AU due to computational requirements [*Agnor et al.*, 1999; *Chambers*, 2001].

Jin et al. [2008] proposed that severe surface density depletions in the proto-planetary disk cause a truncation. Since different angular momentum transport mechanisms are active at different radii in the disk, *Jin et al.* [2008] found a non-monotonic gas surface density for the inner regions of the disk and a surface density minimum at about 1.5 AU. *Izidoro et al.* [2014] performed N-body experiments and deduced that to best fit Mars, a ~75% surface depletion between 1.3 and 2 AU is needed, however the prediction by *Jin et al.* [2008] appear to be more consistent with ~25% and over a narrower annulus. Unlike the hypothesis proposed in *Jin et al.* [2008], *Izidoro et al.* [2014] find that Mars is a stranded embryo scattered out from the interior region of higher surface density, similar to the Grand Tack. Dissimilar to the Grand Tack and like the classical scenario, the asteroid belt is grown from material already in the low surface density region in the *Jin et al.* [2008] and *Izidoro et al.* [2014] models.

3.4.1. 'Grand Tack', Migrating Jupiter and Saturn

A narrow over-packed disk with a truncation of solid material at 1 AU produces ideal outcomes for the Earth/ Mars mass ratio [*Hansen*, 2009], but why such a disk existed needs an explanation. However, a powerful means to re-distribute solid material in the early Solar System are the giant planets. The giant planets must have had a relatively short accretion timescale compared to that of the terrestrial planets (see Section 3.1.1) and so existed to move around terrestrial building block material. Hydrodynamic models and data from extra-solar planetary systems suggest that planets the size of Jupiter can carve an annular gap in a gaseous nebula resulting in their inward migration due to the viscous evolution of the disk [*Ward*, 1980; *Lin and Papaloizou*, 1986]. This migration, type-II migration, can lead to the inward migration of a Jupiter-mass planet on 100,000-year timescales.

A series of works found that the presence of a second massive planet, a Saturn mass planet, can halt and even reverse the inward migration of a Jupiter mass planet [*Masset and Snellgrove*, 2001; *Pierens and Nelson*, 2008; *Pierens et al.*, 2014; *Crida and Morbidelli*, 2007; *Pierens and Raymond*, 2011; *D'Angelo and Marzari*, 2012]. Saturn is found to migrate very close to Jupiter, reaching an exterior 2:3 mean motion resonance with Jupiter, and this proximity causes their gaps to overlap. The net effect is that Saturn's presence allows gas to pass through Jupiter's gap, stopping the inward type II migration. The lower surface density of gas behind Jupiter, owing to the presence of Saturn's gap, then allows for the inner disk to provide a much stronger torque on Jupiter's orbit pushing both planets outward.

The net effect of this inward-then-outward migration scenario is that it provides a possible mechanism to create a truncated disk of solids. *Walsh et al.* [2011] found that if Jupiter reversed its inward migration at 1.5 AU, that is, 'tacked', then the disk of solids would be truncated at 1 AU, however, the violent mechanism causing the truncation and over-packing the disk leaves a more excited initial state than that previously explored by *Hansen* [2009]. Similarly, while *Hansen* [2009] started with 400 similar-sized bodies in the annulus, the *Walsh et al.* [2011] model, dubbed 'the Grand Tack,' started with a bi-modal size distribution of embryos and planetesimals. Despite these differing initial conditions, the terrestrial planets produced were similar including the high Earth/Mars mass ratio as shown in Figure 3.1.

The Grand Tack creates an asteroid belt very differently than the classical models since Jupiter migrates through the asteroid belt twice. Rather than depleting a pre-existing belt via secular resonances and scattering events, the asteroid belt in the Grand Tack is created by the inward scattering of material by the giant planets as they migrate outward, but not all of this material was originally from the outer Solar System. Some of it had been scattered into the outer Solar System from the inner Solar System originally. *Walsh et al.* [2011] hypothesize that material from the outer Solar System corresponds to the broad C-complex of asteroids associated with the carbonaceous chondrites and the material that originated in the inner Solar System but was scattered out and then back in corresponds to the broad S-complex of asteroids associated with the ordinary chondrites. The classical models explain the observed, albeit messy, gradient from S-complex to C-complex in the asteroid belt [*DeMeo and Carry*, 2013, 2014] as a local property of the disk. However, the Grand Tack asserts that this transition occurs over a much larger radial section of the disk (interior and exterior to the giant planets), but the migration of the giant planets naturally re-creates this gradient in the asteroid belt [*Walsh et al.*, 2011, 2012].

This hypothesis has strong implications for delivery of water to Earth since the D/H ratios of carbonaceous chondrites best match Earth's water [*Morbidelli et al.*, 2000]. These outer planetesimals account for a total of 1 to 3% of the total accreted mass for each of the terrestrial planets [*Walsh et al.*, 2011; *O'Brien et al.*, 2014]. As shown in Figure 3.3, they deliver enough water to Earth assuming that they contain ~10% water by mass.

Moreover, unlike the classical models, the mass of C-complex material delivered to the planets is independently calibrated because the initial ratio of C-complex to S-complex material in the protoplanetary disk is known from the ratio that survives in the asteroid belt [*DeMeo and Carry*, 2013, 2014].

The Grand Tack solves each of the outstanding problems identified with the eccentric Jupiter and Saturn model. The giant planets and the terrestrial disk evolve self-consistently in a gas disk, and the Nice model is naturally incorporated. Venus-like, Earth-like, and Mars-like planets are all natural outcomes, and many systems have an architecture similar to the Solar System as shown in Figure 3.4. This is why the Grand Tack alone matches both the concentration statistic and the necessary angular momentum deficit, pre-Nice model in this case, as shown in Figure 3.2.

There are still some open issues regarding the Grand Tack such as the formation of Mercury, which may just be a result of numerical resolution, and some discussed in *Raymond and Morbidelli* [2014].

3.5. EARTH IN THE GRAND TACK MODEL

The Grand Tack model is best able to reproduce the mass-orbit distribution of the terrestrial planets and is the most consistent with the history of the Solar System. This includes reproducing the rapid growth of Mars [*Dauphas and Pourmand*, 2011; *Nimmo and Kleine*, 2007] by acknowledging that Mars is a stranded embryo [*Jacobson and Morbidelli*, 2014]. If Mars is an embryo then we have some information regarding the size of planetary embryos, and if planetesimals are born big, ~100 to 1000 km, as the new story regarding pebble formation suggests, then the only remaining significant free parameter describing the bi-modal mass distribution is the ratio of the total mass of the embryo population to the total mass of the planetesimal population. This ratio has direct consequences for the dynamical friction present in the terrestrial disk: more planetesimal mass means more dynamical friction [*Ida and Makino*, 1993].

Jacobson and Morbidelli [2014] showed that increased dynamical friction also leads to an earlier last giant impact since lower relative velocities accelerate growth. Thus, the later the last giant impact, which is also the Moon-forming impact, then the embryo population must be more massive relative to the planetesimal population. *Jacobson et al.* [2014] showed there is a strong correlation between the time of the last giant impact and the late accreted mass, which is the mass accreted by Earth-like planets after the last giant impact. This makes sense because the mass of planetesimals in the disk decays with time, so the amount of planetesimal mass accreted after a later time is smaller than after an earlier time.

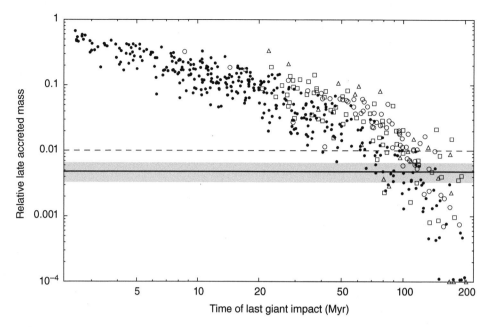

Figure 3.5 The relative late accreted mass for each Earth-like planet formed from the eccentric [empty squares; *O'Brien et al., 2006; Raymond et al., 2009; Fischer and Ciesla, 2014*], circular [empty circles; *O'Brien et al., 2006; Raymond et al., 2009; Fischer and Ciesla, 2014*], and extra-eccentric [empty triangles; *Raymond et al., 2009*] Jupiter and Saturn models and the Grand Tack model [solid points; *Walsh et al., 2011; Jacobson and Morbidelli, 2014; Jacobson et al., 2014; O'Brien et al., 2014*] as a function of the time of the last giant impact. The relative late accreted mass is the mass accreted after the last giant (embryo) impact divided by the final mass of the planet. Earth-like planets from the numerical simulations have masses between 0.5 and 2 $M_\oplus$ and orbits between the current orbits of Mercury and Mars as shown in Figure 1.

This correlation for both the Grand Tack and the classical models is shown in Figure 3.5. Classical models always date the Moon-forming impact later than the Grand Tack.

The geologic record in Earth's mantle provides an estimate of the late accreted mass. From the highly siderophile element record on Earth [*Becker et al.*, 2006; *Walker*, 2009], it appears that ~5 × 10⁻³ Earth masses, $M_\oplus$, of material was delivered to Earth after the Moon-forming event (many lines of evidence restrict this late accreted mass to below 0.01 $M_\oplus$ [*Jacobson et al.*, 2014, also see Morbidelli and Wood in this monograph]). Using the correlation between the late accreted mass and the time of the last giant impact, this late accreted mass estimate dates the Moon-forming impact to ~95 Myr [*Jacobson et al.*, 2014]. Such a late last giant impact requires that most of the mass in the terrestrial disk be in the embryo population rather than the planetesimal population when Jupiter's migration interrupts the oligarchic growth phase [*Jacobson and Morbidelli*, 2014]. This is consistent with models of the oligarchic growth phase that include planetesimal grinding [*Levison et al.*, 2012] as well as a pebble model that rarely creates planetesimals via the streaming instability but efficiently turns planetesimals into embryos via pebble accretion.

3.5.1. The Growth of Earth

We now delve deep into the details of the growth of Earth in the Grand Tack scenario. In the prior sections, we were exploring the overall differences between the terrestrial planet formation models. Here, we focus on specific simulations that produce terrestrial planet systems most like the Solar System. In order to do that we make stricter requirements. We require that a Venus-like planet with mass between 0.4075 and 1.63 $M_\oplus$ be directly interior to an Earth-like planet with a mass between 0.5 and 2 $M_\oplus$ with at least one Mars-like planet exterior to both of those planets with a mass between 0.0535 and 0.114 $M_\oplus$. Furthermore, the Earth-like planet must have a late accreted mass between 0.001 and 0.01 $M_\oplus$. These stringent requirements leave only 10 planetary systems out of the 203 Grand Tack numerical simulations from *Walsh et al.* [2011], *O'Brien et al.* [2014], *Jacobson et al.* [2014], and *Jacobson and Morbidelli* [2014], and all are from the latter two publications since they include the only simulations with a high enough total embryo to planetesimal mass ratio. Three of the ten have two Mars-like planets, but we include them for better statistics.

The growth of these Earth-like planets are shown in Figure 3.6. They all follow a very similar trajectory and

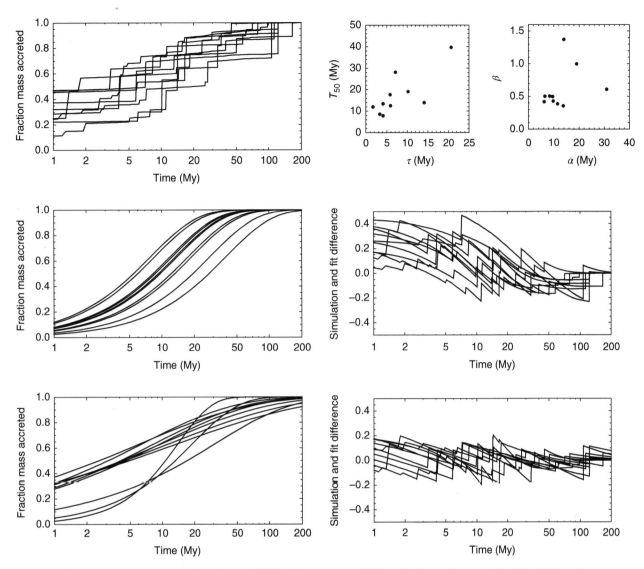

Figure 3.6 Growth curve analysis for Earth-like planets. Upper left graph shows the growth of Earth-like planets as described in Section 3.1, and the left middle and bottom graphs are fits using exponential and Weibull models, respectively. The upper center plot shows the time to reach 50% of the final mass as a function of the exponential fit timescale, and the upper right plot shows the α and β of each of the Weibull fits. The right middle and bottom graphs are the difference between the growth curve and the exponential and Weibull fits, respectively.

grow quickly at first reaching 50% of their growth in $T_{50} \sim 10$ Myr, however, their growth is strongly non-linear. In the past, many have approximated this non-linear growth with an exponential function $M(t) \propto e^{t/\tau}$. The best fits to these simulations have exponential timescales $\tau \sim 3$ to 10 Myr, but we find exponential fits to be inappropriate for the Grand Tack. Figure 3.6 shows exponential fits to the growth profile of each Earth-like planet, and then shows the difference between the actual growth trajectory and those fits. There is a clear underestimate of the growth rate at early times. A much better model is the Weibull model $M(t) \propto e^{(t/\alpha)^{\beta}}$. The best fits to these simulations are shown in Figure 3.6 with $\alpha \sim 10$ Myr and $\beta \sim 0.5$.

This model has no systematic offset; however, even this model can be off by 20%.

3.5.2. Composition of Earth and the Other Terrestrial Planets

In the Grand Tack, the inner Solar System is thoroughly radially mixed by the inward migration of Jupiter [*O'Brien et al.*, 2014]. This naturally produces Venus-like and Earth-like planets that are compositionally very similar. Both planets contain multiple embryos and many planetesimals from throughout the disk. Figure 3.7 shows their compositions as a cumulative function of semi-major axis.

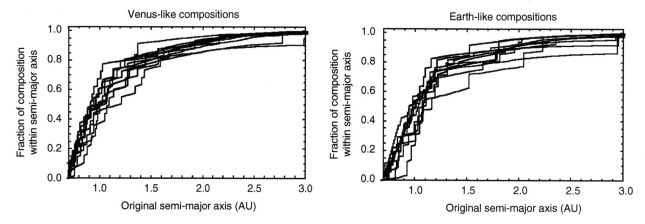

Figure 3.7 The composition curves for Venus-like and Earth-like planets. These curves correspond to the terrestrial planetary systems described in Section 3.1. The composition curves show the cumulative mass in the final planet as a function of semimajor axis. Some mass is accreted from exterior of 3 AU.

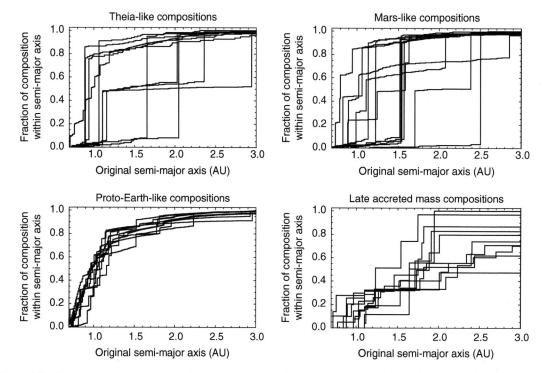

Figure 3.8 The composition curves for the Theia-like, Mars-like, and proto-Earth-like planets and the late accreted mass composition. These curves correspond to the terrestrial planetary systems described in Section 3.1. The composition curves show the cumulative mass in the final planet as a function of semimajor axis. Some mass is accreted from exterior of 3 AU.

There are three different compositional regions made clear. First, ~75% of the mass originates from 0.7 AU, the inner edge of the terrestrial disk, to ~1.1 AU, the location of the 3:2 inner mean motion resonance with the location of Jupiter's 'tack' at 1.5 AU. The mass from within this region is remarkably linearly representative. About ~22 to 25% of each planet's mass originates from exterior of 1.1 AU to the outer edge of the terrestrial disk at about 4 AU. This mass is not delivered linearly with semi-major axis, but with slightly more mass delivered from the inner section than the outer. This is because material scattered by Jupiter further out in the disk was less likely to be scattered all the way into the 1 AU region where the proto-Earth and proto-Venus were growing. The final ~1 to 3% of mass comes from the outer disk exterior to the giant planets [*O'Brien et al.*, 2014].

The feeding zones of the proto-Earth before the last giant (Moon-forming) impact is shown for each simulation in Figure 3.8. It very much resembles the final Earth-like planet, which makes sense because Theia, the last giant impactor, and the late accreted mass, which are also shown in Figure 3.8, contribute only a small amount of mass. The late accreted mass appears to sample the entire inner disk. Unlike the final Earth-like planet, the composition of the late accreted mass doesn't reflect the location of Jupiter's tack.

Theia is usually a stranded embryo like Mars, although sometimes Theia is much larger and looks very similar to the proto-Earth. The feeding zones of Mars-like planets are also shown in Figure 3.8, and these reveal that while Mars-like planets often come from the outer disk, they do not always. Similarly, while Theia often comes from the inner terrestrial disk near 1 AU, it can come from the outer terrestrial disk. Some Mars-like planet compositions are composed of different embryos and so are unlikely to match the Hf-W growth constraints.

3.6. CONCLUSION AND DISCUSSION

The growth of Earth cannot be isolated from the formation of the rest of the Solar System. Although the number of constraints regarding the history of the Solar System are few, they are powerful. The mass and orbit distribution of the terrestrial planets and the evidence for the late heavy bombardment rule out the classical terrestrial plant formation models. Trunfcated disk models reproduce the mass and orbit distribution of the terrestrial planets, particularly the Earth/Mars mass ratio. Of these models, only the Grand Tack is completely self-consistent and consistent with a late heavy bombardment via the Nice model.

The growth of Earth in the Grand Tack model follows a Weibull accretion curve to within 20%. During its growth, it samples ~75% of its mass linearly from the inner terrestrial disk edge to the location of the 3:2 inner mean motion resonance with the location of Jupiter's tack. The remainder of its mass comes predominately from the rest of the terrestrial disk but biased toward the regions exterior but closes to 1.1 AU. Plenty of water is delivered to Earth via C-complex asteroids from the outer Solar System. The total mass of these asteroids delivered into the inner Solar System is calibrated by the ratio of S-complex to C-complex asteroids in the Main Belt.

From the highly siderophile record on Earth, the age of the Moon is estimated to be ~95 Myr [*Jacobson et al.*, 2014]. Since this age estimate is high, the ratio of the total mass in the embryo population to the total mass in the planetesimal population must also be high [*Jacobson and Morbidelli*, 2014]. Since Mars is a stranded embryo [*Nimmo and Kleine*, 2007; *Dauphas and Pourmand*, 2011], the mass of individual embryos is also known [*Jacobson and Morbidelli*, 2014]. These two facts strongly constrain the structure of the terrestrial disk at the time of Jupiter's inward migration. Even though this could be explained by planetesimal grinding, it could also be the result of pebble processes.

Pebble processes are likely the key to growing past the boulder barrier and creating planetesimals via the streaming instability. They also likely contribute a significant amount of mass to embryos via pebble accretion. These processes are likely to leave unique compositional gradients and signatures in the disk. Combining these gradients with models of the composition of Earth and observations of Earth's composition will be a powerful test of these hypotheses.

REFERENCES

Agnor, C. B., R. M. Canup, and H. F. Levison (1999), On the Character and Consequences of Large Impacts in the Late Stage of Terrestrial Planet Formation, *ICARUS*, *142*(1), 219–237.

Alexander, C. M. O. D., J. N. Grossman, D. S. Ebel, and F. J. Ciesla (2008), The Formation Conditions of Chondrules and Chondrites, *Science*, *320*(5), 1617.

Allègre, C. J., G. Manhès, and C. Göpel (2008), The major differentiation of the Earth at ~4.45 Ga, *Earth and Planetary Science Letters*, *267*(1-2), 386–398.

Armitage, P. J. (2014), Lecture notes on the formation and early evolution of planetary systems, pp. 1–71.

Becker, H., M. F. Horan, R. J. Walker, S. Gao, J. P. Lorand, and R. L. Rudnick (2006), Highly siderophile element composition of the Earth's primitive upper mantle: Constraints from new data on peridotite massifs and xenoliths, *Geochimica et Cosmochimica Acta.*, *70*(17), 4528–4550.

Bitsch, B., A. Crida, A. S. Libert, and E. Lega (2013), Highly inclined and eccentric massive planets. I. Planet-disc interactions, *Astronomy and Astrophysics*, *555*, 124.

Blum, J. and G. Wurm (2008), The Growth Mechanisms of Macroscopic Bodies in Protoplanetary Disks, *Annual Review of Astronomy and Astrophysics*, *46*(1), 21–56.

Bottke, W. F., D. Nesvorný, D. Vokrouhlický, and A. Morbidelli (2010), The Irregular Satellites: The Most Collisionally Evolved Populations in the Solar System, *The Astronomical Journal*, *139*(3), 994–1014.

Bottke, W. F., D. Vokrouhlický, S. Marchi, T. Swindle, E. R. D. Scott, and J. R. Weirich (2014), The Evolution of Giant Impact Ejecta and the Age of the Moon, *45th Lunar and Planetary Science Conference*, *45*, 1611.

Brasser, R., K. J. Walsh, and D. Nesvorný (2013), Constraining the primordial orbits of the terrestrial planets, *Monthly Notices of the Royal Astronomical Society*, *433*(4), 3417–3427.

Briceño, C., A. K. Vivas, N. Calvet, L. Hartmann, R. Pacheco, D. Herrera, L. Romero, P. Berlind, G. Sánchez, J. A. Snyder, and P. Andrews (2001), The CIDA-QUEST Large-Scale Survey of Orion OB1: Evidence for Rapid Disk Dissipation in a Dispersed Stellar Population, *Science*, *291*(5), 93–97.

Bryden, G., M. Rożyczka, D. N. C. Lin, and P. Bodenheimer (2000), On the Interaction between Protoplanets and Protostellar Disks, *The Astrophysical Journal*, *540*(2), 1091–1101.

Carballido, A., S. Fromang, and J. Papaloizou (2006), Midplane sedimentation of large solid bodies in turbulent protoplanetary discs, *Monthly Notices of the Royal Astronomical Society*, *373*(4), 1633–1640.

Carporzen, L., B. P. Weiss, L. T. Elkins-Tanton, D. L. Shuster, D. Ebel, and J. Gattacceca (2011), From the Cover: Magnetic evidence for a partially differentiated carbonaceous chondrite parent body, in *Proceedings of the National Academy of Sciences*, pp. 6386–6389.

Castillo-Rogez, J. C., D. L. Matson, C. Sotin, T. V. Johnson, J. I. Lunine, and P. C. Thomas (2007), Iapetus' geophysics: Rotation rate, shape, and equatorial ridge, *ICARUS*, *190*(1), 179–202.

Castillo-Rogez, J. C., T. V. Johnson, M. H. Lee, N. J. Turner, D. L. Matson, and J. Lunine (2009), 26Al decay: Heat production and a revised age for Iapetus, *ICARUS*, *204*(2), 658–662.

Chambers, J. E. (2001), Making More Terrestrial Planets, *ICARUS*, *152*(2), 205–224.

Chambers, J. E. (2007), On the stability of a planet between Mars and the asteroid belt: Implications for the Planet V hypothesis, *ICARUS*, *189*(2), 386–400.

Chambers, J. E. (2010), Planetesimal formation by turbulent concentration, *ICARUS*, *208*(2), 505–517.

Chambers, J. E. (2013), Late-stage planetary accretion including hit-and-run collisions and fragmentation, *ICARUS*, *224*(1), 43–56.

Chambers, J. E. (2014), Giant planet formation with pebble accretion, *ICARUS*, *233*, 83–100.

Chambers, J. E. and P. M. Cassen (2002), The effects of nebula surface density profile and giant-planet eccentricities on planetary accretion in the inner solar system, *Meteoritics & Planetary Science*, *37*, 1523–1540.

Chapman, C. R., B. A. Cohen, and D. H. Grinspoon (2007), What are the real constraints on the existence and magnitude of the late heavy bombardment?, *ICARUS*, *189*(1), 233–245.

Cohen, B. A., T. D. Swindle, and D. A. Kring (2000), Support for the Lunar Cataclysm Hypothesis from Lunar Meteorite Impact Melt Ages, *Science*, *290*(5), 1754–1756.

Crida, A. and A. Morbidelli (2007), Cavity opening by a giant planet in a protoplanetary disc and effects on planetary migration, *Monthly Notices of the Royal Astronomical Society*, *377*(3), 1324–1336.

Crida, A., Z. Sándor, and W. Kley (2008), Influence of an inner disc on the orbital evolution of massive planets migrating in resonance, *Astronomy and Astrophysics*, *483*(1), 325–337.

Ćuk, M. (2012), Chronology and sources of lunar impact bombardment, *ICARUS*, *218*(1), 69–79.

Ćuk, M., B. J. Gladman, and S. T. Stewart (2010), Constraints on the source of lunar cataclysm impactors, *ICARUS*, *207*(2), 590–594.

Cuzzi, J. N., A. R. Dobrovolskis, and J. M. Champney (1993), Particle-gas dynamics in the midplane of a protoplanetary nebula, *ICARUS*, *106*, 102.

Cuzzi, J. N., R. C. Hogan, and K. Shariff (2008), Toward Planetesimals: Dense Chondrule Clumps in the Protoplanetary Nebula, *The Astrophysical Journal*, *687*(2), 1432–1447.

D'Angelo, G. and F. Marzari (2012), Outward Migration of Jupiter and Saturn in Evolved Gaseous Disks, *The Astrophysical Journal*, *757*(1), 50.

Dauphas, N. and A. Pourmand (2011), Hf-W-Th evidence for rapid growth of Mars and its status as a planetary embryo, *Nature*, *473*(7), 489–492.

DeMeo, F. E. and B. Carry (2013), The taxonomic distribution of asteroids from multi-filter all-sky photometric surveys, *ICARUS*, *226*(1), 723–741.

DeMeo, F. E. and B. Carry (2014), Solar System evolution from compositional mapping of the asteroid belt, *Nature*, *505*(7), 629–634.

Dominik, C., J. Blum, J. N. Cuzzi, and G. Wurm (2007), Growth of Dust as the Initial Step Toward Planet Formation, *Protostars and Planets V*, pp. 783–800.

Dubrulle, B., G. Morfill, and M. Sterzik (1995), The dust sub-disk in the protoplanetary nebula, *ICARUS*, *114*, 237–246.

Elkins-Tanton, L. T., B. P. Weiss, and M. T. Zuber (2011), Chondrites as samples of differentiated planetesimals, *Earth and Planetary Science Letters*, *305*(1), 1–10.

Fernandez, J. A. and W. H. Ip (1984), Some dynamical aspects of the accretion of Uranus and Neptune—The exchange of orbital angular momentum with planetesimals, *Icarus (ISSN 0019-1035)*, *58*, 109–120.

Fischer, R. A. and F. J. Ciesla (2014), Dynamics of the terrestrial planets from a large number of N-body simulations, *Earth and Planetary Science Letters*, *392*, 28–38.

Goldreich, P. and W. R. Ward (1973), The Formation of Planetesimals, *Astrophysical Journal*, *183*, 1051–1062.

Gomes, R. S. (2003), The origin of the Kuiper Belt high-inclination population, *ICARUS*, *161*(2), 404–418.

Gomes, R. S., H. F. Levison, K. Tsiganis, and A. Morbidelli (2005), Origin of the cataclysmic Late Heavy Bombardment period of the terrestrial planets, *Nature*, *435*(7), 466–469.

Greenberg, R., J. F. Wacker, W. K. Hartmann, and C. R. Chapman (1978), Planetesimals to planets: Numerical simulation of collisional evolution, *ICARUS*, *35*(1), 1–26.

Guillot, T., S. Ida, and C. W. Ormel (2014), On the filtering and processing of dust by planetesimals 1. Derivation of collision probabilities for non-drifting planetesimals, *arXiv.org*, p. 7328.

Haisch, K. E. J., E. A. Lada, and C. J. Lada (2001), Disk Frequencies and Lifetimes in Young Clusters, *The Astrophysical Journal*, *553*(2), L153–L156.

Halliday, A. N. (2008), A young Moon-forming giant impact at 70-110 million years accompanied by late-stage mixing, core formation and degassing of the Earth, *Philosophical Transactions of the Royal Society A: Mathematical, Physical and Engineering Sciences*, *366*(1883), 4163–4181.

Hansen, B. M. S. (2009), Formation of the Terrestrial Planets from a Narrow Annulus, *The Astrophysical Journal*, *703*(1), 1131–1140.

Hopkins, P. F. (2014), Jumping the Gap: The Formation Conditions and Mass Function of Pebble-Pile Planetesimals, *arXiv.org*, p. 2458.

Ida, S. and D. N. C. Lin (2008), Toward a Deterministic Model of Planetary Formation. IV. *Effects of Type I Migration, The Astrophysical Journal*, *673*(1), 487–501.

Ida, S. and J. Makino (1992), N-body simulation of gravitational interaction between planetesimals and a protoplanet. I. Velocity distribution of planetesimals, *ICARUS*, *96*, 107–120.

Ida, S. and J. Makino (1993), Scattering of planetesimals by a protoplanet—Slowing down of runaway growth, *ICARUS*, *106*, 210.

Ikoma, M. and H. Genda (2006), Constraints on the Mass of a Habitable Planet with Water of Nebular Origin, *The Astrophysical Journal*, *648*(1), 696–706.

Iwasaki, K., H. Emori, K. Nakazawa, and H. Tanaka (2002), Orbital Stability of a Protoplanet System under a Drag Force Proportional to the Random Velocity, *Publications of the Astronomical Society of Japan*, *54*, 471–479.

Izidoro, A., N. Haghighipour, O. C. Winter, and M. Tsuchida (2014), Terrestrial Planet Formation in a Protoplanetary Disk with a Local Mass Depletion: A Successful Scenario for the Formation of Mars, *The Astrophysical Journal*, *782*(1), 31.

Jacobsen, S. B. (2005), The Hf-W Isotopic System and the Origin of the Earth and Moon, *Annual Review of Earth and Planetary Sciences*, *33*, 531–570.

Jacobson, S. A. and A. Morbidelli (2014), Lunar and terrestrial planet formation in the Grand Tack scenario, *Phil. Trans. R. Soc. A*, *372*, 0174.

Jacobson, S. A., A. Morbidelli, S. N. Raymond, D. P. O'Brien, K. J. Walsh, and D. C. Rubie (2014), Highly siderophile elements in Earth's mantle as a clock for the Moon-forming impact, *Nature*, *508*(7494), 84–87.

Jansson, K. and A. Johansen (2014), Formation of pebble-pile planetesimals, *arXiv.org*, p. 2535.

Jin, L., W. D. Arnett, N. Sui, and X. Wang (2008), An Interpretation of the Anomalously Low Mass of Mars, *The Astrophysical Journal*, *674*(2), L105–L108.

Johansen, A. and P. Lacerda (2010), Prograde rotation of protoplanets by accretion of pebbles in a gaseous environment, *Monthly Notices of the Royal Astronomical Society*, *404*, 475–485.

Johansen, A., J. S. Oishi, M.-M. Mac Low, H. H. Klahr, T. Henning, and A. Youdin (2007), Rapid planetesimal formation in turbulent circumstellar disks, *Nature*, *448*(7), 1022–1025.

Johansen, A., J. Blum, H. Tanaka, C. Ormel, M. Bizzarro, and H. Rickman (2014), The multifaceted planetesimal formation process, *arXiv.org*, p. 1344.

Kita, N. T., G. R. Huss, S. Tachibana, Y. Amelin, L. E. Nyquist, and I. D. Hutcheon (2005), Constraints on the Origin of Chondrules and CAIs from Short-lived and Long-Lived Radionuclides, *Chondrites and the Protoplanetary Disk*, *341*, 558.

Kley, W. (2000), The Orbital Evolution of Planets in Disks, *Disks*, *219*, 69.

Kley, W. and G. Dirksen (2006), Disk eccentricity and embedded planets, *Astronomy and Astrophysics*, *447*(1), 369–377.

Kobayashi, H. and N. Dauphas (2013), Small planetesimals in a massive disk formed Mars, *ICARUS*, *225*(1), 122–130.

Kobayashi, H. and H. Tanaka (2010), Fragmentation model dependence of collision cascades, *ICARUS*, *206*(2), 735–746.

Kokubo, E. and S. Ida (1995), Orbital evolution of protoplanets embedded in a swarm of planetesimals, *ICARUS*, *114*, 247–257.

Kokubo, E. and S. Ida (1996), On Runaway Growth of Planetesimals, *ICARUS*, *123*(1), 180–191.

Kokubo, E. and S. Ida (1998), Oligarchic Growth of Protoplanets, *ICARUS*, *131*(1), 171–178.

Kretke, K. A. and H. F. Levison (2014), Challenges in Forming the Solar System's Giant Planet Cores via Pebble Accretion, *arXiv.org*, p. 4430.

Lambrechts, M. and A. Johansen (2012), Rapid growth of gas-giant cores by pebble accretion, *Astronomy and Astrophysics*, *544*, 32.

Lambrechts, M. and A. Johansen (2014), Forming the cores of giant planets from the radial pebble flux in protoplanetary discs, *Astronomy and Astrophysics*, *572*, A107.

Lambrechts, M., A. Johansen, and A. Morbidelli (2014), Separating gas-giant and ice-giant planets by halting pebble accretion, *Astronomy and Astrophysics*, *572*, A35.

Laskar, J. (1997), Large scale chaos and the spacing of the inner planets, *Astronomy and Astrophysics*, *317*, L75–L78.

Lécuyer, C. (1998), The hydrogen isotope composition of seawater and the global water cycle, *Chem. Geol.*, *145*, 249–261.

Levison, H. F. and A. Morbidelli (2003), The formation of the Kuiper belt by the outward transport of bodies during Neptune's migration, *Nature*, *426*(6), 419–421.

Levison, H. F., A. Morbidelli, C. Van Laerhoven, R. S. Gomes, and K. Tsiganis (2008), Origin of the structure of the Kuiper belt during a dynamical instability in the orbits of Uranus and Neptune, *ICARUS*, *196*(1), 258–273.

Levison, H. F., E. Thommes, and M. J. Duncan (2010), Modeling the Formation of Giant Planet Cores. I. Evaluating Key Processes, *The Astronomical Journal*, *139*(4), 1297–1314.

Levison, H. F., A. Morbidelli, K. Tsiganis, D. Nesvorný, and R. S. Gomes (2011), Late Orbital Instabilities in the Outer Planets Induced by Interaction with a Self-gravitating Planetesimal Disk, *The Astronomical Journal*, *142*(5), 152.

Levison, H. F., M. J. Duncan, and D. M. Minton (2012), Modeling Terrestrial Planet Formation with Full Dynamics, Accretion, and Fragmentation, *AAS/Division for Planetary Sciences Meeting Abstracts*, *44*.

Levison, H. F., K. Kretke, and K. J. Walsh (2014), The Formation of Terrestrial Planets from the Direct Accretion of Pebbles, *AAS/Division for Planetary Sciences Meeting Abstracts*, *46*.

Lin, D. N. C., and J. Papaloizou (1986), On the tidal interaction between protoplanets and the protoplanetary disk. III - Orbital migration of protoplanets, *The Astrophysical Journal*, *309*, 846–857.

Lissauer, J. J. (1987), Timescales for planetary accretion and the structure of the protoplanetary disk, *ICARUS*, *69*, 249–265.

Malhotra, R. (1995), The Origin of Pluto's Orbit: Implications for the Solar System Beyond Neptune, *Astronomical Journal v. 110*, *110*, 420.

Mamajek, E. E. (2009), Initial Conditions of Planet Formation: Lifetimes of Primordial Disks, in *EXOPLANETS AND DISKS: THEIR FORMATION AND DIVERSITY: Proceedings of the International Conference. AIP Conference Proceedings*, pp. 3–10, Department of Physics and Astronomy, University of Rochester, Rochester, NY 14627-0171.

Marchi, S., W. F. Bottke, B. A. Cohen, K. Wünnemann, D. A. Kring, H. Y. McSween, M. C. de Sanctis, D. P. O'Brien, P. Schenk, C. A. Raymond, and C. T. Russell (2013), High-velocity collisions from the lunar cataclysm recorded in asteroidal meteorites, *Nature Geoscience*, *6*(4), 303–307.

Marchi, S., W. F. Bottke, L. T. Elkins-Tanton, M. Bierhaus, K. Wuennemann, A. Morbidelli, and D. A. Kring (2014), Widespread mixing and burial of Earth's Hadean crust by asteroid impacts, *Nature*, *511*(7), 578–582.

Marty, B. (2012), The origins and concentrations of water, carbon, nitrogen and noble gases on Earth, *Earth and Planetary Science Letters*, *313*, 56–66.

Masset, F. and M. Snellgrove (2001), Reversing type II migration: resonance trapping of a lighter giant protoplanet, *Monthly Notices of the Royal Astronomical Society*, *320*(4), L55–L59.

Morbidelli, A. and A. Crida (2007), The dynamics of Jupiter and Saturn in the gaseous protoplanetary disk, *ICARUS*, *191*(1), 158–171.

Morbidelli, A., J. E. Chambers, J. I. Lunine, J.-M. Petit, F. Robert, G. B. Valsecchi, and K. E. Cyr (2000), Source regions and time scales for the delivery of water to Earth, *Meteoritics & Planetary Science*, *35*, 1309–1320.

Morbidelli, A., H. F. Levison, K. Tsiganis, and R. S. Gomes (2005), Chaotic capture of Jupiter's Trojan asteroids in the early Solar System, *Nature*, *435*(7041), 462–465.

Morbidelli, A., K. Tsiganis, A. Crida, H. F. Levison, and R. S. Gomes (2007), Dynamics of the Giant Planets of the Solar System in the Gaseous Protoplanetary Disk and Their Relationship to the Current Orbital Architecture, *The Astronomical Journal*, *134*(5), 1790–1798.

Morbidelli, A., W. F. Bottke, D. Nesvorný, and H. F. Levison (2009), Asteroids were born big, *ICARUS*, *204*(2), 558–573.

Morbidelli, A., R. Brasser, R. S. Gomes, H. F. Levison, and K. Tsiganis (2010), Evidence from the Asteroid Belt for a Violent Past Evolution of Jupiter's Orbit, *The Astronomical Journal*, *140*(5), 1391–1401.

Morbidelli, A., S. Marchi, W. F. Bottke, and D. A. Kring (2012a), A sawtooth-like timeline for the first billion years of lunar bombardment, *Earth and Planetary Science Letters*, *355*, 144–151.

Morbidelli, A., J. I. Lunine, D. P. O'Brien, S. N. Raymond, and K. J. Walsh (2012b), Building Terrestrial Planets, *Annual Review of Earth and Planetary Sciences*, *40*, 251–275.

Morbidelli, A., H. S. Gaspar, and D. Nesvorný (2014), Origin of the peculiar eccentricity distribution of the inner cold Kuiper belt, *ICARUS*, *232*, 81–87.

Morishima, R., M. W. Schmidt, J. Stadel, and B. Moore (2008), Formation and Accretion History of Terrestrial Planets from Runaway Growth through to Late Time: Implications for Orbital Eccentricity, *The Astrophysical Journal*, *685*(2), 1247–1261.

Muralidharan, K., P. Deymier, M. Stimpfl, N. H. de Leeuw, and M. J. Drake (2008), Origin of water in the inner Solar System: A kinetic Monte Carlo study of water adsorption on forsterite, *ICARUS*, *198*(2), 400–407.

Murray-Clay, R., K. Kratter, and A. N. Youdin (2011), Core Accretion at Wide Separations: The Critical Role of Gas, *AAS/Division for Extreme Solar Systems Abstracts*, *2*, 0804.

Nagasawa, M., D. N. C. Lin, and E. Thommes (2005), Dynamical Shake-up of Planetary Systems. I. *Embryo Trapping and Induced Collisions by the Sweeping Secular Resonance and Embryo-Disk Tidal Interaction*, *The Astrophysical Journal*, *635*(1), 578–598.

Nakagawa, Y., C. Hayashi, and K. Nakazawa (1983), Accumulation of planetesimals in the solar nebula, *ICARUS*, *54*, 361–376.

Nesvorný, D., D. Vokrouhlický, and A. Morbidelli (2007), Capture of Irregular Satellites during Planetary Encounters, *The Astronomical Journal*, *133*(5), 1962–1976.

Nesvorný, D., A. N. Youdin, and D. C. Richardson (2010), Formation of Kuiper Belt Binaries by Gravitational Collapse, *The Astronomical Journal*, *140*(3), 785–793.

Nimmo, F. and T. Kleine (2007), How rapidly did Mars accrete? Uncertainties in the Hf W timing of core formation, *ICARUS*, *191*(2), 497–504.

O'Brien, D. P., A. Morbidelli, and H. F. Levison (2006), Terrestrial planet formation with strong dynamical friction, *ICARUS*, *184*(1), 39–58.

O'Brien, D. P., K. J. Walsh, A. Morbidelli, S. N. Raymond, and A. M. Mandell (2014), Water Delivery and Giant Impacts in the 'Grand Tack' Scenario, *ICARUS*, *223*, 74–84.

Okuzumi, S., H. Tanaka, H. Kobayashi, and K. Wada (2012), Rapid Coagulation of Porous Dust Aggregates outside the Snow Line: A Pathway to Successful Icy Planetesimal Formation, *The Astrophysical Journal*, *752*(2), 106.

Ormel, C. W. and H. H. Klahr (2010), The effect of gas drag on the growth of protoplanets. Analytical expressions for the accretion of small bodies in laminar disks, *Astronomy and Astrophysics*, *520*, 43.

Owen, T. and A. Bar-Nun (1995), Comets, Impacts, and Atmospheres, *ICARUS*, *116*(2), 215–226.

Papaloizou, J. C. B. (2003), Disc-Planet Interactions: Migration and Resonances in Extrasolar Planetary Systems, *Celestial Mechanics and Dynamical Astronomy*, *87*(1), 53–83.

Papaloizou, J. C. B., R. P. Nelson, and F. Masset (2001), Orbital eccentricity growth through disc-companion tidal interaction, *Astronomy and Astrophysics*, *366*, 263–275.

Pfalzner, S., M. Steinhausen, and K. Menten (2014), Short Dissipation Times of Proto-planetary Disks: An Artifact of Selection Effects?, *The Astrophysical Journal Letters*, *793*(2), L34.

Pierens, A. and R. P. Nelson (2008), Constraints on resonant-trapping for two planets embedded in a protoplanetary disc, *Astronomy and Astrophysics*, *482*(1), 333–340.

Pierens, A. and S. N. Raymond (2011), Two phase, inward-then-outward migration of Jupiter and Saturn in the gaseous solar nebula, *Astronomy and Astrophysics*, *533*, 131.

Pierens, A., S. Raymond, D. Nesvorný, and A. Morbidelli (2014), Outward migration of Jupiter and Saturn in 3:2 or 2:1 resonance in radiative disks: implications for the Grand Tack and Nice models, *arXiv.org*, p. 543.

Pollack, J. B., O. Hubickyj, P. Bodenheimer, J. J. Lissauer, M. Podolak, and Y. Greenzweig (1996), Formation of the Giant Planets by Concurrent Accretion of Solids and Gas, *ICARUS*, *124*(1), 62–85.

Raymond, S. N. and A. Morbidelli (2014), The Grand Tack model: a critical review, *eprint arXiv:1409.6340*.

Raymond, S. N., T. R. Quinn, and J. I. Lunine (2004), Making other earths: dynamical simulations of terrestrial planet formation and water delivery, *ICARUS*, *168*(1), 1–17.

Raymond, S. N., D. P. O'Brien, A. Morbidelli, and N. A. Kaib (2009), Building the terrestrial planets: Constrained accretion in the inner Solar System, *ICARUS*, *203*(2), 644–662.

Raymond, S. N., E. Kokubo, A. Morbidelli, R. Morishima, and K. J. Walsh (2013), Terrestrial Planet Formation at Home and Abroad, *arXiv.org*, p. 1689.

Ryder, G. (1990), Lunar samples, lunar accretion and the early bombardment of the moon, *EOS (ISSN 0096-3941)*, *71*, 313.

Ryder, G. (2002), Mass flux in the ancient Earth-Moon system and benign implications for the origin of life on Earth, *Journal of Geophysical Research (Planets)*, *107*(E), 5022.

Safronov, V. S. (1972), Evolution of the protoplanetary cloud and formation of the earth and planets., *Evolution of the protoplanetary cloud and formation of the earth and planets*.

Safronov, V. S. and E. V. Zvjagina (1969), Relative sizes of the largest bodies during the accumulation of planets, *ICARUS*, *10*(1), 109–115.

Snellgrove, M. D., J. C. B. Papaloizou, and R. P. Nelson (2001), On disc driven inward migration of resonantly coupled planets with application to the system around GJ876, *Astronomy and Astrophysics*, *374*, 1092–1099.

Strom, R. G., R. Malhotra, T. Ito, F. Yoshida, and D. A. Kring (2005), The Origin of Planetary Impactors in the Inner Solar System, *Science*, *309*(5), 1847–1850.

Taylor, D. J., K. D. McKeegan, and T. M. Harrison (2009), Lu-Hf zircon evidence for rapid lunar differentiation, *Earth and Planetary Science Letters*, *279*(3), 157–164.

Tera, F., D. A. Papanastassiou, and G. J. Wasserburg (1974), Isotopic evidence for a terminal lunar cataclysm, *Earth and Planetary Science Letters*, *22*(1), 1–21.

Thommes, E., M. Nagasawa, and D. N. C. Lin (2008), Dynamical Shake-up of Planetary Systems. II. N-Body Simulations of Solar System Terrestrial Planet Formation Induced by Secular Resonance Sweeping, *The Astrophysical Journal*, *676*(1), 728–739.

Touboul, M., T. Kleine, B. Bourdon, H. Palme, and R. Wieler (2007), Late formation and prolonged differentiation of the Moon inferred from W isotopes in lunar metals, *Nature*, *450*(7), 1206–1209.

Tsiganis, K., R. S. Gomes, A. Morbidelli, and H. F. Levison (2005), Origin of the orbital architecture of the giant planets of the Solar System, *Nature*, *435*(7), 459–461.

Villeneuve, J., M. Chaussidon, and G. Libourel (2009), Homogeneous Distribution of 26Al in the Solar System from the Mg Isotopic Composition of Chondrules, *Science*, *325*(5), 985.

Wada, K., H. Tanaka, T. Suyama, H. Kimura, and T. Yamamoto (2008), Numerical Simulation of Dust Aggregate Collisions. II. Compression and Disruption of Three-Dimensional Aggregates in Head-on Collisions, *The Astrophysical Journal*, *677*(2), 1296–1308.

Wada, K., H. Tanaka, T. Suyama, H. Kimura, and T. Yamamoto (2009), Collisional Growth Conditions for Dust Aggregates, *The Astrophysical Journal*, *702*(2), 1490–1501.

Wada, K., H. Tanaka, S. Okuzumi, H. Kobayashi, T. Suyama, H. Kimura, and T. Yamamoto (2013), Growth efficiency of dust aggregates through collisions with high mass ratios, *Astronomy and Astrophysics*, *559*, A62.

Walker, R. J. (2009), Highly siderophile elements in the Earth, Moon and Mars: Update and implications for planetary accretion and differentiation, *Chemie der Erde - Geochemistry*, *69*, 101–125.

Walsh, K. J. and H. F. Levison (2014), Formation of the Terrestrial Planets from an Annulus, *AAS/Division of Dynamical Astronomy Meeting*, *45*.

Walsh, K. J. and A. Morbidelli (2011), The effect of an early planetesimal-driven migration of the giant planets on terrestrial planet formation, *Astronomy and Astrophysics*, *526*, 126.

Walsh, K. J., A. Morbidelli, S. N. Raymond, D. P. O'Brien, and A. M. Mandell (2011), A low mass for Mars from Jupiter's early gas-driven migration, *Nature*, *475*(7), 206–209.

Walsh, K. J., A. Morbidelli, S. N. Raymond, D. P. O'Brien, and A. M. Mandell (2012), Populating the asteroid belt from two parent source regions due to the migration of giant planets—"The Grand Tack," *Meteoritics & Planetary Science*, *47*(1), 1941–1947.

Ward, W. R. (1980), Scanning Secular Resonances: a Cosmogonical Broom?, *LUNAR AND PLANETARY SCIENCE XI*, *11*, 1199–1201.

Weidenschilling, S. J. (1977), Aerodynamics of solid bodies in the solar nebula, *Monthly Notices of the Royal Astronomical Society*, *180*, 57–70.

Weidenschilling, S. J. (1980), Dust to planetesimals—Settling and coagulation in the solar nebula, *ICARUS*, *44*, 172–189.

Weidenschilling, S. J. (2011), Initial sizes of planetesimals and accretion of the asteroids, *ICARUS*, *214*(2), 671–684.

Weidenschilling, S. J. and J. N. Cuzzi (1993), Formation of planetesimals in the solar nebula, *Protostars and planets III*, pp. 1031–1060.

Weiss, B. P., J. Gattacceca, S. Stanley, P. Rochette, and U. R. Christensen (2010), Paleomagnetic Records of Meteorites and Early Planetesimal Differentiation, *Space Science Reviews*, *152*(1), 341–390.

Wetherill, G. W. (1985), Occurrence of giant impacts during the growth of the terrestrial planets, *Science (ISSN 0036-8075)*, *228*, 877–879.

Wetherill, G. W. (1991), Why Isn't Mars as Big as Earth?, *Abstracts of the Lunar and Planetary Science Conference*, *22*, 1495.

Wetherill, G. W. and G. R. Stewart (1989), Accumulation of a swarm of small planetesimals, *ICARUS*, *77*, 330–357.

Wetherill, G. W. and G. R. Stewart (1993), Formation of planetary embryos—Effects of fragmentation, low relative velocity, and independent variation of eccentricity and inclination, *ICARUS*, *106*, 190.

Whipple, F. L. (1972), On certain aerodynamic processes for asteroids and comets, *From Plasma to Planet*, p. 211.

Yin, Q., S. B. Jacobsen, K. Yamashita, J. Blichert-Toft, P. Télouk, and F. Albarède (2002), A short timescale for terrestrial planet formation from Hf–W chronometry of meteorites, *Nature*, *418*(6901), 949–952.

Youdin, A. N. (2010), From Grains to Planetesimals, *EAS Publications Series*, *41*, 187–207.

Youdin, A. N. and F. H. Shu (2002), Planetesimal Formation by Gravitational Instability, *The Astrophysical Journal*, *580*(1), 494–505.

Zhou, J.-L., D. N. C. Lin, and Y.-S. Sun (2007), Post-oligarchic Evolution of Protoplanetary Embryos and the Stability of Planetary Systems, *The Astrophysical Journal*, *666*(1), 423–435.

Zsom, A., C. W. Ormel, C. Güttler, J. Blum, and C. P. Dullemond (2010), The outcome of protoplanetary dust growth: pebbles, boulders, or planetesimals? II. Introducing the bouncing barrier, *Astronomy and Astrophysics*, *513*, 57.

4

Late Accretion and the Late Veneer

Alessandro Morbidelli[1] and Bernard J. Wood[2]

ABSTRACT

The concept of Late Veneer has been introduced by the geochemical community to explain the abundance of highly siderophile elements in the Earth's mantle and their approximate chondritic proportions relative to each other. However, in the complex scenario of Earth accretion, involving both planetesimal bombardment and giant impacts from chondritic and differentiated projectiles, it is not obvious what the "Late Veneer" actually corresponds to. In fact, the process of differentiation of the Earth was probably intermittent, and there was presumably no well-defined transition between an earlier phase where all metal sunk into the core and a later phase in which the core was a closed entity separated from the mantle. In addition, the modelers of Earth formation have introduced the concept of "Late Accretion," which refers to the material accreted by our planet after the Moon-forming event. Characterizing Late Veneer, Late Accretion, and the relationship between the two is the major goal of this chapter.

4.1. INTRODUCTION

Late Veneer, Late Accretion, and Late Heavy Bombardment are concepts introduced by different scientific communities to address the very last stage of Earth formation when our planet acquired the final bit of its mass, after the end of the core-mantle differentiation process, and/or the giant impact that gave origin to the Moon. Often, however, the same names are used to indicate quite different processes. Moreover, it is quite common to identify the Late Veneer/Accretion/Bombardment with the origin of volatiles (including water) on our planet, which is not necessarily correct.

Therefore, the goal of this chapter is to define these concepts, and describe the processes that they refer to and their mutual relationships. We start in Section 4.2 with the Late Veneer, as defined by the geochemical community. In Section 4.3 we briefly describe the mode of formation of the Earth, which leads in a natural way to define the concept of Late Accretion. Section 4.4 will discuss the relationships between Late Veneer and Late Accretion. Finally in Section 4.5 we will come to the issue of the delivery of volatiles to the Earth and its chronology.

4.2. THE LATE VENEER AS DEFINED IN GEOCHEMISTRY

The concept of a "Late Veneer" was developed from the observed abundance patterns of siderophile (iron-loving) elements in the silicate Earth. Figure 4.1 shows the relative abundances of a wide range of elements [*McDonough and Sun*, 1995] in the primitive upper mantle (otherwise known as bulk silicate Earth [BSE]) as a function of their condensation temperature from a gas of

[1]*CNRS Laboratoire Lagrange, Observatoire de la Côte d'Azur, Université Cote d'Azur, CNRS, Nice, France*
[2]*Department of Earth Sciences, University of Oxford, UK*

The Early Earth: Accretion and Differentiation, Geophysical Monograph 212, First Edition.
Edited by James Badro and Michael Walter.
© 2015 American Geophysical Union. Published 2015 by John Wiley & Sons, Inc.

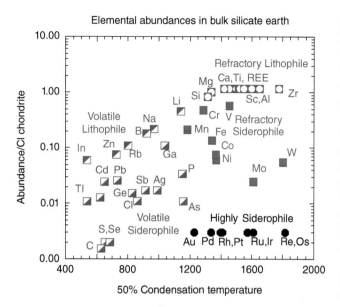

Figure 4.1 Abundance relative to CI chondrites and Mg of elements in the bulk silicate Earth, as a function of condensation temperature. The latter is defined as the temperature at which 50% of the element is in solid form, at the pressure conditions on the Minimum Mass Solar Nebula [*Lodders*, 2003].

solar composition at a total pressure of 10^{-4} bar [*Lodders*, 2003]. Refractory lithophile elements, which condense at high temperatures (e.g., Ca, Ti, Sc, Zr,REE) are inapproximately the same ratios one to another as in the CI chondrite class of primitive meteorites. This gives us a reference level for all other elements. Refractory siderophile elements are depleted in BSE due to their partitioning into the core. To a first approximation one can make a mass balance between the BSE and the "missing" contents (assuming chondritic ratios in bulk Earth) of these siderophile elements and obtain the mass of the core and its elemental concentrations (dominated by Fe). Additionally, as can be seen in Figure 4.1, those elements with low condensation temperatures, which were volatile in the solar nebula, are depleted in BSE relative to the chondritic reference. This is due, as discussed in Section 4.5, to the Earth being dominated by high temperature materials that condensed and accreted in the inner solar system. Note that the same relationship between siderophile and lithophile elements observed in refractory elements applies to the more volatile elements; volatile siderophile elements are more depleted in bulk silicate Earth than volatile lithophile elements.

Measurements of the concentrations of the noble metals (Re, Os, Ir, Ru, Pt, Rh, Pd, Au) in mantle peridotites and igneous rocks generated by partial melting of the mantle [*Chou*, 1978; *Chou et al.*, 1983] demonstrated that these "highly siderophile" elements (HSE) are, as expected, even more depleted in BSE than the other important siderophile elements shown in Figure 4.1. Experimental measurements show, however, that despite their strong depletions, these elements are much more abundant in silicate Earth than would be expected from simple equilibrium between mantle and core [*Holzheid et al.*, 2000; *Kimura, et al.*, 1974; *Mann, et al.*, 2012; *O'Neill*, 1991]. In other words, they are orders of magnitude more siderophile than indicated by their relative abundances in BSE. Partition coefficients of elements depend on temperature and pressure [*Righter et al.*, 2008] so that, in principle, it is possible that the values measured in laboratory experiments do not correspond to the conditions existing on Earth during core formation. However, HSEs are in approximately chondritic ratio one to another in the primitive upper mantle [*Chou*, 1978], and it is highly unlikely that conditions of temperature and pressure exist to make the partition coefficients of the eight HSEs equal to each other. In fact, no combination of pressure, temperature, and oxygen fugacity appears to be capable of reproducing the mantle concentrations of the HSEs by metal-silicate equilibrium [*Mann et al.*, 2012].

Thus, the most plausible explanation for the HSEs' overabundance in BSE and their chondritic relative abundances is the addition of material of approximate chondritic composition to the upper mantle after core segregation had ceased [*Chou*, 1978; *Kimura, Lewis, and Anders*, 1974]. These are the origins of the concept of the "Late Veneer," a small amount (<1% Earth mass) of chondrite-like material added to a convecting silicate Earth and never re-equilibrated with the metal in the core [*Anders*, 1968; *Turekian and Clark*, 1969]. From the abundance of HSEs in BSE, the mass of the mantle and the range of HSE contents in the various types of chondritic meteorites, the modern estimate for the Late Veneer mass is $4.86 +/- 1.63 \times 10^{-3} M_E$, where M_E is the Earth's mass [*Walker*, 2009; *Jacobson et al.*, 2014].

It is important to emphasize that, as far as the highly siderophile elements are concerned, the late chondritic veneer has been completely homogenized into the mantle by convection. This process of mixing might have taken a long time to be completed. Indeed, the HSE content in komatiites decreases with their increasing age, for komatiites older than 3 Gy (see Fig. 4.2) [*Maier et al.*, 2009; *Wilbold et al.*, 2011], with one exception pointed out in *Touboul et al.* [2012]. It seems unlikely that this trend could be due to a slow delivery of HSEs to the Earth, because the terrestrial bombardment should have decayed much faster than is implied by this HSE vs. age trend (see Section 4.3). So, the most likely interpretation is that it took about 1.5 billion years to completely mix the mantle and bring the HSEs from near the surface to the deeper parts of the mantle from where the old komatiites are derived.

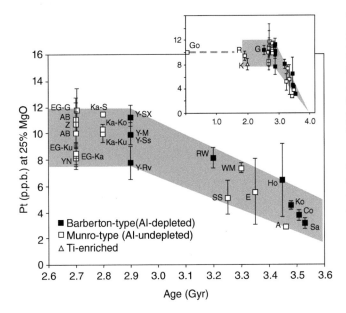

Figure 4.2 The abundance of HSEs as a function of age for the Earth's oldest komatiites. From *Maier et al.* [2009].

4.3. LATE ACCRETION MASS AS DEFINED IN ACCRETION MODELS

The detailed description of the process of terrestrial planet growth is the object of Chapter 3. Here it is enough to emphasize that terrestrial planets are believed to grow from a disk of planetesimals (small bodies resembling the asteroids) and planetary embryos (bodies with a mass intermediate between the mass of the Moon and the mass of Mars, which formed very rapidly, before the disappearance of gas from the system). In view of its short accretion timescale, Mars itself could be a stranded planetary embryo [*Kleine et al.*, 2004b; *Halliday et al.*, 2005; *Dauphas and Pourmand*, 2011; *Jacobson and Morbidelli*, 2014].

Therefore, the growth of a planet like the Earth is characterized by a sequence of giant impacts from planetary embryos and smaller impacts from planetesimals. In this contest, we define "Late Accretion" as the tail of the accretion history of a planet due to the planetesimal bombardment that happens after the last giant impact has occurred. For the Earth, the last giant impact is presumably the one that generated the Moon (see Section 4.4.3), so that late accretion occurs after lunar formation.

Because the dynamical lifetime of planetesimals is finite (planetesimals are removed by collisions with the planets, ejection to hyperbolic orbit, collisions with the Sun, or mutual collisions that grind them into small debris), the number of planetesimals in the system decays roughly exponentially with time. Hence the bombardment rate of the planets is expected to decay in the same fashion. From numerical experiments, the half-life of the process is from 10 to 50 Myr [*Bottke et al.*, 2007; *Morbidelli et al.*, 2012]. Unfortunately, geological activity on the Earth has erased any clear evidence of impacts occurring over the first billion years or so. Instead, a lot of information is preserved in the lunar ancient crater record.

From lunar craters we learn that the size distribution of the projectiles striking the Moon was similar to that of today's main belt asteroids [*Strom et al.*, 2005]. The so-called impact basins ("craters" with diameters larger than 300 km) trace the impacts of the largest projectiles (up to ~200 km in diameter for the projectile of the SPA basin). The last of the basin-forming events occurred relatively late compared to the time of formation of the first solids in the Solar System (the Calcium-Aluminium Inclusions, which formed 4.567 Gyr ago) [*Bouvier and Wadhwa*, 2010]. The Imbrium basin, which, from stratigraphy analysis, appears to be the second from last basin in the chronological sequence [*Wilhelms*, 1987] formed 3.85 Gyr ago according to the dating of lunar samples returned by the Apollo program [*Stoffler and Ryder*, 2001]. The Orientale basin (the last basin on the Moon) presumably did not occur long after Imbrium, given that crater counts on the ejecta blankets of both basins are similar.

Forming basins so late as a tail of an exponentially decaying process would require that the initial population of large planetesimals was very numerous [*Bottke et al.*, 2007]. However, this is inconsistent with the small amount of mass accreted by the Moon since its formation, which is on the order of $5 \times 10^{-6} M_E$ as deduced from HSEs abundance in the lunar mantle [*Day et al.*, 2010] and in the crust [*Ryder*, 2002] and the total number of lunar basins [*Neumann et al.*, 2013]. This argues strongly in favor of a surge in the frequency of impacts sometime prior to the formation of Imbrium [*Bottke et al.*, 2007]. Several other pieces of evidence point to such a surge, often called lunar Cataclysm or Late Heavy Bombardment. (The latter term, however, is sometimes associated with the entire Late Accretion process, so we will not use it in this review to avoid confusion.) For instance, impact ages on lunar samples [*Tera et al.*, 1974] and lunar meteorites [*Cohen et al.*, 2000] indicate that the impact rates were lower before ~4.1 Gyr than in the subsequent 3.9–3.7 Gyr period.

A surge in the bombardment rate, associated with an increase of the velocity distribution of the projectiles, is consistent with a sudden change in the orbital configuration of the giant planets, which would have dislodged part of the asteroid population from the main belt, refurbishing the population of planet-crossing objects

[*Bottke et al.*, 2012]. Such a change is needed, sometime during the history of the Solar System, to reconcile the current orbits of the giant planets with those that these planets should have had when they emerged from the phase of the gas-dominated proto-planetary disk [*Morbidelli et al.*, 2007; *Morbidelli*, 2013].

Observational constraints [*Marchi et al.*, 2012] and theoretical modeling [*Bottke et al.*, 2012] suggest that about 10 basins formed on the Moon during the cataclysm, from Nectaris to Orientale, and that the cataclysm started about 4.1 Gyr ago [see *Morbidelli et al.*, 2012 for a review]. The total number of basins on the Moon, detected by the GRAIL mission through gravitational anomalies [*Neumann et al.*, 2013] is ~40. Thus, the cataclysm accounts for about 1/4 to 1/3 of the total mass (~$5 \times 10^{-6} M_E$) that the Moon accreted since its formation. From all of these constraints, a timeline of the lunar bombardment can be derived [*Morbidelli et al.*, 2012].

Knowing the impact history of the Moon, one can then scale it to the Earth, using the ratio of gravitational cross sections. Monte Carlo simulations also allow assessments of the effects of small number statistics related to the largest impactors [*Marchi et al.*, 2014]. The results change vastly depending on whether one limits the size frequency distribution (SFD) of the impactors to the size of Ceres (the largest asteroid in the main belt today, ~900km in diameter) or extrapolates it to larger sizes.

If the projectile size distribution is limited to Ceres-size, given the constraint on the mass delivered to the Moon, the total amount of mass that the Earth acquires during Late Accretion is less than $10^{-3} M_E$, significantly smaller than the chondritic mass estimated for the Late Veneer (see Section 4.2). But if planetesimals larger than Ceres are permitted, the same constraint on the mass delivered to the Moon allows a significant fraction of the simulations (~12%) to bring a mass to the Earth consistent with the mass of the Late Veneer (see Fig. 4.3 and *Bottke et al.*, 2010; *Raymond et al.*, 2013). About 8% of the simulations bring a somewhat larger mass. However, only 0.5% of the simulations deliver more than $0.01 M_E$ to the Earth.

4.4. RELATIONSHIP BETWEEN LATE VENEER AND LATE ACCRETION

As we have seen above, the Late Veneer mass and the Late Accretion mass are defined from two different concepts. The amount of Late Veneer mass is well constrained, but it may be questionable if all of it was delivered to the Earth after the Moon-forming event. The estimate of the late accretion mass is quite uncertain, as it

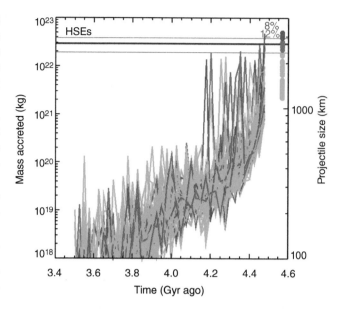

Figure 4.3 The bombardment of the Earth obtained in 50 Monte Carlo simulations calibrated on the bombardment history of the Moon. Prior to 4.1 Gyr ago, the asteroid-like SFD of the impactors is extrapolated up to D = 4000 km projectiles. Since 4.1 Gyr ago, the projectile SFD is limited at Ceres' size. Each curve shows the mass accreted in each 25 Myr time-bin. The cumulative accreted mass is indicated by the yellow dots on the right-hand side of the panel. The horizontal band marks the Late Veneer mass and its uncertainty. Red curves (12% of the runs) cumulatively carry a mass consistent with the Late Veneer mass, green curves (80%) carry a smaller mass, and blue curves (8%) a larger mass. Adapted from *Marchi et al.* [2014].

depends on the assumed size distribution of projectiles when scaling the lunar impact crater record to the Earth. It would be very useful if we could equate with confidence the Late Accretion mass to the Late Veneer mass. To assess whether this equation is reasonable, we address two key questions below.

4.4.1. Can the Late Accretion Mass Be Significantly Smaller Than the Late Veneer Mass?

To answer this question we need to consider whether any of the HSE content of bulk silicate Earth could have accreted prior to the Moon-forming event. The Moon is believed to have been formed during the last giant impact on Earth. Because of its large size (at least more massive than one-fourth of Mars), the projectile is expected to have been differentiated. Numerical simulation of the Moon-forming process [*Canup and Asphaug*, 2001; *Reufer et al.*, 2012; *Cuk and Stewart*, 2012; *Canup*, 2012] show that the core of the projectile merges with the core of the Earth and that the temperature of the Earth is raised above the global magma ocean threshold. However, it is unclear which fraction of the terrestrial mantle has a chance to equilibrate with the projectile's core in this process. Also, it is unclear whether there is still metallic iron in the Earth's mantle at this stage that can take advantage of the new magma ocean phase to percolate into the core. Obviously, if a substantial fraction of pre-existing mantle HSEs had not been depleted during the episode of core growth associated with the Moon-forming event, the late veneer mass deduced from the current HSE abundance would overestimate the mass accreted after the Moon's formation, that is, the Late Accretion mass.

If we consider the abundances of the refractory siderophile elements Ni and Co in BSE (Fig. 4.1), then it has been known for nearly 20 years that these can only be explained, under conditions of metal-silicate equilibrium, by core segregation at very high pressures [*Li and Agee*, 1996; *Thibault and Walter*, 1995]. When the weakly siderophile elements V, Cr, and Nb are also considered, the most likely explanation of the core formation process is that it began under strongly reducing conditions at low pressure and temperature and that the Earth became more oxidized as it grew, and the pressures and temperatures of core segregation increased [*Wade and Wood*, 2005; *Wood et al.*, 2008]. These observations led Mann *et al.* [2012] to determine the pressure, temperature, and oxygen fugacity effects on the metal-silicate partitioning of the HSEs (Re, Ir, Ru, Pt, Rh, Pd). Their data demonstrate that these elements all become less siderophile with increasing temperature and to some extent with increasing pressure and that by a long extrapolation one might infer that the mantle concentrations of Pt, Rh, and Pd would approach those observed at equilibrium conditions of about 60 GPa and 3560 K. Nevertheless, the metal-silicate partition coefficients of Ru, Re, and Ir would remain 1–2 orders of magnitude too high to explain the mantle concentrations of these elements even under these extreme conditions.

Since equilibrium between core and mantle would not generate the observed abundance pattern of HSEs in BSE, the only way in which the current concentrations could contain some "pre-Moon-formation" component is if it were a result of partial disequilibrium during the Moon-forming event. For example, a small amount of metal left behind by inefficient core segregation would contain the HSEs in chondritic relative proportions because these elements are so strongly partitioned into the metal. This would be the case, for instance, if the Earth's hemisphere opposite to the impact point of the Moon-forming projectile remained largely unaffected by the collision, as suggested by the recent simulations in *Cuk and Stewart* [2012] (where the mantle in the opposite hemisphere becomes the lower mantle of the final Earth, although its temperature is very high so that it could still undergo metal-silicate segregation).

Alternatively, internal production of Fe^{3+} in the mantle by, for example, disproportionation of Fe^{2+} to Fe^{3+} plus Fe^0 in the perovskite field [*Frost et al.*, 2004; *Kyser et al.*, 1986] could lead to the oxygen fugacity of the upper mantle exceeding the conditions of iron metal stability. Any additional metal added(including the HSEs) would then be oxidized and mixed back into the silicate part of Earth rather than being segregated to the core. In this way the giant impact associated with the Moon-forming event would have delivered HSEs to the Earth's mantle, rather than depleting them.

Tungsten isotopes provide a potential means of independently estimating both the mass of the Late Veneer and of Late Accretion. ^{182}W was produced in the early Earth by decay of ^{182}Hf with a half-life of 8.9 Myr. This extinct radioactive system, in which W is siderophile and Hf is lithophile, is extensively used to estimate the accretionary timescales of planetary bodies [*Yin et al.*, 2002].

A Late Veneer that provided ~0.5% of an Earth mass with HSEs in chondritic form should have provided the same amount of W with chondritic $^{182}W/^{184}W$, i.e., with ε_W of −1.9. Given that siderophile W was strongly partitioned into the core during the principal accretionary phase and depleted by a factor ~16 in the mantle relative to chondrites, this amount of chondritic W is sufficient to have significantly altered the ε_W of BSE, decreasing it by about 0.25 in ε_W units. Measurements of the W isotopic compositions of 3.8 Gyr age rocks from Isua, Greenland [*Willbold et al.*, 2011] show that these have slightly higher $^{182}W/^{184}W$ (ε_W of 0.13 ± 0.04) than modern day silicate Earth (−0.01 ± 0.02). This result is consistent with the Isua rocks reflecting the composition of BSE prior to the Late Veneer (this interpretation should be validated by checking that Isua rocks are HSE poor), while the modern Earth exhibits homogenization of the Late Veneer with the pre-existing mantle (see Fig. 4.2). *Willbold et al.* (2011) show that this shift in the W isotopic composition of BSE, if attributed to mixing-in of the Late Veneer, would lead to a mass of the Late Veneer of between 0.002 and 0.009 of the mass of bulk silicate Earth, depending on the nature of the chondritic material added in the Late Veneer. Thus, W-isotopes yield an estimate for the mass of the Late Veneer, which is consistent with all HSEs in silicate Earth arriving with the Late Veneer. This argument however does not require that this mass was delivered after the Moon-forming event.

For this purpose it is interesting to look at the differences in W isotope composition between the Earth and the Moon. The Earth and the Moon have essentially identical isotope compositions for many elements: O [*Wiechert et al.*, 2001], Ti [*Zhang et al.*,

2012], Si [*Georg et al.*, 2007], Cr [*Lugmair and Shykolyukov*, 1998]. Whatever the origins of these identical isotopic compositions, it is reasonable to assume that, at formation, the Moon and the Earth had identical W-isotope composition as well. However, if the HSE abundances are diagnostic of late accretion, the Earth should have subsequently accreted much more chondritic material than the Moon. Thus, its W-isotope composition should now be different from that of the Moon; the Moon should now have ε_w ~0.2. Until recently, the W-isotope composition difference between the Earth and its satellite was not resolved: $\varepsilon_w^{Moon}=0.1$ +/−0.1 [*Touboul et al.*, 2007].

Recently, however, *Kleine et al.* [2014] reported to have achieved a more precise determination: $\varepsilon_w^{Moon}=0.29$ +/−0.08 (the value reported in the reference is 0.17, but it has been later refined to 0.29 [*Kleine*, private communication]). *Touboul et al.* [2015] reported a value of 0.21 +/−0.05. Originally *Touboul et al.* [2007, 2012] interpreted any difference in W-isotopic composition between the Moon and the Earth as an indication that the Moon formed before all [182]Hf was extinct. However, it now turns out that the Hf/W ratios in the Earth and Moon mantles are the same [*König et al.*, 2011] so the W-isotope composition of the two bodies should not change relative to each other over time, whatever the age of the Moon. Thus, the difference reported by *Kleine et al.* is more easily explained if a chondritic mass consistent with the Late Veneer mass was delivered to the Earth after the Moon-forming event.

If W suggests that the Late Accretion mass cannot be significantly smaller than the Late Veneer mass, two arguments suggest that nevertheless a fraction of the HSEs should pre-date the closure of Earth's core and/or the moon-forming event.

A first argument comes from sulfur isotopes. As can be seen in Figure 4.1, S is depleted in BSE due to its volatility in the solar nebula and also because it is highly siderophile, even though it is not conventionally named as one of the HSEs. To a first approximation, the S content of BSE (250 ppm) [*McDonough and Sun*, 1995] is consistent with an addition of 0.5% of CI chondrite to the Earth in the Late Veneer. Thus, its abundance is consistent with the Late Veneer mass. However, recent measurements of the S-isotopic composition of BSE [*Labidi et al.*, 2013] have shown that the dominant mantle end-member of S sampled by MORB has $\delta^{34}S$ of −1.9 ‰ rather than the ~0.0, expected for chondritic sulfur.

Measurements of the S isotopic compositions of carbonaceous, ordinary and enstatite chondrites range from −0.26 to +0.49 ‰ [*Gao and Thiemens*, 1993] so the mantle end-member is compositionally distinct from conventionally recognized planetary "building-blocks." Based on limited experimental data on S isotope partitioning between silicate and metal, *Labidi et al.* [2013] showed that the difference

between chondrites and the mantle end-member is consistent with the presence of sulfur in the mantle that had equilibrated with the metal of the core. The isotopic measurements are consistent with the amount of "Late-Veneer" sulfur in the mantle being about 40% of the total, although this figure has very large uncertainties. The conclusion is that some fraction of highly siderophile, volatile S was present in the mantle before the Late Veneer was added and that it is probable that the same applies to other highly siderophile elements as well.

A second argument is brought by the analysis of Kostomuksha komatiites [*Touboul et al.*, 2012], for which combined [182]W, [186,187]Os, and [142,143]Nd isotopic data indicate that their mantle source underwent metal-silicate fractionation well before 30 Myr of Solar System history. Surprisingly, Kostomuksha komatiites appear to come from a mantle source that had a HSE content that was 80% of that of the present-day mantle. Thus, the fraction of the current HSE mantle budget that predates the Moon-forming event would be $0.8 \times f$, where f is the fraction of the total mantle mass represented by the source of the Kostomuksha komatiites. Unfortunately, f is not known even at the order of magnitude level

In summary, we can exclude the Late Accretion mass being order(s) of magnitude less than the Late Veneer chondritic mass deduced from HSE abundances in the terrestrial mantle. For instance we can exclude that, after the Moon-forming event, the Earth accreted only 0.0001 M_E, which would be the mass expected by scaling the mass accreted by the Moon ($5 \times 10^{-6} M_E$) according to the gravitational cross-sections. This supports the scenario of stochastic accretion proposed by *Bottke et al.* [2010] (i.e., that the projectile size distribution extended to sizes significantly larger than Ceres). (See the right panel of Fig. 4.3.) However, it is possible and plausible that a fraction of the HSEs, perhaps as much as ~50%, pre-date the Moon-forming event. It may appear surprising that the fraction of the current HSEs budget that pre-dates the Moon-forming event is small, given that only a portion of the Earth's mantle is expected to have equilibrated with the core of the Moon-forming projectile. A possible explanation is that the HSEs have been sequestered into the core due to the exsolution of iron sulfide during the global magma ocean phase [*O'Neil*, 1991; Rubie et al., 2015].

4.4.2. Can the Late Accretion Mass Be Significantly Larger Than the Late Veneer Mass?

If the impactors delivering most of the mass during Late Accretion were very big (larger than 1000 km in diameter), as advocated by *Bottke et al.*, 2010, they were presumably differentiated objects. Thus, most of their HSEs were sequestered in their cores. The idea proposed in *Bottke et al.* is that the cores of these objects equilibrated with the terrestrial mantle, the metallic iron was oxidized and therefore it did not join the terrestrial core, so that the projectile's HSEs were delivered to the Earth's mantle. However, if only a fraction X of the projectiles' cores equilibrated with the mantle, while the remaining 1-X fraction plunged directly into the Earth core, the HSE mantle budget would require a Late Accretion mass M_{LA} equal to M_{LV}/X, where M_{LV} is the Late Veneer mass. Thus, *Albarede et al.* [2013] suggested that, if X is small, M_{LA} can be much larger than 5×10^{-3} M_E.

An upper bound on the late-accreted mass of Earth can be derived from the fact that the silicate parts of Earth and Moon have very similar isotopic compositions. If a large mass had been added to Earth in a few large projectiles after the Moon-forming event, the chemical nature of Late Accretion would have been dominated by a single parent body composition (rather than an average of many projectile properties) and then the Earth and the Moon would be more different than they are observed to be. Below, we examine the isotopic systems of O and Ti.

Wiechert et al. [2001] found that Earth and the Moon lay on the same O isotope fractionation line:

$$\Delta^{17}O_{Moon} - \Delta^{17}O_{Earth} = 0.0008 + / - 0.001\%$$

where $\Delta^{17}O$ measures the distance from the terrestrial fractionation line on an oxygen three-isotope diagram. All reported uncertainties are 1σ. Recently, *Herwartz* [2013] announced that they resolved a difference between the two bodies: $\Delta^{17}O_{Moon} - \Delta^{17}O_{Earth} = 0.0012\%$. If this measurement is correct and if we assume that Earth and Moon equilibrated oxygen isotope ratios at the time of the Moon-forming event, knowing that the Earth received more late-accreted mass than the Moon, we conclude that the dominant nature of the projectiles during Late Accretion must have had a carbonaceous chondrite (excluding CI), enstatite or HED meteoritic composition, because these are the only meteoritic compositions that are situated below or on the terrestrial fractionation line in the oxygen-isotope diagram. Then we can constrain how much late-accreted mass could be added to Earth from carbonaceous CV, CO, CK, CM, CR, and CH meteoritic compositions using their measured oxygen isotope compositions. We find 0.4+/−0.3%, 0.3+/−0.2%, 0.3+/−0.2%, 0.5+/−0.5%, 0.8+/−0.7%, and 0.8+/−0.7% Earth mass, respectively. For HEDs and enstatite chondrites, oxygen isotopes do not provide very useful constraints, because these meteorite groups are close to the terrestrial fractionation line.

Further constraints are provided by the relative Ti isotope composition of the Earth and the Moon. The Moon-Earth difference is $\epsilon^{50}Ti_{Moon} - \epsilon^{50}Ti_{Earth} = -0.04 +/- 0.02$ [*Zhang et al.*, 2012]. HEDs and, to a lesser extent, enstatite chondrites, can be excluded as the dominant component of Late Accretion because they have negative

titanium isotope compositions ($\epsilon^{50}Ti_{enstatite} = -0.23 +/- 0.09$ and $\epsilon^{50}Ti_{HED} = -1.24 +/- 0.03$), and any large addition of this composition would result in a positive $\epsilon^{50}Ti$ measurement difference between the Moon and Earth. A source of uncertainty in this analysis appears from the possibility that the Moon-Earth difference is positive, because its 2σ uncertainty encompasses the zero value. Considering this possibility and assuming that $\epsilon^{50}Ti_{Moon} - \epsilon^{50}Ti_{Earth} = 0.01$ (a 2.5σ deviation), we can still constrain that at most $0.008 M_E$ of HED composition could have been added to the Earth; however, we cannot place such a similarly strict constraint on a projectile of enstatite chondrite composition.

In order to constrain the mass that the Earth could have accreted from differentiated enstatite chondrite projectiles, we turn again to the difference in W- isotope ratio between the Moon and the Earth. If only a fraction X of the cores of the projectiles equilibrated with the mantle, a fraction 1-X of the projectiles' mantles have to have been delivered as an achondritic contribution to the terrestrial mantle. Because enstatite chondrites are very reduced, the mantle of differentiated enstatite chondrite objects is expected to be very depleted in W, but with a highly radiogenic isotopic ratio (i.e., large positive ϵ_W). Eucrite and Aubrite meteorites could be used as a proxy of what the mantle of a differentiated enstatite-chondrite projectile could look like (ϵ_W equal to 22 and 11, respectively). This radiogenic contribution would increase ϵ_W of the BSE relative to the Moon, making the Moon negative in ϵ_W relative to the final terrestrial standard (which sets the zero value by definition). But we have seen before that the Moon has a positive $\epsilon_W = 0.29$ relative to the BSE. This effectively limits the amount of mass delivered by differentiated enstatite chondrite projectiles.

Figure 4.4 shows this constraint in graphical form. Depending on the assumed composition of the projectile mantle (Eucrite-like, left panel or Aubrite like, right panel), the figure shows as a gray area the set of parameters X,Y (Y being the fraction of HSEs in the BSE delivered after the Moon-forming event) consistent with the measured Moon-Earth difference in ϵ_W and its uncertainty. The diagonal dashed blue lines show the total mass delivered to the Earth.

As one can see, also in this case the amount of material delivered to the Earth should not exceed 0.01 M_E.

4.4.3. Summary and Implications for Moon Formation

The discussions in Sections 4.4.1 and 4.4.2 suggest that the Late Veneer mass and the Late Accretion mass are the same within a factor of 2. In particular, the Earth should not have accreted more than 1% of its mass after the Moon-forming event. This result has two implications.

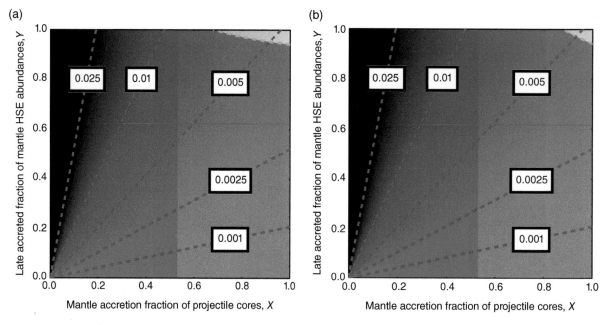

Figure 4.4 In the parameter space X (fraction of the projectile core that equilibrates with the mantle) and Y (fraction of HSEs in the silicate Earth delivered as Late Accretion), the gray region represents the parameter space consistent with a value of ϵ_W (Moon-Earth) = 0.29 +/− 0.08 [*Kleine et al.*, 2014]. The left panel has been computed for a Eucrite-like composition of the mantle of the differentiated projectile and the right panel for an Aubrite-like composition (in terms of W- content and ϵ_W). The diagonal dashed blue lines show the total mass accreted by the Earth. Adapted from *Jacobson et al.* [2014] using the new value of the lunar ϵ_W and a mass concentration of W in the Earth's mantle that is 1/16 of chondritic.

First, the Moon should have formed during the last giant impact that brought mass to the Earth. In fact, the short accretion timescale of Mars suggests that planetary embryos had masses on the order of 0.5–1 Mars mass. Thus, each accretional giant impact of embryos on Earth should have delivered at least 5–10% of an Earth mass to our planet. If the Moon-forming giant impact had not been the last one, our planet would have accreted several percent of its mass during the subsequent giant impact, which is excluded by the arguments presented in Section 4.4.2.

Second, the Moon formed relatively late. *Jacobson et al.* [2014] showed with computer simulations that there exists a correlation between the timing of the last giant impact and the mass that the planet acquires afterward (Fig. 4.5). From this correlation one deduces that the Moon should have formed from 65 to 120 Myr after formation of the first solids for to Earth to accrete a Late Veneer of 4.86 +/− 1.63 $\times 10^{-3} M_E$. If the Late Accretion mass is 0.01 M_E, the Moon should have formed no earlier than 50 Myr, within 1σ uncertainty. This is in agreement with the date of the Moon-forming collision using some radioactive chronometers [e.g., *Allegre et al.*, 2008 using the Pb-Pb chronometer; *Halliday*, 2008 using the Rb–Sr chronometer], but it is well known that other radioactive chronometers are discordant [e.g., *Yin and Antognini*, 2008 for the Hf-W chronometer; *Taylor et al.*, 2009 for the Lu-Hf chronometer].

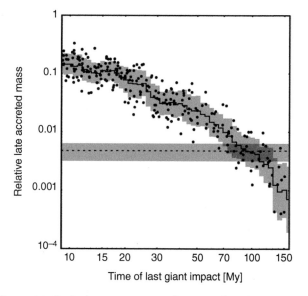

Figure 4.5 Each dot represents a planet produced in a terrestrial planet formation simulation, indicating the time of the last giant impact and the amount of mass (normalized by the final planet mass) accreted from planetesimals after that impact. The distribution of points shows a clear correlation whose running mean and standard deviation are illustrated by the solid staircase-like line and the surrounding gray band, respectively. The Late Veneer mass and its uncertainty are depicted as a horizontal dark gray band. Adapted from *Jacobson et al.* [2014].

4.5. LATE VENEER AND THE ORIGIN OF EARTH'S VOLATILES

An important issue that has arisen recently is the relationship between Late Veneer and the origin of terrestrial volatiles, including moderately volatile elements such as S, Pb, and Ag (Fig. 4.1). Cosmochemically, the simplest model for the origin of volatiles is that the Earth accreted from volatile-depleted materials that changed little in bulk composition as accretion progressed (homogeneous accretion). An alternative hypothesis [*Albarède*, 2009] is that the Earth first formed completely devoid of volatile constituents. *Albarède* suggests that the Earth's complement of volatile elements could reasonably have been established later, by addition of volatile-rich material after the Moon-forming event.

There are several chemical and isotopic arguments in favor of a view intermediate between the homogeneous accretion model and the extreme heterogeneous accretion model of *Albarède* [2009]. Firstly, although the silicate Earth is substantially depleted in volatile elements relative to CI chondrites (Fig. 4.1), it is clearly most strongly depleted in those volatile elements that are also siderophile (e.g., S, Se, Au, Ge, C). This suggests that these elements have partitioned into the core, implying that volatile elements were added to Earth during the principal phase of accretion and core segregation, and not solely as part of a Late Veneer. S-isotopic evidence indicating that some fraction of terrestrial S equilibrated with metal of the core [*Labidi et al.*, 2013] is further evidence for a pre-Late Veneer volatile accretion to the Earth.

A second argument against the accretion of volatile elements solely in the Late Veneer is that such a model can hardly be mass-balanced. For example, the concentration of ^{204}Pb in CI chondrites (the most volatile rich meteorites) is 42 ppb [*Palme and O'Neill*, 2003]. In contrast, the silicate Earth contains 2.5 ppb of ^{204}Pb (Fig. 4.1). Supplying this as a Late Veneer of CI chondrite composition would add 6% to the mass of the primitive mantle and deliver Re, Ag, S, C, and water at levels that are 5, 7, 11, 13, and 5 to 20 times greater, respectively, than found in bulk silicate Earth. Atmospheric losses may reduce the final abundances but is unlikely in the required proportions. Delivering all terrestrial Pb during the Late Veneer would also generate a Moon-Earth difference in W isotopic composition of $\sim -1.3\varepsilon_w$ and of 0.5‰ for $\Delta^{18}O$, significantly different from observations [*Touboul et al.*, 2007; *Wiechert et al.*, 2001]. These inconsistencies are made even worse if the Late Veneer were more volatile-depleted than CI chondrites. For example, Re-Os systematics indicate that the Late Veneer may have had a composition similar to that of H chondrites [*Drake and Righter*, 2002]. In this case, the ^{204}Pb mass balance requires addition of $\sim 60\%$ of the mass of the Earth's mantle.

Schönbächler et al. [2010] provide a complement to these arguments against the Late Veneer as the sole source of the volatile elements. These authors show that the silicate Earth is identical to CI chondrites in Ag isotopes. Since ^{107}Ag was produced in the early Solar System by decay of ^{107}Pd ($t_{1/2} = 6.5$ My) and Pd is much more siderophile than Ag, it is relatively straightforward to calculate the conditions under which the $^{107}Ag/^{109}Ag$ of BSE can be the same as that of CI chondrites. The relatively refractory Pd is generally regarded as having been added to Earth throughout accretion, while moderately volatile Ag could have been added throughout accretion (in a homogeneous accretion scenario) or as part of the Late Veneer. The end-member homogeneous accretion model can be excluded [*Schönbächler et al.*, 2010] because the apparent age of segregation of Pd from Ag would be, in a 2-stage model, much older than that indicated by the Hf-W system, i.e., 9 Myr instead of 30 Myr. More protracted accretion with a certain fraction of disequilibrium between metal and silicate does not lead to concordance between Hf-W and Pd-Ag "ages" of silicate Earth. The most plausible way of bringing agreement between these two short-lived isotope systems is if the bulk of Earth's Ag had been added after ^{107}Pd was extinct (i.e., more than 30 Myr after the origin of the Solar System). Reducing by dilution the Ag- isotope ratio resulting from Pd decay, however, requires the delivery of 13% of the mass of the Earth in volatile rich material. This, however, would provide too much Ag relative to the current mantle composition; thus, this delivery should have happened while core formation was still ongoing (i.e., not in the Late Veneer phase), which resulted in a large fraction of the added Ag being segregated to the core.

A final argument against the possibility of accretion of volatiles during the sole phase of Late Veneer comes from molybdenum-ruthenium isotopes [*Dauphas et al.*, 2004]. In differentiated and bulk primitive meteorites, the isotopic anomalies of these two elements correlate with one another, although new data [*Chen et al.*, 2010; *Burkhardt et al.*, 2011] seem to make the correlation less sharp. Molybdenum is only moderately siderophile, thus most of the amount of Mo presently in the mantle was delivered before the completion of core formation. In contrast, ruthenium is highly siderophile so that nearly all of the mantle Ru was delivered in the Late Veneer [see, however, *Rubie et al.*, 2015 for an opposite view]. The fact silicate Earth lies on the Mo-Ru cosmic correlation supports the idea that the Earth accreted quite homogeneously. For instance, if the Earth's mantle had obtained Mo from enstatite chondrites (dry accretion) and Ru from CM chondrites (wet Late Veneer) then the silicate Earth should be at $\varepsilon^{92}Mo = 0$ and $\varepsilon^{102}Ru = -1.2$ (i.e., completely off the correlation), unlike what is observed.

In summary, although some elementary ratios argue for a volatile-rich Late Veneer [*Wang and Becker*, 2013], it is unlikely that a dominant fraction of Earth's volatiles could be delivered by the Late Veneer.

4.6. CONCLUSIONS

Late Veneer and Late Accretion are related to different concepts. The first is the delivery of chondritic material that, once on our planet, avoided any metal-silicate segregation. The second is the delivery of material after the last giant impact on Earth. We have shown that Late Veneer and Late Accretion are probably not the same. Some of the HSEs in the Earth's mantle possibly pre-date the Moon-forming event, and late-accreted differentiated projectiles probably did not deliver their full HSE budget to the Earth's mantle. Nevertheless, we estimate that the masses delivered as Late Veneer and Late Accretion are probably within a factor of 2 of each other.

The small amount of mass accreted by the Earth after the Moon-forming event (<1% of the Earth's mass) implies that the Moon formed during the last accretional giant impact on our planet and that it formed relatively late, approximately 100 Myr after the first solar system solids. It also argues against the possibility that the entire volatile element budget of the Earth was acquired during the Late Veneer. Geochemical and isotopic arguments also rule out this hypothesis.

The large difference in late accreted masses on the Earth ($\sim 5 \times 10^{-3}$ M_E) and on the Moon ($5 \times 10^{-6} M_E$) suggests that the post-Moon-formation accretion process was dominated by a few large impactors, which, by virtue of the large ratio of gravitational cross sections, hit our planet but missed our satellite [*Bottke et al.*, 2010; *Raymond et al.*, 2013].

ACKNOWLEDGMENTS

A. M. was supported by the European Research Council (ERC) Advanced Grant "ACCRETE" (contract number 290568). The authors wish to thank R. J. Walker and an anonymous reviewer for their constructive reports as well as D. Rubie and S. Jacobson for enriching discussions.

REFERENCES

Albarède, F. (2009) Volatile accretion history of the terrestrial planets and dynamic implications. *Nature*, 461, 1227–1233.

Albarède, F., C. Ballhaus, J. Blichert-Toft, C. T. Lee, B. Marty, F. Moynier, and Q. Z. Yin (2013), Asteroidal impacts and the origin of terrestrial and lunar volatiles. *Icarus*, 222, 44–52.

Allegre, C. J., G. Manhes, and C. Gopel (2008), The major differentiation of the Earth at ~4.45 Ga. *Earth and Planetary Science Letters*, 267, 386–398.

Anders, E. (1968), Chemical Processes in Early Solar System, as Inferred from Meteorites. Accounts Chem Res, *1*(10):289–298, doi: 10.1021/Ar50010a001.

Bottke, W. F., H. F. Levison, D. Nesvorny, and L. Dones (2007), Can planetesimals left over from terrestrial planet formation produce the lunar Late Heavy Bombardment? *Icarus*, 190, 203–223.

Bottke, W. F., R. J. Walker, J. M. D. Day, D. Nesvorny, and L. Elkins-Tanton (2010), Stochastic Late Accretion to Earth, the Moon, and Mars. *Science*, 330, 1527.

Bottke, W. F., D. Vokrouhlicky, D. Minton, D. Nesvorny, A. Morbidelli, R. Brasser, B. Simonson, and H. F. Levison (2012), An Archaean heavy bombardment from a destabilized extension of the asteroid belt. *Nature*, 485, 78–81.

Bouvier, A. and M. Wadhwa (2010), The age of the Solar System redefined by the oldest Pb-Pb age of a meteoritic inclusion. *Nature Geoscience*, 3, 637–641.

Burkhardt, C., T. Kleine, F. Oberli, A. Pack, B. Bourdon, and R. Wieler (2011), Molybdenum isotope anomalies in meteorites: Constraints on solar nebula evolution and origin of the Earth. *Earth and Planetary Science Letters*, 312, 390–400.

Canup, R. M. (2012), Forming a Moon with an Earth-like Composition via a Giant Impact. *Science*, 338, 1052.

Canup, R. M. and E. Asphaug (2001), Origin of the Moon in a giant impact near the end of the Earth's formation. *Nature*, 412, 708–712.

Chen, J. H., D. A. Papanastassiou, and G. J. Wasserburg (2010), Ruthenium endemic isotope effects in chondrites and differentiated meteorites. *Geochimica et Cosmochimica Acta.*, 74, 3851–3862.

Chou, C. L. (1978), Fractionation of siderophile elements in the earth's upper mantle. In: *Lunar and Planetary Science Conference*, vol. 9. pp. 219–230.

Chou, C. L., D. M. Shaw, and J. H. Crocket (1983), Siderophile trace elements in the Earth's oceanic crust and upper mantle. *Journal of Geophysical Research*, 88:A507–A518.

Cohen, B. A., T. D. Swindle, and D. A. Kring (2000), Support for the Lunar Cataclysm Hypothesis from Lunar Meteorite Impact Melt Ages. *Science*, 290, 1754–1756.

Cuk, M. and S. T. Stewart (2012), Making the Moon from a Fast-Spinning Earth: A Giant Impact Followed by Resonant Despinning. *Science*, 338, 1047.

Dauphas, N. and A. Pourmand (2011), Hf-W-Th evidence for rapid growth of Mars and its status as a planetary embryo. *Nature*, 473, 489–492.

Dauphas, N., A. M. Davis, B. Marty, and L. Reisberg (2004), The cosmic molybdenum-ruthenium isotope correlation. *Earth and Planetary Science Letters*, 226, 465–475.

Day, J. M. D., R. J. Walker, O. B. James, and I. S. Puchtel (2010), Osmium isotope and highly siderophile element systematics of the lunar crust. *Earth and Planetary Science Letters*, 289, 595–605.

Drake, M. J. and K. Righter (2002), Determining the composition of the Earth. *Nature*, 416, 39–44.

Frost, D. J., C. Liebske, F. Langenhorst, C. A. McCammon, R. G. Trønnes, and D. C. Rubie (2004), Experimental evidence for the existence of iron-rich metal in the Earth's lower mantle. *Nature*, 428(6981), 409–412.

Gao, X. and M. H. Thiemens (1993), Variations of the Isotopic Composition of Sulfur in Enstatite and Ordinary Chondrites. *Geochimica et Cosmochimica Acta.*, 57(13):3171–3176, doi:10.1016/0016-7037(93)90301-C.

Georg, R. B., A. N. Halliday, E. A. Schauble, and B. C. Reynolds (2007), Silicon in the Earth's core. *Nature*, *447*, 1102–1106.

Halliday, A. N. (2008), A young Moon-forming giant impact at 70–110 million years accompanied by late-stage mixing, core formation and degassing of the Earth. Royal Society of London Philosophical Transactions Series A, *366*, 4163–4181.

Halliday, A. N., B. J. Wood, and T. Kleine (2005), Runaway Growth of Mars and Implications for Core Formation Relative to Earth. AGU Fall Meeting Abstracts, *3*.

Herwartz, D. (2013), Differences in the D17O between Earth, Moon and enstatite chondrites. Royal Society/Kavli Institute Meeting on "The Origin of the Moon— Challenges and Prospects," 25–26 September, Chicheley Hall, Buckinghamshire, UK.

Holzheid, A., P. Sylvester, H. S. C. O'Neill, D. C. Rubie, and H. Palme (2000), Evidence for a late chondritic veneer in the Earth's mantle from high-pressure partitioning of palladium and platinum. *Nature*, *406*(6794), 396–399, doi:10.1038/35019050.

Jacobson, S. A. and A. Morbidelli (2014), Lunar and Terrestrial Planet Formation in the Grand Tack Scenario. Proceedings of the Royal Society A., in press.

Jacobson, S. A., A. Morbidelli, S. N. Raymond, D. P. O'Brien, K. J. Walsh, and D. C. Rubie (2014), Highly siderophile elements in Earth's mantle as a clock for the Moon-forming impact. *Nature*, *508*, 84–87.

Kimura, K., R. S. Lewis, and E. Anders (1974), Distribution of Gold and Rhenium between Nickel-Iron and Silicate Melts– Implications for Abundance of Siderophile Elements on Earth and Moon. *Geochimica et Cosmochimica Acta.*, *38*(5):683–701, doi:10.1016/0016-7037(74)90144-6.

Kleine, T., K. Mezger, H. Palme, and C. Munker. (2004a), The W isotope evolution of the bulk silicate Earth: constraints on the timing and mechanisms of core formation and accretion. *Earth and Planetary Science Letters*, *228*(1–2), 109–123.

Kleine, T., K. Mezger, C. Munker, H. Palme, and A. Bischoff (2004b), 182Hf- 182W isotope systematics of chondrites, eucrites, and martian meteorites: Chronology of core formation and early mantle differentiation in Vesta and Mars. *Geochimica et Cosmochimica Acta.*, *68*, 2935–2946.

Kleine, T., T. S. Kruijer, and P. Sprung (2014), Lunar 182W and the Age and Origin of the Moon. Lunar and Planetary Science Conference, *45*, 2895.

Konig, S., C. Munker, S. Hohl, H. Paulick, A. R. Barth, M. Lagos, J. Pfander, and A. Buchl (2011), The Earth's tungsten budget during mantle melting and crust formation. *Geochimica et Cosmochimica Acta.*, *75*, 2119–2136.

Kyser, T. K., J. R. Oneil, and I. S. E. Carmichael (1986), Possible nonequilibrium oxygen isotope effects in mantle nodules, an alternative to the Kyser-O'Neil-Carmichael o18/o16 geothermometer. *Contributions to Mineralogy and Petrology*, *93*(1), 120–123.

Labidi, J., P. Cartigny, and M. Moreira (2013), Non-chondritic sulphur isotope composition of the terrestrial mantle. *Nature*, *501*(7466):208–211, doi:10.1038/Nature12490

Li, J. and C. B. Agee (1996), Geochemistry of mantle-core differentiation at high pressure. *Nature*, *381*, 686–689.

Lodders, K. (2003), Solar system abundances and condensation temperatures of the elements. *Astrophys J.*, *591*, 1220–1247.

Lugmair, G. W. and A. Shukolyukov (1998), Early solar system timescales according to 53Mn–53Cr systematics. *Geochimica et Cosmochimica Acta.*, *62*, 2863–2886.

Maier, W. D., S. J. Barnes, I. H. Campbell, M. L. Fiorentini, P. Peltonen, S. J. Barnes, and R. H. Smithies (2009), Progressive mixing of meteoritic veneer into the early Earth's deep mantle. *Nature*, *460*, 620–623.

Mann, U., D. J. Frost, D. C. Rubie, H. Becker, and A. Audetat (2012), Partitioning of Ru, Rh, Pd, Re, Ir and Pt between liquid metal and silicate at high pressures and high temperatures - Implications for the origin of highly siderophile element concentrations in the Earth's mantle. *Geochimica et Cosmochimica Acta.*, *84*, 593– 613, doi:10.1016/J.Gca.2012.01.026

Marchi, S., W. F. Bottke, D. A. Kring, and A. Morbidelli (2012), The onset of the lunar cataclysm as recorded in its ancient crater populations. *Earth and Planetary Science Letters*, *325*, 27–38.

Marchi, S., W. F. Bottke, L. T. Elkins-Tanton, M. Bierhaus, K. Wuennemann, A. Morbidelli, and D. A. Kring (2014), Widespread mixing and burial of Earth's Hadean crust by asteroid impacts. Submitted.

McDonough, W. F. and S.-s. Sun (1995), The composition of the Earth. *Chemical Geology*, *120*(3–4), 223–253.

Morbidelli, A. (2013), Dynamical Evolution of Planetary Systems. Planets, Stars and Stellar Systems. Volume 3: Solar and Stellar Planetary Systems, *63*.

Morbidelli, A., K. Tsiganis, A. Crida, H. F. Levison, and R. Gomes (2007), Dynamics of the Giant Planets of the Solar System in the Gaseous Protoplanetary Disk and Their Relationship to the Current Orbital Architecture. *The Astronomical Journal*, *134*, 1790–1798.

Morbidelli, A., S. Marchi, W. F. Bottke, and D. A. Kring (2012), A sawtooth-like timeline for the first billion years of lunar bombardment. *Earth and Planetary Science Letters*, *355*, 144–151.

Neumann, G. A. and 10 colleagues (2013), The Inventory of Lunar Impact Basins from LOLA and GRAIL. Lunar and Planetary Science Conference, *44*, 2379.

O'Neill, H. S. (1991), The Origin of the Moon and the Early History of the Earth–a Chemical-Model .2. The Earth. *Geochimica et Cosmochimica Acta.*, *55*(4):1159– 1172, doi:10.1016/0016-7037(91)90169-6.

Palme, H., H. S. C. O'Neill (2003), Cosmochemical Estimates of Mantle Composition. In: R. W. Carlson (Ed.) *The Mantle and Core, vol 2.* Elsevier, Amsterdam, pp 1–38.

Raymond, S. N., H. E. Schlichting, F. Hersant, and F. Selsis (2013), Dynamical and collisional constraints on a stochastic late veneer on the terrestrial planets. *Icarus*, *226*, 671–681.

Reufer, A., M. M. M. Meier, W. Benz, and R. Wieler (2012), A hit-and-run giant impact scenario. *Icarus*, *221*, 296–299.

Righter, K., M. Humayun, and L. Danielson (2008), Partitioning of palladium at high pressures and temperatures during core formation. *Nature Geoscience*, *1*(5), 321– 323, doi:10.1038/Ngeo180.

Rubie, D.C., D. J. Frost, U. Mann, Y. Asahara, F. Nimmo, K. Tsuno, P. Kegler, A. Holzheid, and H. Palme (2011), Heterogeneous accretion, composition and core-mantle differentiation of the Earth. *Earth and Planetary Science Letters*, *301*, 31– 42.

Rubie, D. C., V. Laurenz, S. A. Jacobson, A. Morbidelli, H. Palme, and D. J. Frost (2015), The Hadean matte, magma ocean solidification and Earth's late veneer, *Abstract of the 2015 Goldschmidt conference.*

Ryder, G. (2002), Mass flux in the ancient Earth-Moon system and benign implications for the origin of life on Earth. *Journal of Geophysical Research (Planets), 107,* 5022.

Schönbächler, M., R. W. Carlson, M. F. Horan, T. D. Mock, and E. H. Hauri (2010), Heterogeneous accretion and the moderately volatile element budget of Earth. *Science, 328,* 884–887.

Stoffler, D. and G. Ryder (2001), Stratigraphy and Isotope Ages of Lunar Geologic Units: Chronological Standard for the Inner Solar System. *Space Science Reviews, 96,* 9–54.

Strom, R.G., R. Malhotra, T. Ito, F. Yoshida, and D. A. Kring (2005), The Origin of Planetary Impactors in the Inner Solar System. *Science, 309,* 1847–1850.

Taylor, D.J., K. D. McKeegan, T. M. Harrison, and E. D. Young (2009), Early differentiation of the lunar magma ocean. New Lu-Hf isotope results from Apollo 17. *Geochimica et Cosmochimica Acta Supplement, 73,* 1317.

Tera, F., D. A. Papanastassiou, and G. J. Wasserburg (1974), Isotopic evidence for a terminal lunar cataclysm. *Earth and Planetary Science Letters, 22,* 1–21.

Thibault, Y. and M. J. Walter (1995), The influence of pressure and temperature on the metal-silicate partition coefficients of nickel and cobalt in a model C1 chondrite and implications for metal segregation in a deep magma ocean. *Geochimica Et Cosmochimica Acta., 59*(5), 991–1002.

Touboul, M., T. Kleine, B. Bourdon, H. Palme, and R. Wieler (2007), Late formation and prolonged differentiation of the Moon inferred from W isotopes in lunar metals. *Nature, 450,* 1206–1209.

Touboul, M., I. S. Puchtel, and R. J. Walker (2012), 182W Evidence for Long-Term Preservation of Early Mantle Differentiation Products. *Science, 335,* 1065–1069.

Touboul, M., I. S. Puchtel, and R. J. Walker (2015), Tungsten isotopic evidence for disproportional late accretion to the Earth and Moon. *Nature, 520,* 530–533.

Turekian, K. K. and S. P. Clark (1969), Inhomogeneous Accumulation of Earth from Primitive Solar Nebula. *Earth and Planetary Science Letters, 6*(5), 346–348, doi:10.1016/0012-821x(69)90183-6.

Wade, J. and B. J. Wood (2005), Core formation and the oxidation state of the Earth. *Earth and Planetary Science Letters, 236,* 78–95.

Walker, R. J. (2009), Highly siderophile elements in the Earth, Moon and Mars: Update and implications for planetary accretion and differentiation. *Chemie der Erde Geochemistry, 69,* 101–125.

Walker, R. J., M. F. Horan, J. W. Morgan, H. Becker, J. N. Grossman, and A. E. Rubin (2002), Comparative 187Re–187Os systematics of chondrites: Implications regarding early solar system processes. *Geochimica et Cosmochimica Acta., 66,* 4187–4201.

Wang, Z. C. and H. Becker (2013), Ratios of S, Se and Te in the silicate Earth require a volatile-rich late veneer. *Nature, 499*(7458):328, doi:10.1038/Nature12285.

Wiechert, U., A. N. Halliday, D.-C. Lee, G. A. Snyder, L. A. Taylor, and D. A. Rumble (2001), Oxygen isotopes and the Moon-forming giant impact. *Science, 294,* 345–348.

Wilhelms, D.E. (1987), The geologic history of the Moon. (U.S. Geol. Surv. Prof. Pap., 1348).

Willbold, M., T. Elliott, and S. Moorbath (2011), The tungsten isotopic composition of the Earth's mantle before the terminal bombardment. *Nature, 477*(7363), 195–U191, doi:10.1038/Nature10399.

Wood, B. J., A. Halliday, and M. Rehkamper (2010), Volatile accretion history of the Earth. *Nature, 467,* E6–E7.

Wood, B.J., J. Wade, and M. R. Kilburn (2008), Core formation and the oxidation state of the Earth: Additional constraints from Nb, V and Cr partitioning. *Geochimica et Cosmochimica Acta., 72,* 1415–1426.

Yin, Q. Z., S. B. Jacobsen, K. Yamashita, J. Blichert-Toft, P. Telouk, and F. Albarede (2002), A short timescale for terrestrial planet formation from Hf-W chronometry of meteorites. *Nature, 418*(6901), 949–952.

Yin, Q. Z. and J. Antognini (2008), Isotopic and elemental constraints on the first 100 Myr of Earth history. *Geochimica et Cosmochimica Acta Supplement, 72,* 1061.

Zhang, J., N. Dauphas, A. M. Davis, I. Leya, and A. Fedkin (2012), The proto-Earth as a significant source of lunar material. *Nature Geoscience, 5,* 251–255.

5

Early Differentiation and Core Formation: Processes and Timescales

Francis Nimmo[1] and Thorsten Kleine[2]

ABSTRACT

The Earth's core formed via a series of high-energy collisions with already-differentiated objects, likely resulting in several distinct magma ocean epochs. The cores of these impactors probably underwent only limited emulsification and moderate (~50%) isotopic re-equilibration with the target mantle during the collision. Later impactors likely originated from more distant regions of the inner Solar System and were plausibly more volatile-rich and more oxidized than earlier impactors. Short-lived isotopes, especially the hafnium-tungsten (Hf-W) system, provide the strongest constraints on the timescale of accretion and core formation. These short-lived isotopes and dynamical models provide a mutually self-consistent, albeit approximate, chronology. Terrestrial core formation took more than 30 Myr but less than about 200 Myr to complete.

5.1. INTRODUCTION

The Earth—and the other terrestrial planets—are differentiated bodies, a metallic core overlain by a silicate mantle. But the nebula from which these planets ultimately formed originally consisted mainly of undifferentiated dust grains. The aim of this chapter is to review how and when the process of differentiation is thought to have taken place.

In the first half of this chapter we focus on "how"; how did the planets ultimately grow, and by what mechanisms did core formation take place? In the second half, we focus on "when"; what evidence do we have for the timescales of melting and core formation? Although the main focus of this chapter is the Earth, the processes of accretion and differentiation are general. As a result, considerable insight can be gained by looking at the evidence from both meteorites and other terrestrial bodies, and we do so briefly in this chapter.

The contents of this chapter are closely related to others in this volume. In particular, the processes by which the elemental and stable isotopic composition of the core were established are examined in Chapter 6. We do not focus on silicate differentiation (mantle melting and crust formation), which is treated in more detail in Chapter 8. Nor do we discuss the final addition of material to the mantle (the "late veneer") after core formation effectively ceased (see Chapter 4).

[1]*Department Earth and Planetary Sciences, University of California, Santa Cruz, California, USA*
[2]*Institut für Planetologie, Westfälische Wilhelms-Universität Münster, Münster, Germany*

The Early Earth: Accretion and Differentiation, Geophysical Monograph 212, First Edition.
Edited by James Badro and Michael Walter.

This chapter covers similar ground to various earlier reviews. An old but still useful treatment of the physics of core formation is given by *Stevenson* [1990]; an accessible survey of the earliest Earth is by *Zahnle et al.* [2007]. A comprehensive review with a similar scope to this one may be found in *Rubie et al.* [2015].

5.1.1. How Does Core Formation Occur?

In one sense, core formation is a very simple process. It is energetically favorable for the densest components of a planet (the metals) to sink to the center. However, this redistribution of mass requires deformation of some kind to occur. Neither metals nor silicates are particularly deformable at low temperatures, and as a result, core formation can only occur at elevated temperatures (Section 5.1.1.3). As we discuss in Section 5.1.1.2, the extent to which planetary bodies are heated as they grow depends on the details of the accretion process. Our current understanding of terrestrial planet accretion is summarized in Section 5.1.1.1.

An important conclusion, at least for the Earth, is that many of the bodies that formed the Earth were themselves already differentiated. As a result, the idea of there being a single instant of core formation is too simple; in reality, the delivery of a metallic component to the deep interior happened many times, and potentially by several different mechanisms. To the extent that it implies a single event, "core formation" is thus a misnomer. We nonetheless continue to employ it as useful shorthand for the processes by which metallic materials are transported to depth.

5.1.1.1. Overview of Accretion in the Terrestrial Planet Zone

The processes of planetary accretion and core growth are inextricably linked. Planets grow via collisions, which both deliver core material and produce conditions under which differentiation is favored (see below). As a result, we will begin with a brief overview of accretion. More detailed reviews may be found in *Chambers* [2010], *Morbidelli et al.* [2012], and Chapter 3.

Conventionally, terrestrial planet accretion is thought to occur in four stages. The initial growth from micron-sized dust grains to kilometer-sized bodies is not well understood, because of the tendency of meter-scale bodies to break themselves apart. Nonetheless, this process must have occurred rapidly–in perhaps 10^3 years–to avoid loss of the dust grains via gas drag. Kilometer-scale bodies are large enough to start perturbing their neighbors' orbits, at which point a second stage, that of runaway growth, ensues. In this stage, larger bodies are more effective in focusing impacts than their smaller neighbors, and so grow more rapidly, becoming yet more effective and so

on. This process slows once the local feeding zone is exhausted, and transitions into the more orderly third stage of oligarchic growth [*Kokubo and Ida*, 1998]. At a few Myr after Solar System formation, the region around 1AU will have contained tens of Moon- to Mars-sized "embryos" embedded within a cloud of surviving smaller planetesimals. At about the same time, early stellar activity will have removed any remaining gas and dust not already incorporated into larger objects. The presence of gas is important because it can drive planetary migration. For instance, early migration of Jupiter and Saturn may have sculpted the protoplanetary disk from which the terrestrial planets subsequently formed, potentially explaining the small mass of Mars [*Walsh et al.*, 2011].

The final stage of terrestrial planet accretion involves growth via collisions with embryos; typical timescales for this process are tens of Myr [e.g., *Agnor et al.*, 1999]. From the point of view of the early Earth, this stage is the most important, because it involves the largest transfers of mass and energy (see below). Because of the relatively small number of bodies involved, this process can be modeled relatively easily. Unfortunately, however, the process is stochastic, and so there are many different growth scenarios that could have yielded the early Earth.

Figure 5.1 shows the typical growth history of an Earth-like body (bold line), from the numerical simulations of *Raymond et al.* [2006]. Growth occurs mainly via half-a-dozen or so collisions with comparable-sized objects; although there is a steady background mass accumulation via small impacts, the total mass fraction contributed by this process is small. This kind of growth history can be crudely approximated by continuous growth at an exponentially-decreasing rate (thin line). Although this kind of analytical model is advantageous for modeling the isotopic effects of core formation (see Section 5.2.3 below), it does not capture the discontinuous, stochastic nature of the real accretion process.

Beyond the discontinuous, stochastic nature of Earth's growth, there are two other relevant characteristics of accretion. First, the "feeding zone" from which the proto-Earth accretes material expands with time [*O'Brien et al.*, 2006, 2014; *Bond et al.*, 2010]. As a result, material delivered later in the accretion process tends to have originated at greater distances from the Sun. This result is potentially important for volatile content and partitioning behavior of siderophile elements during core formation (see Section 5.2.3.3). Second, in reality only roughly one in every two collisions actually result in the two bodies merging [*Kokubo and Genda*, 2010, *Chambers*, 2013]; other collisions are either "hit-and-run" events [*Asphaug et al.*, 2006] or (more rarely) result in net mass loss (erosion) of the target body. Hit-and-run events prolong the duration of the final stages of accretion. Erosive events are important because they can change the bulk composition

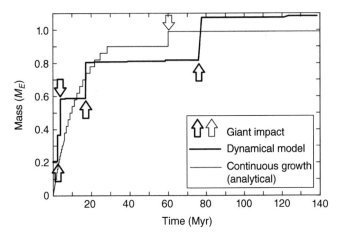

Figure 5.1 Bold line shows a typical growth curve from an N-body simulation [run2a of *Raymond et al.*, 2006]; arrows indicate giant impacts. The thin line assumes growth at an exponentially decaying rate (equation 5.4) with a timescale τ of 10 Myr and a later final Moon-forming impact. Exponential growth models with longer e-folding time-scales have been used to reproduce the Hf-W and U-Pb isotope systematics of the Earth [*Halliday*, 2004; *Rudge et al.*, 2010] under the assumption of incomplete re-equilibration during impacts (Section 5.2.4.1). M_E is one Earth mass.

(and isotopic signature) of the target bodies [*Dwyer et al.*, 2015]; for instance, the Earth may have a non-chondritic bulk composition (see Chapter 2) thanks to preferential impact removal of incompatible elements contained in the crust [*O'Neill and Palme*, 2008]. We discuss this issue further below.

5.1.1.2. Thermal State of Accreting Bodies

Since core formation requires elevated temperatures, we need to consider possible heat sources available during accretion. There are two principal sources. The first is the decay of short-lived radioisotopes, the most important of which is ^{26}Al with a half-life of ~0.7 Myr; ^{60}Fe with a half-life of ~2.6 Myr was probably not present in sufficient quantities to be important [*Tang and Dauphas*, 2012]. The heat conduction timescale for a silicate sphere of radius R is given by $R^2/\pi^2\kappa$ where κ is the thermal diffusivity; the resulting timescale is ~0.7 Myr $(R/15$ km$)^2$. Consequently, an asteroid that grows to a diameter much greater than 30 km within the first Myr or so will be unable to conduct away the heat produced by ^{26}Al and will thus experience heating. The exact amount of heating will depend on the details of the growth process, but ^{26}Al is enormously energetic. The total energy released by ^{26}Al decay is sufficient to heat typical chondritic materials by about 4000 K [*Rubie et al.*, 2015]. As a result, early-formed bodies will certainly experience widespread melting (and there is now abundant evidence that this actually happened as discussed below).

The second energy source is the release of gravitational energy. Crudely speaking, an impactor's kinetic energy is (largely) converted into heating the target, with the depth to which this heat is buried depending on the impactor size. As long as the impactors are not too small, the heat

will be buried sufficiently deep that cooling via radiation is ineffective, and as a result the body heats up. In the simplest scenario, this effect results in an inverted temperature profile (later impactors deliver more energy), but for Earth-sized bodies, this effect is likely to be overwhelmed by the deep, heterogeneous heating and mixing caused by giant impacts.

Assuming all the gravitational energy is converted to heat, the globally-averaged temperature increase ΔT due to accretion of a planet of mass M is given by $\Delta T = 3GM/5RC_p \approx 35,000$ K $(M/M_E)^{2/3}$, where M_E is the mass of the Earth and C_p is the specific heat capacity [e.g., *Rubie et al.*, 2015]. This highly simplified calculation neglects the likely large lateral and radial variations in temperature due to giant impacts [e.g., *Canup*, 2004] and the subsequent redistribution of material via rebound and relaxation [*Tonks and Melosh*, 1993]. Nonetheless, it serves to illustrate that Earth- and Mars-sized ($M = 0.1$ M_E) bodies are likely to have undergone extensive melting, whereas for an object the size of Vesta ($M = 5 \times 10^{-5}$ M_E) gravitational heating is insignificant.

An additional source of heat is further gravitational potential energy release as iron sinks toward the center of the planet [*Stevenson*, 1990]. This heat-source is small compared to the total impact energy but can serve to accelerate the iron transport process because the heat is deposited locally, reducing local viscosities [e.g., *Ricard et al.*, 2009; *Sramek et al.*, 2010].

5.1.1.2.1. Terrestrial Magma Oceans

It seems inescapable that much of the Earth's mantle was molten during late-stage accretion. However, the depth and duration of these melting episodes are less clear. Impacts tend to deposit less heat at depth than in the near-surface, while the mantle melting temperature increases quite steeply with depth (~1 K/km) [e.g., *Rubie et al.*, 2015]. As a result, it is possible that the lowermost mantle never experienced complete melting. A more detailed discussion of mantle melting and re-freezing may be found in chapters 7 and 8.

The melting history of the mantle depends very strongly on magma ocean lifetimes. If the magma ocean lifetime is long compared to the interval between giant impacts (~10 Myr) then complete mantle melting is more likely than if each magma ocean freezes before the next impact. Unfortunately, magma ocean lifetimes are not well understood. Radiative cooling of an exposed convecting magma ocean is rapid (~1 kyr) [*Solomatov*, 2000]. On the other hand, if a surface conductive lid develops, or a thick atmosphere is present [*Abe and Matsui*, 1986; *Zahnle et al.*, 2007], cooling timescales can be ~100 Myr, which is a big difference. Tidal dissipation could also have extended the lifetime of a partially-molten layer [*Zahnle et al.*, 2007]. Unlike the Moon, where a buoyant Al-rich plagioclase crust developed, any conductive lid on the Earth would be dense (because Al partitions into garnet at higher pressures). Such a lid would have a tendency to founder and/or be disrupted by impacts, both of which would negate its insulating properties. As a result, in the absence of an atmosphere, relatively rapid magma ocean cooling appears likely. This being the case, the Earth probably experienced several magma ocean epochs, with relatively rapid re-freezing following each giant impact event. The apparently high mantle ^{3}He/^{22}Ne ratio has been attributed to fractionation as the result of multiple magma ocean episodes [*Tucker and Mukhopadhyay*, 2014].

5.1.1.3. Core Formation Mechanisms

Core formation involves several distinct mechanisms by which a dense metallic phase may be transported to the deep interior: percolation, diking, diapirism, and direct delivery via impacts. More thorough treatments are given in *Stevenson* [1990] and *Rubie et al.* [2015].

The mechanism involving the lowest levels of stress is percolation. Because Fe (and even more so Fe-S) melts at lower temperatures than silicates, undifferentiated small bodies heated internally will tend to develop metallic melts dispersed within a solid silicate matrix. The metal-silicate density contrast will cause the metal to percolate downward. The characteristic timescale for the two phases to separate is the so-called compaction timescale, originally derived for silicate melt percolation [*McKenzie*, 1984]. Because of the high density and low viscosity of molten iron, core formation by percolation is expected to be rapid [*McCoy et al.*, 2006]. However, at least in the absence of shear stresses [*Yoshino et al.*, 2003], percolation requires an interconnected melt network to exist. This in turn requires either the so-called dihedral angle between melt and solid to be <60° or the melt fraction to exceed a critical value (typically a few percent). The dihedral angle is likely >60° in the upper mantle; for the lower mantle it is not yet clear what the dihedral angle is [*Terasaki et al.*, 2007; *Shih et al.*, 2013]. In any event, if percolation were the dominant mechanism in the upper mantle, one would expect to see a few percent metallic iron stranded there, which is not observed.

Percolation involves distributed flow through a solid matrix. However, if sufficiently large bodies of liquid metal develop, the associated stresses may become large enough to permit other transport mechanisms to operate. Although percolation in theory could be important in the lower mantle, the likely accumulation of large iron bodies (e.g., at the base of a magma ocean) implies that other transport mechanisms will dominate. As argued above, iron transport in the upper mantle also takes place by other mechanisms (e.g., sinking through a magma ocean). Lastly, percolation *sensu stricto* is unlikely to operate over an extended depth range, as it requires that the temperature remain above the metallic solidus but below the silicate solidus. In short, percolation is not expected to be an important iron transport mechanism for the proto-Earth.

If the silicates are relatively cold and brittle, transport may occur via fluid-filled cracks, analogous to dikes [*Stevenson*, 1990]. As long as the stresses are sufficient to overcome the fracture toughness of the surrounding material, the rate-limiting process is fluid drag on the crack walls, and as a result downward delivery of iron by this mechanism is very rapid.

Alternatively, if the silicates are warmer they will deform by viscous creep rather than brittle failure. In this case a dense body of iron will act like a diapir, sinking through the deformable silicates. The rate of sinking depends strongly on the size of the diapir and the viscosity of the surrounding material, but it can be rapid (~kyrs) [*Ricard et al.*, 2009]. The deformation of this material can lead to heating, which enhances the rate at which the diapir sinks [*Ricard et al.*, 2009; *Sramek et al.*, 2010; *Samuel et al.*, 2010]. This picture is relevant to small impactor cores, which will be decelerated effectively by the mantle and subsequently exhibit laminar flow. It is not relevant to the largest impactors, which will transit the mantle without slowing appreciably (see below).

The above mechanisms all assume an initially solid (or mostly solid) background matrix. However, temperatures may become sufficiently elevated that the silicates also melt (Section 5.1.1.2). In this case, any pre-existing metal will sink at a rate controlled by the viscosity of the molten silicates, which is comparable to that of water [*Rubie et al.*, 2003]. As a result, this sinking process is very rapid, despite the potentially vigorous convection taking place in the molten mantle. One can thus envisage situations in which a pile-up of metal occurs at the base of a local melt pool (for small impacts) or a global magma ocean. Subsequent deeper transport then occurs via one of the other mechanisms discussed above.

During the final stages of accretion, metallic material is delivered as pre-existing cores contained within impactors of comparable size to the target, striking the target at many km/s. The stresses involved in these giant impacts are so large that the material strength of the target is irrelevant. The entire mantle behaves effectively like a fluid. As a result, simulations show that the impactor core merges with the target core on the free-fall timescale of roughly an hour [e.g., *Canup*, 2004]. The high speed and low viscosity of iron make this kind of process enormously turbulent, something that cannot be adequately captured by numerical models. This process, which probably is most relevant for defining the geochemical and isotopic signature of core formation, is very different from the laminar models of diapir descent described above. Diapirism may occur after small impacts deliver metal to the base of a magma ocean, but it is not an appropriate description of the giant impacts that likely deliver the bulk of the metal content of the Earth's core.

Because of the difficulty in modeling turbulent flow during impacts, the extent to which the impactor core mixes and equilibrates with the target silicates is very poorly understood. Unfortunately, the extent of equilibration is crucial in determining the extent to which metals partition into the mantle and the isotopic consequences of this partitioning. The mixing process is certainly scale-dependent: impactor cores that are small compared to the mantle thickness probably re-mix effectively, while larger ones do not. This issue is discussed in more detail in Section 5.2.3.1.

5.1.1.4. Composition of the Earth's Core

The composition of the Earth's core is important because it is known (to some extent), and thus provides a constraint on how accretion occurred. For instance, as will be seen, different trajectories in oxygen fugacity make quite different predictions about core composition. The core also has an indirect effect, in that core formation is likely to have removed most siderophile elements from the mantle (see Section 5.2.3.3). This process is central to the Hf-W chronometer discussed below. A recent review of core composition appears in *Rubie et al.* [2015], and only a brief summary is given here.

Although the Earth's core is mainly Fe+Ni, it has long been known that the seismically-determined core density is less than that of a simple Fe-Ni alloy. For the outer core the difference is 6 to 10%, with a somewhat smaller difference for the inner core [e.g., *Alfe et al.*, 2002]. Thus, the core must contain one or more light elements, with commonly-cited suspects including sulfur, silicon, oxygen, and carbon [e.g., *Poirier*, 1994].

The Earth's dynamo is thought to be at least partly driven by compositional convection due to expulsion of light element(s) from the inner core as it solidifies [e.g., *Nimmo*, 2015]. *Alfe et al.* [2002] used first principles computations to argue that O, but not S or Si, is excluded from crystalline iron, and therefore, that O must make up at least part of the light element budget. They concluded that the outer core contains 10+/2.5% molar S or Si and 8+/–2.5% molar O, while the inner core contains 8.5+/–2.5% molar S/Si.

Whether or not a particular element partitions into the core depends on the pressure, temperature, and oxygen fugacity conditions at the relevant time [e.g., *Tsuno et al.*, 2013]. Measurement of partitioning behavior at the high P,T conditions associated with core formation is experimentally challenging [e.g., *Siebert et al.*, 2013]; moreover, the oxygen fugacity is likely to have evolved as core formation proceeded [e.g., *Wade and Wood*, 2005; *Rubie et al.*, 2011; *Siebert et al.*, 2013], further complicating analysis. Some calculations favor Si as the dominant light element [*Rubie et al.*, 2011; *Ricolleau et al.*, 2011], while others prefer O [*Siebert et al.*, 2013]; S is less popular because of its high volatility [*McDonough*, 2003]. A comparison of experimental and seismically-derived velocities suggest that core concentrations of O are relatively low [*Huang et al.*, 2011] and that S and/or Si are more important [*Morard et al.*, 2013]. A core possessing at least some Si is also consistent with small differences in stable Si isotopes between Earth's mantle and chondrites [*Georg et al.*, 2007], although this difference may also arise from processes within the solar nebula [*Dauphas et al.*, 2014]. At present, the identity of the light element(s) in the Earth's core remains an open question.

One further trace element of potential importance to the core is potassium, because radioactive decay of this element can help drive a long-lived dynamo and influences the long-term temperature evolution of the core [e.g., *Nimmo*, 2015]. K does not appear to partition efficiently into metal, at least under moderate P,T conditions [*Bouhifd et al.*, 2007; *Corgne et al.*, 2007], and the deficiency of K relative to U and Th in the Earth's mantle compared to chondrites is readily explained by potassium's greater volatility. Ultimately, geoneutrino studies [e.g., *Araki et al.*, 2005] should directly constrain how much (if any) potassium the core contains.

5.1.1.5. Lessons from Other Bodies

So far, we have treated the processes involved in core formation theoretically. Fortunately, however, in most cases there is observational evidence for the processes described. Here we will briefly discuss pertinent observations from bodies other than the Earth. For the Earth, we summarize the likely processes operating in Section 5.1.1.6 below, and discuss the chronology of core formation in more detail in Section 5.2.

Abundant evidence exists for differentiation and core formation in asteroids. Perhaps most obviously, magmatic iron meteorites and metallic parent bodies (such as 16 Psyche) require core formation to have occurred, while differentiated achondrites are depleted in siderophile elements [e.g., *Mittlefehldt et al.*, 1998], indicating loss of metal due to core formation. The size of the parent bodies in which this differentiation took place is roughly constrained by cooling rates estimated from exsolution textures [e.g., *Yang and Goldstein*, 2006]. Typical diameters are 30–100 km. Some rapid cooling rates have been attributed to bodies in which the silicate mantle was stripped by a giant impact [*Yang et al.*, 2007], in which case the pre-impact body could have been larger. Even in this case, however, gravitational heating would still be insufficient to cause melting (see above). On the other hand, these sizes are sufficient to cause melting due to ^{26}Al decay if the bodies formed early enough. The Hf-W chronometer (see below) suggests that the parent bodies of magmatic iron meteorites formed and differentiated within ~2 Myr after formation of the earliest known Solar System solids, calcium-aluminium inclusions, or CAIs [e.g., *Kruijer et al.*, 2014a], and were therefore certainly subject to ^{26}Al heating.

On the other hand, the existence of chondritic meteorites implies that not all asteroids underwent differentiation. Because of the rapid decay of ^{26}Al, bodies that accreted more than 2–3 Myr after Solar System formation are unlikely to have experienced enough heating to undergo differentiation. Chondrites, therefore, either derive from very small bodies or from bodies that accreted more than ~2 Myr after CAI formation. A recent suggestion is that some chondrites are samples from the surface of differentiated asteroids [*Elkins-Tanton et al.*, 2011].In any event, relatively small changes in the rate of growth can have very dramatic effects on the thermal evolution of a body, and thus whether or not it underwent differentiation.

Interestingly, some classes of meteorites appear to have been caught in the act of differentiation. Hand-specimen size samples of acapulcoites and lodranites show evidence for localized formation of metal melts, which, however, did not fully segregate [*McCoy et al.*, 2006]. Hf-W isotope systematics suggest that such bodies accreted 1.5–2 Myr after CAI formation [*Touboul et al.*, 2009], consistent with moderate heating by ^{26}Al decay.

As noted above, gravitational energy alone is probably sufficient to cause differentiation and core formation on Mars. However, the growth of Mars (as measured by the Hf-W isotopic system, Section 5.2.1) may have been sufficiently rapid that ^{26}Al also played a role [*Dauphas and Pourmand*, 2011], although the Hf-W age of core formation

in Mars is debated [*Mezger et al.*, 2013; *Nimmo and Kleine*, 2007]. As noted above, there are dynamical scenarios that permit this kind of rapid growth to occur. From our point of view, Mars is important because it likely represents the kind of precursor body from which the Earth was built. Early core formation in these precursor bodies could have potentially important consequences for the Hf-W isotope systematics of the Earth's mantle (see below).

The Moon probably formed as the result of a giant impact onto the proto-Earth. The small core size and volatile-depleted nature of the Moon is a consequence of this unusual mode of formation. Although many dynamical aspects of this event remain uncertain (such as impactor size, mass fraction of impactor material in the Moon) [*Cuk and Stewart*, 2012; *Canup*, 2012], the isotopic similarities between the Earth and the Moon are most easily accounted for if the Moon is largely derived from the Earth's mantle [*Zhang et al.*, 2012] or if both the Earth and impactor formed at similar heliocentric distances from a homogeneous inner disk reservoir [*Wiechert et al.*, 2001; *Dauphas et al.*, 2014]. The Earth-Moon W isotope similarity is not easily explained in either case, however, because the different mixing proportions of impactor material in the Moon and bulk silicate Earth should lead to W isotope heterogeneities [*Touboul et al.*, 2007; *Kruijer et al.*, 2015], unless very specific impactor and proto-Earth compositions and giant impact conditions are invoked [*Dauphas et al.*, 2014]. The Moon-forming impact is generally assumed to be the last large impact that the Earth experienced, and as such its formation marks the effective end of the main phase of core formation. We discuss this issue further below in Section 5.2.5. The Moon is also relevant because it may provide a record of the Earth's mantle prior to the waning stage of accretion.

5.1.1.6. Summary: Core Formation Processes on Earth

Core formation on Earth was a multi-stage process, with the majority of the metal being delivered to the proto-Earth by giant impacts with already-differentiated bodies. These impacts were highly energetic and resulted in massive melting of significant fractions of the mantle. Although the evidence is not conclusive, it is likely that the resulting magma ocean lifetimes were short compared to the interval between large impacts. These multiple melting episodes likely allowed mantle heterogeneities to develop and persist.

For the largest impacts, the impactor core's passage through the mantle was rapid (~hours) and emulsification and mixing was probably restricted; as a result, chemical or isotopic re-equilibration with the silicates will have been limited. Smaller impactor cores may have undergone more complete emulsification, resulting in ponding of metallic material at the base of the magma ocean and subsequent downward transport by diapirism or diking. Figure 5.2 summarizes our physical picture of how core

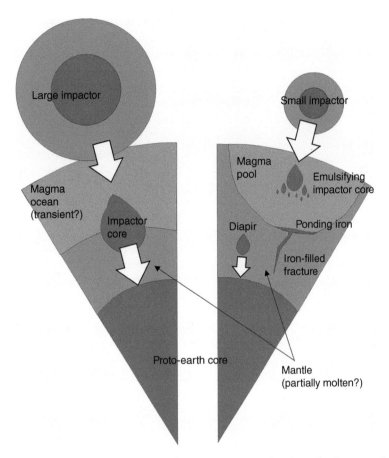

Figure 5.2 Summary picture of the distinct metal segregation mechanisms for large and smaller impacts. Giant impacts probably result in widespread mantle melting and a magma ocean of uncertain but potentially short duration. Such large impacts probably result in minimal emulsification and mixing of the impactor core. If the mantle is initially solid, a small impact will generate a melt pool at the base of which metal will accumulate.

formation proceeds following both large and smaller impacts. In the next section, we discuss the timescales of the relevant processes.

5.2. WHEN DOES IT OCCUR?

There are various unstable isotopes that provide potential chronometers of core formation (and thus accretion). The most useful of these is the hafnium-tungsten (Hf-W) system, which is described in some detail in Sections 5.2.1 to 5.2.3. Other chronometers, principally the palladium-silver (Pd-Ag), uranium-lead (U-Pb), and iodine-xenon (I-Xe) systems, also provide some constraints, and are discussed briefly in Section 5.2.4.

5.2.1. Principles of Hf-W System

A recent and thorough review of the Hf-W system may be found in *Kleine et al.* [2009]. A good overview of theoretical models may be found in *Jacobsen* [2005].

The principles of the system are as follows. The short-lived, now-extinct radionuclide ^{182}Hf decays to ^{182}W with a half-life, $t_{1/2}$, of 8.9 Myr (comparable to the timescale of accretion). Since both Hf and W are refractory, there presumably has been little Hf/W fractionation in the solar nebula, and so the Hf/W ratio of bulk planetary bodies can be assumed to be approximately the same as in chondrites. During core formation, the chondritic Hf/W ratio of an originally undifferentiated body is internally fractionated, because most W will have been removed from the mantle, which will still retain its full complement of Hf. If core formation occurred during the effective lifetime of ^{182}Hf (i.e., ca. 6 half-lives, ~50 Myr), then the mantle will develop excess ^{182}W, the magnitude of which depends on the timing of core formation and the Hf/W ratio of the mantle. On the other hand, if core formation occurred after extinction of ^{182}Hf, then no in-growth of mantle ^{182}W will occur. Thus, the amount of excess ^{182}W in the mantle (or the ^{182}W deficit in a core) can be used to establish the timing of core formation. Additional fractionation may occur during silicate melting, where W is more incompatible than Hf, potentially leading to ^{182}W variations unrelated to core formation (see Section 5.2.2) [*Kleine et al.*, 2004; *Foley et al.*, 2005].

The excess ^{182}W of a sample can be expressed as relative to a CHondritic Uniform Reservoir (CHUR), in which case it is defined as follows:

$$\Delta\varepsilon_W(t) = \left[\frac{\left(^{182}W/^{184}W\right)_{sample}(t)}{\left(^{182}W/^{184}W\right)_{CHUR}(t)} - 1 \right] \times 10^4 \quad (5.1)$$

Note that $\Delta\varepsilon_w(t)$ is time-variable, because $(^{182}W/^{184}W)_{CHUR}$ varies due to ^{182}Hf decay; also note that for a chondritic material $\Delta\varepsilon_w(t)$ is always zero by definition.

Measured W isotope compositions are usually expressed as ε_w, which is the deviation of the present-day ^{182}W/^{184}W ratio of a sample from that of the terrestrial standard:

$$\varepsilon_W = \left[\frac{\left(^{182}W/^{184}W\right)_{sample}}{\left(^{182}W/^{184}W\right)_{standard}} - 1 \right] \times 10^4 \quad (5.2)$$

Note that for the present-day $\Delta\varepsilon_w = \varepsilon_w + 1.9$, because the present-day ε_w of chondrites is -1.9 [*Kleine et al.*, 2002, 2004; *Schönberg et al.*, 2002; *Yin et al.*, 2002].

Consider an initially chondritic undifferentiated body that undergoes instantaneous core formation at t_{cf} (We will refer to this as a "two-stage" model.) The timing of core formation t_{cf} is given by [e.g., *Kleine et al.*, 2009]

$$t_{cf} = \frac{1}{\lambda} \cdot \ln\left[\frac{1.59 \cdot f^{Hf/W}}{\Delta\varepsilon_W} \right] \quad (5.3)$$

Here $f^{Hf/W}$ is the Hf/W fractionation factor defined as $f^{Hf/W} = [(^{180}Hf/^{184}W)_{sample}/(^{180}Hf/^{184}W)_{CHUR}] - 1$, where the numerical quantity inside the square brackets in equation (5.3) arises from the difference in present-day $\varepsilon_w = -1.9$ and initial $\varepsilon_w = -3.49$ of chondrites [*Burkhardt et al.*, 2012; *Kruijer et al.*, 2014b] This equation demonstrates that determining a time of core formation requires knowledge of the ^{182}W anomaly and Hf/W ratio characteristic of the entire silicate portion (or the entire metal core) of a differentiated planet. Equation (5.3) also makes clear that for a given Hf/W ratio of the mantle a larger ^{182}W anomaly implies an earlier core formation. It also makes the role of core-mantle partitioning clearer. If tungsten partitions more strongly into the core, then $f^{Hf/W}$ is larger and the inferred core formation timescale, for a given ^{182}W anomaly, is later.

The two-stage model time (equation 5.3) is useful for assessing the relative speeds of accretion of different bodies. However, for Earth-sized bodies it should not be taken to represent the actual timing of accretion or core formation, which are both protracted, multi-stage processes. At best, it represents the earliest age at which core formation could have ceased. More realistic core formation models and their consequences for the Hf-W isotope systematics will be discussed in more detail below (see Section 5.2.3).

5.2.2. Hf-W Isotope Systematics of the Earth, Moon, Mars, and Differentiated Asteroids

The Hf-W isotope systematics of differentiated asteroids and terrestrial planets are reasonably well established. The Hf-W data show the expected systematics for core formation during the lifetime of ^{182}Hf. Magmatic iron meteorites as samples of the metallic core of differentiated protoplanets exhibit ^{182}W deficits relative to chondrites, whereas samples derived from the silicate portion of differentiated bodies are characterized by ^{182}W excesses (Fig. 5.3).

Eucrites, angrites, and martian meteorites exhibit variable ^{182}W excesses, which at least in part result from Hf/W fractionation by silicate melting processes during the lifetime of ^{182}Hf [*Kleine et al.*, 2004; *Foley et al.*, 2005; *Kleine et al.*, 2012; *Touboul et al.*, 2015b]. This makes estimating the ε_w of the bulk mantle of these bodies difficult. For Mars there is abundant evidence for early mantle differentiation from variations in ^{142}Nd/^{144}Nd (from the decay of short-lived ^{146}Sm) [e.g., *Debaille et al.*, 2007]. Some shergottites, however, exhibit near-chondritic ^{142}Nd/^{144}Nd and can thus be used to estimate the ^{182}W anomaly of the bulk martian mantle prior to mantle differentiation [*Kleine et al.*, 2004; *Foley et al.*, 2005]. This approach leads to $\varepsilon_w = 0.45\pm0.15$ for bulk silicate Mars [*Kleine et al.*, 2009]. For the eucrite and angrite parent bodies, the ε_w of the bulk mantle can be determined from the Hf-W isotopic evolution of eucrites and angrites, respectively, which sampled the mantle at different times and, hence, provide a record of the ε_w of the mantle over time. For the angrite parent body at least two mantle reservoirs exist, which probably acquired their distinct Hf/W ratios as a result of metal-silicate fractionation [*Kleine et al.*, 2012]. In this case, there is no single Hf/W and ε_w characteristic of the entire mantle.

Terrestrial and lunar samples show less variable ε_w, but recently small ^{182}W excesses have been reported for Archean samples from Isua, Greenland [*Willbold et al.*, 2011] and the Nuvvuagittuq Greenstone Belt, Quebec, Canada [*Touboul et al.*, 2014], and for some komatiites [*Touboul et al.*, 2012]. These small ^{182}W excesses either reflect the ε_w of the bulk silicate Earth prior to (complete) addition of the late veneer, or result from early differentiation processes within the mantle. Either way, for determining an Hf-W age of core formation in the Earth,

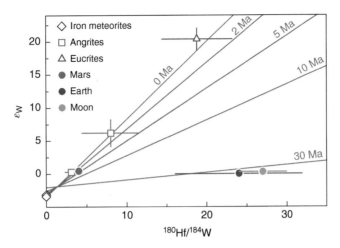

Figure 5.3 Hf-W systematics of differentiated planetary bodies. Data sources as in Table 5.1. Note that chondrites and, hence, undifferentiated bulk planetary objects, plot at $\varepsilon_W = -1.9$. Diagonal solid lines indicate different two-stage core formation times (equation 5.3). Longer accretion timescales result in a smaller tungsten anomaly for a given value of $^{180}Hf/^{184}W$.

the ε_W of the bulk silicate Earth prior to addition of the late veneer should be used. Mass balance indicates that this value should be approximately 0.15–0.40 higher than that of the modern Earth's mantle, in good agreement with values measured for some Isua samples and komatiites [*Willbold et al.*, 2011; *Touboul et al.*, 2012, 2014]. Lunar samples show variable and large apparent ^{182}W excesses, but these largely result from cosmic ray-induced neutron capture reactions [e.g., *Kleine et al.*, 2009]. Two recent studies have shown that samples unaffected by neutron capture show small but well-resolved ^{182}W excesses of between ~0.2 and ~0.3 ε_W [*Kruijer et al.*, 2015; *Touboul et al.*, 2015a], in good agreement with the estimated W isotope composition of the pre-late veneer bulk silicate Earth.

Owing to the different incompatibilities of Hf and W during silicate melting, the Hf/W ratio characteristic of the bulk silicate portion of a differentiated planetary body cannot be measured directly. Instead it must be inferred from the ratio of W to a refractory lithophile element of similar incompatibility (such as U and Th) and by assuming chondritic ratios of these refractory lithophile elements relative to Hf (see Table 5.1). Figure 5.3 shows that the bulk silicate portions of the Earth, Moon, and eucrite parent body are characterized by higher Hf/W ratios than those of Mars and the angrite parent body. This highlights the fact that the partitioning of W into metal can vary depending on the conditions of core formation and suggests that core formation in Mars and the angrite parent body occurred under more oxidizing conditions than on the other three bodies (note that the Moon inherited its high Hf/W from the Earth's mantle).

Table 5.1 summarizes the calculated two-stage model ages (equation 5.3) for core formation in the Earth, Moon, Mars, and the parent bodies of magmatic iron meteorites. For the eucrite and angrite parent bodies, the two-stage model ages are very imprecise because the ε_W and Hf/W of the bulk mantle are poorly constrained. For these bodies, the time of core formation is best determined from the intersection of the Hf-W isotope evolution line of the mantle with the chondritic evolution line [*Kleine et al.*, 2012; *Touboul et al.*, 2015b]. As discussed above, the model timescales of the Earth and Moon and probably also Mars should not be taken literally, because core formation at least in these bodies is not a single event. Nonetheless, it serves to show that the Earth experienced a protracted accretion history, compared to bodies like Mars or the parent bodies of differentiated meteorites, all of which have much shorter two-stage timescales [*Kleine et al.*, 2009].

5.2.3. Multi-stage Core Formation During Accretion of the Earth

As outlined above, assuming instantaneous core formation is not appropriate for Earth-sized bodies, in which core formation was a stochastic, multi-stage process driven by multiple giant impacts. Rather than assuming instantaneous core formation, a better approximation is to model core formation as a continuous process occurring during protracted accretion [e.g., *Halliday*, 2004; *Kleine et al.*, 2004; *Jacobsen*, 2005]. This requires an appropriate model for the growth of the Earth and assumptions regarding the degree of re-equilibration of newly accreted material with the mantle of proto-Earth.

To track the isotopic evolution of the Earth, its growth is often assumed to occur at an exponentially decreasing rate (see Fig. 5.1), such that

Table 5.1 Values represent those of the bulk silicate portions or in the case of the iron meteorites of the bulk metal core. For the Earth, ε_W has been calculated by subtracting the late veneer (i.e., the pre-late veneer value of the BSE is given) [Kruijer et al., 2015]. Note that for the angrite parent body two mantle reservoirs with distinct ^{180}Hf/^{184}W and ε_W exist [Kleine et al., 2012]. The ^{180}Hf/^{184}W of the bulk silicate Earth is the average of estimates based on Ta/W [König et al., 2011], Th/W [Newsom et al., 1996; Arevalo and McDonough, 2008], and U/W ratios [Arevalo et al., 2008]. The ^{180}Hf/^{184}W of the bulk silicate Moon was calculated using Th/W and U/W ratios [Palme and Rammensee, 1981; Münker, 2010], and for both the Earth and Moon chondritic Th/Hf and U/Hf ratios from Dauphas and Pourmand [2011] were used. For derivation of ε_W values for the Earth and Moon, see the text. ^{180}Hf/^{184}W for the bulk mantle of Mars is from Dauphas and Pourmand [2011], and ε_W is from Kleine et al. [2009]. Values for eucrites are from Touboul et al. [2015b], for angrites from Kleine et al. [2012], and for iron meteorites from Kruijer et al. [2013, 2014]. $\Delta\varepsilon_W$ and $f^{Hf/W}$ are both relative to chondrites ($\varepsilon_W = -1.9$; ^{180}Hf/^{184}W = 1.35). For the Earth, Moon, Mars, and iron meteorites, the two-stage core formation time t_{cf} is calculated from equation 5.3. For eucrites and angrites, the two-stage model age is very uncertain, because the ^{180}Hf/^{184}W and ε_W characteristic for the bulk silicate portion of these bodies is only poorly constrained. For these two bodies the time of core formation is most precisely determined, however, through the intersection of the Hf-W isotope evolution of the mantle with the chondritic evolution line [see Kleine et al., 2012]. Note that the ^{180}Hf/^{184}W values for the Earth and Moon are slightly different from previously accepted values (which were used in preparing Figures 5.4 and 5.5) due to a recent revision of the chondritic Hf/Th and Hf/U ratios [Dauphas and Pourmand, 2011].

	^{180}Hf/^{184}W	$f^{Hf/W}$	ε_W	$\Delta\varepsilon_W$	t_{cf} (Myr)
Earth	24±8	17±6	0.16–0.38	2.06–2.28	~32
Moon	27±3	19±2	0.27±0.04	2.17±0.11	~34
Mars	4.0±0.5	2.0±0.4	0.45±0.15	2.35±0.18	~4
Eucrites	19±4	13±3	21.4±1.7	22.3±1.7	< ~1
Angrites	3.1±0.8	1.3±0.6	0.3±0.5	2.2±0.5	< ~2
	8.0±3.6	4.9±2.7	6.2±2.1	8.1±2.1	< ~2
Iron meteorites	~0	−1	−3.15±0.07	−1.25±0.12	~3.1
			−3.40±0.03	−1.5±0.1	~0.7

$$M(t)/M_E = 1 - e^{-t/\tau} \qquad (5.4)$$

where τ is the mean-life of accretion, corresponding to the time to reach 63% of the Earth's mass. Assuming continuous core formation and full metal-silicate equilibration, this model requires a mean-life of accretion of ~11 Myr to generate the observed $\Delta\varepsilon_W$ [Yin et al., 2002]. In the specific case of the exponential model, this value of τ corresponds to an end of accretion at ~30 Myr (as given by time to reach 95% of Earth's mass in the exponential model), similar to the two-stage model age [Jacobsen et al., 2005; Rudge et al., 2010]. Note that in case of incomplete equilibration, a longer mean-life of accretion is required to generate the observed $\Delta\varepsilon_W$ of the bulk silicate Earth (see below).

5.2.3.1. Re-equilibration During Core Formation in the Earth

In reality, core formation occurred neither as a continuous influx of metallic material, nor as an instantaneous single differentiation event. Instead, it occurred stochastically, when individual impactors (already differentiated) collided with the proto-Earth (Section 5.1.1.1). This being the case, both the pre-history of the impactors and the extent of re-equilibration between metal and silicates during impact are important.

Consider a collision between two differentiated bodies. If there is no re-equilibration between silicates and metal,

then the final ^{182}W anomaly will simply be the mass-weighted average of the two initial anomalies. The earlier isotopic history of the bodies is thus preserved, and the ^{182}W anomaly of Earth's mantle would say nothing about the timescale of core formation within the Earth. On the other hand, if there is complete re-equilibration between the entirety of the Earth's mantle and the metal core of the newly accreted body, the earlier isotopic history of the bodies is erased, and only then would the resulting ^{182}W anomaly entirely reflect core formation in the Earth. Thus, re-equilibration results in the reduction of the ^{182}W excess in the mantle and as such has a very strong effect on the final ^{182}W anomaly observed. Consequently, higher degrees of re-equilibration require faster accretion to generate the same ^{182}W anomaly, and make the final outcome less sensitive to the pre-collision isotopic characteristics of the bodies.

Figure 5.4 shows the effects of incomplete re-equilibration on the calculated growth timescale in the exponential accretion model (equation 5.4). Here k is the fraction of impactor core material re-equilibrated. As can be seen, the inferred time to complete 90% of Earth formation varies strongly, depending on the value of k chosen. High degrees of re-equilibration with rapid growth yield the same isotopic anomaly as lower degrees of re-equilibration and slower growth. Unfortunately, the actual degree of re-equilibration is both poorly understood and probably time-variable (see below), resulting in significant uncertainty in the actual core formation timescale. Nonetheless, for the Hf-W system, this model suggests that the bulk of Earth formation certainly took more than 30 Myr to complete, and probably was mostly complete within ~100 Myr, which is approximately consistent with dynamical models (Figure 5.1; *Jacobson et al.*, 2014). Figure 5.4 also plots similar constraints from the U-Pb system, which are discussed further in Section 5.2.4.1.

Figure 5.4 demonstrates that determining the timescale of accretion and core formation in the Earth requires knowledge of the degree of re-equilibration during impact. Unfortunately, theoretical understanding of the extent of re-equilibration during impacts remains an unsolved problem. Isotopic re-equilibration takes place over lengthscales of ~1 cm, based on likely iron transport timescales and diffusivities [*Rubie et al.*, 2003]. Thus, for extensive re-equilibration to occur, the incoming impactor core must be emulsified down to cm-scale droplets, a change in scale of 8 orders of magnitude. Presumably both buoyancy- and shear-instabilities develop at the interface between the

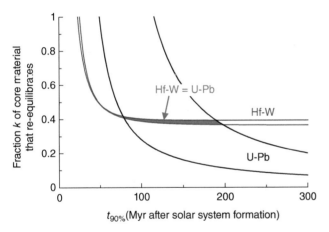

Figure 5.4 Tradeoff between re-equilibration fraction k and growth time for a continuous core formation model (equation 5.4) assuming exponentially decaying accretion [reproduced from *Kleine and Rudge*, 2011]. Here $t_{90\%}$ represents the time to 90% of the final mass in the exponential growth model, which would be equivalent to the time of the Moon-forming impact (see Section 5.2.5). To produce the same present-day tungsten anomaly of the BSE, higher degrees of re-equilibration require more rapid growth. Also shown are results of the U-Pb isotope system in the same accretion and core formation model (see Section 5.2.4.1). There is a region of overlap at k~0.4, where both systems agree. For Hf-W the uncertainty arising from uncertainty in Hf/W of the BSE is shown; for U-Pb the uncertainty arising from the unknown bulk Pb isotope composition of the Earth is shown [see *Rudge et al.*, 2010].

descending impactor and the surrounding (probably molten) mantle [*Dahl and Stevenson*, 2010]. A turbulent cascade develops, reducing the characteristic size of the liquid metal blobs. The key question is therefore how far this process proceeds before the impactor and target core merge.

This kind of question is very hard to answer numerically because of the high degree of turbulence and large range in length-scales involved [*Samuel*, 2012]. Laboratory experiments provide more insight [*Deguen et al.*, 2011; *Deguen*, 2014], and suggest that impactor cores with diameters comparable to the thickness of the magma ocean will not undergo efficient equilibration, while smaller cores will. For a typical body with a (molten) mantle half the radius of the planet, this criterion implies that impactors larger than roughly one-eighth the target mass will not experience efficient equilibration. Such impacts are a common feature of the final stages of accretion (Figure 5.1). Scaling arguments based on turbulent cascades suggest equilibration may be ineffective for impactor cores with radii >100 km [*Dahl and Stevenson*, 2010]. The effective k, integrated over all impactors, is then only about 0.1. If magma oceans are short-lived features, that again militates against extensive equilibration, because metal transport through a solid mantle tends to require iron bodies with lengthscales >> 1 cm (Section 5.1.1.3). Evidently complete equilibration is an unrealistic assumption, but current uncertainties in the correct value of k to employ are large, and it certainly depends on the scale of the impactor.

A further complication, which is usually ignored, is that only a fraction of the target mantle may initially equilibrate with the incoming core material. This process has the same effect as incomplete core equilibration on the subsequent isotopic evolution of the bulk mantle [*Sasaki and Abe*, 2007, *Morishima et al.*, 2013]. For bodies like Mars with a low $f^{Hf/W}$, this effect of incomplete mantle equilibration is unimportant [*Morishima et al.*, 2013], but for larger values of $f^{Hf/W}$, like the Earth's, it may be significant. Further study of this topic is required.

Determining a precise timescale for the formation of Earth's core is hampered by the tradeoff between equilibration factor and accretion rate. The Hf-W isotope systematics of the Earth alone, therefore, cannot precisely constrain the timescale of accretion and core formation. Nevertheless, Hf-W chronometry provides a strict lower limit of ~30 Myr after Solar System formation for the completion of Earth's accretion and core formation. There are at least three additional sources of information, which in principle can help to obtain a more detailed picture of the Earth's accretion. One, described in Section 5.2.3.2, is to use dynamical models to fix the accretion timescale, and then use $\Delta\varepsilon_W$ to determine the degree of re-equilibration required. Another approach (Section 5.2.3.3) is to track both Hf-W isotope evolution and siderophile element partitioning in a given accretion model. Finally, the Hf-W chronometer can be combined with other chronometers having somewhat different characteristics, to derive an accretion scenario that matches all constraints (Section 5.2.4).

5.2.3.2. Combining Isotope Evolution Calculations with Dynamical Models

Various groups have started to incorporate isotopic evolution calculations into dynamical models of the final stages of terrestrial planet accretion [*Nimmo and Agnor*, 2006; *Morishima et al.*, 2013; *Kobayashi and Dauphas*, 2013]. If one assumes that these models provide an accurate description of how the Earth actually accreted, then the W isotopic evolution of the growing bodies can be calculated and the extent of re-equilibration determined by matching the Earth's observed ^{182}W anomaly.

An example of carrying out isotopic calculations on accretion simulations is shown in Figure 5.5. Across these eight simulations [*O'Brien et al.*, 2006], the time at which a body exceeds 90% of its final mass ranges from 4 to 63 Myr, and the final giant impact can occur as early as 14 Myr and as late as 232 Myr. Although these two extremes are not consistent with isotopic constraints on Earth's accretion timescale (see above), this range of values nevertheless serves to illustrate the highly stochastic nature of the accretion process.

Figure 5.5 plots the average mantle tungsten anomaly across all Earth-mass bodies as a function of equilibration fraction k, where k is assumed constant throughout.

As expected, lower degrees of equilibration result in higher tungsten anomalies, and vice versa. On average, Earth-like tungsten anomalies are obtained for intermediate degrees of equilibration ($k = 0.3$–0.7).

An advantage of this approach is that it tracks the isotopic history of each body; as a result, there is no need (unlike analytical models) to assume that all bodies follow an identical isotopic evolution. It is also easy to explore the consequences of allowing k or the partitioning behavior of Hf and W to vary as a function of time, radial position, or body size. Perhaps the biggest disadvantage of this approach is that accretion is stochastic so that large numbers of simulations need to be run [e.g., *Fischer and Ciesla*, 2014], and then only probabilistic statements can be made.

This approach is also only as reliable as the accretion models themselves. For instance, recent accretion models have started to investigate the effects of imperfect accretion (Section 5.1.1.1), which can change both the overall accretion timescale and the bulk chemistry of surviving bodies. Preliminary investigations [*Dwyer et al.*, 2015] suggest that the latter effect is relatively minor for Earth-mass planets, because of the averaging effects of multiple large impacts. However, the overall increase in accretion timescale requires k to decrease slightly (to 0.2 –0.6) to match the observed tungsten anomaly of the Earth's mantle.

Overall, results based on accretion simulations suggest that the Earth accreted material that (on average) underwent intermediate degrees of re-equilibration ($k\sim0.5$). In reality, larger impactors probably suffered less re-equilibration and smaller ones more, but the details are highly uncertain.

5.2.3.3. How Did the Partitioning Behavior of Siderophile Elements Evolve During Accretion?

In the models discussed up to this point, it is assumed that the fractionation of Hf from W during core formation is time-independent (i.e., the partition coefficients D_{Hf} and D_W are constant). This is a convenient simplification because D_{Hf} and D_W can be calculated, given the known mantle Hf/W ratio and iron:silicate ratio of the Earth. However, reality is clearly more complicated because the partition coefficient of W varies with pressure (P), temperature (T), and oxygen fugacity (fO_2), and thus will have changed as accretion proceeded. For instance, the relevant partition coefficients apply at the point when iron-silicate equilibration ceases; thus, for small liquid iron blobs sinking through a magma ocean, the relevant P,T conditions are those at the base of the magma ocean. Since the depth of the magma ocean probably increased as the Earth grew, the P,T conditions of metal-silicate equilibration and with them the partitioning behavior of W will have also changed. Furthermore, D_W also is

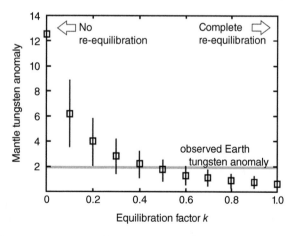

Figure 5.5 Mean mantle tungsten anomaly $\Delta\varepsilon_W$ as a function of equilibration factor k based on tracking the isotopic evolution of bodies in a suite of N-body accretion models. Constant tungsten partitioning coefficients are assumed. Reproduced from *Nimmo et al.* [2010].

strongly dependent on oxygen fugacity, so that W becomes less siderophile under oxidizing conditions [*Cottrell et al.*, 2009; *Wade et al.*, 2013]. For instance, the much lower Hf/W ratio of bulk silicate Mars compared to the Earth's mantle (Table 5.1) probably reflects core formation under more oxidizing conditions.

The characterization of partition coefficients at the relevant P,T and fO_2 conditions is an ongoing challenge [e.g., *Siebert et al.*, 2013]. The result is non-uniqueness. Various groups have made different assumptions about the evolution of P,T and fO_2, all of which are capable of matching the inferred mantle siderophile concentrations. Some groups favor progressively oxidizing conditions [e.g., *Wade and Wood*, 2005; *Rubie et al.*, 2011], while others advocate oxidized conditions throughout [*Siebert et al.*, 2013]. These different scenarios make quite different predictions about the nature of the light elements in the core, and can thus in theory be tested. Further discussion of this topic may be found in Chapter 6.

Obviously, any successful accretion and core formation model must not only satisfy the Hf-W isotope systematics but also the observed siderophile element depletions in the Earth's mantle. This makes combining siderophile element and Hf-W isotope systematics a promising approach for more tightly constraining the core formation history of the Earth. Such a combined approach may in particular help to assess the degree of re-equilibration during metal segregation within the Earth. However, *Rudge et al.* [2010] showed that the siderophile element and Hf-W isotope systematics of the Earth's mantle are equally consistent with 100% re-equilibration as well as with ~40% re-equilibration during core formation, depending on the growth history assumed. This result may be compared with the 30 to 70% range in re-equilibration factors derived from dynamical models (Fig. 5.5). Ultimately, although it is clearly essential to consider siderophile element partitioning and Hf-W isotope systematics together, this approach so far only provides limited constraints on the degree of metal-silicate re-equilibration, the most important parameter for Hf-W chronometry of core formation in the Earth.

5.2.4. Other Chronometers

Although the Hf-W system represents the best single core formation chronometer, it does suffer from limitations. An alternative approach to obtain better constraints on the timing of core formation is to employ multiple isotopic systems. For the Earth, the three most relevant systems are the U-Pb, Pd-Ag, and I-Xe. Both U-Pb and Pd-Ag are sensitive to core formation, but interpretation is complicated by other factors, and the I-Xe system is sensitive to degassing, which is often assumed to occur as a result of late-stage giant impacts.

5.2.4.1. Evidence from Combined Hf-W and U-Pb Isotope Systematics

The U-Pb isotope system was the first to provide a precise age for the Earth [*Patterson*, 1956]. The principle of this system to date core formation in the Earth is as follows. In a plot of $^{207}Pb/^{204}Pb$ *vs.* $^{206}Pb/^{204}Pb$, all estimates for the Pb isotope composition of the bulk silicate Earth plot to the right of the geochron (Fig. 5.6), indicating that the Earth's mantle underwent a major U/Pb fractionation event some time after Solar System formation. Most estimates for the Pb isotope composition of the BSE return ages of between ~50 and ~150 Myr after the start of the Solar System [see summary in *Rudge et al.*, 2010 and *Wood and Halliday*, 2010]. All the Pb-Pb ages have in common that they are younger than the Hf-W model age for core formation in the Earth. The disparate Pb-Pb and Hf-W ages of the Earth's core have been interpreted to reflect disequilibrium during core formation [*Halliday*, 2004; *Kleine et al.*, 2004; *Rudge et al.*, 2010; *Kleine and Rudge*, 2011]. It is noteworthy that the Hf-W and U-Pb isotope systematics provide consistent $t_{90\%}$ timescales of 80–200 Myr for a degree of re-equilibration of ~40% (Fig. 5.4), in reasonable agreement with the range obtained from dynamical models.

The apparent discrepancy between the Hf-W and Pb-Pb model ages may alternatively reflect the fact that Earth's accretion is not correctly described by the exponential accretion model (cf. Fig. 5.1). Using a more general growth model shows that the Hf-W and Pb-Pb

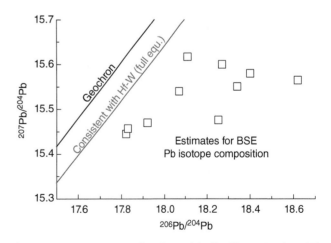

Figure 5.6 Comparison of estimated bulk silicate Earth (BSE) lead isotope ratios with those expected for a zero-age reservoir as given by the Geochron. Blue line indicates loci of data points for which the timing obtained from Hf-W and U-Pb isotope systematics in the exponential growth model (equation 5.4) would be consistent, assuming full equilibration. For estimates of BSE compositions see *Rudge et al.* [2010].

ages of the Earth's core can be reconciled if the Earth grew rapidly initially, with $t_{63\%}$ between 0.4 and 2.5 Myr, followed by a much more protracted accretion period terminated by a 'late' giant impact at ~100 Myr [*Rudge et al.*, 2010].

Although the Hf-W and Pb-Pb ages of the Earth can reasonably well be brought into agreement (Fig. 5.4), the significance of Pb-Pb age is debated, and it has been argued that the Pb-Pb age does not date core formation but the arrival of late veneer some time after core formation [*Albarede*, 2009]. In this case, core formation would have occurred earlier than given by the Pb-Pb age. However, this model would require large additions (~4%) of primitive chondritic material to the Earth after the cessation of core formation, inconsistent with the ~0.5% mass of the late veneer derived from abundances of highly siderophile elements in the Earth's mantle [*Wood et al.*, 2010]. To overcome this problem, *Albarede et al.* [2013] proposed that during addition of the late veneer most of the metal entered Earth's core, thereby allowing for a larger mass to be added.

However, as argued above, core formation in the Earth almost certainly involved some disequilibrium, in which case the Hf-W and Pb-Pb ages for core formation are in good agreement (Fig. 5.4). Thus, there is no need to invoke more complicated models in which the two systems date distinct events. Moreover, an apparent disparity between the Hf-W and Pb-Pb ages only arises if complete re-equilibration and a certain growth model (the exponential model) are assumed. Given the stochastic nature of accretion and the likelihood of disequilibrium during giant impacts, these two assumptions are probably not valid. Thus, the most straightforward interpretation of the Hf-W and Pb-Pb systematics is that both provide constraints on the timing of core formation. The strength of using both systems lies in the fact that they date different stages of core formation because of their very different half-lives [*Rudge et al.*, 2010; *Kleine and Rudge*, 2011]. Whereas Hf-W as a short-lived nuclide system mainly constrains the early stages of accretion and core formation; therefore, it provides little information on the end of core formation. This is because the Hf-W system becomes extinct after *ca.* 60 Myr and so the W isotope composition of the mantle is mainly set by very early Hf/W fractionation. In contrast, the long-lived Pb-Pb system provides little information on the early stages of accretion and core formation but mainly constrains the end of core formation. This is because the present-day Pb isotope composition of the mantle mainly records the final U/Pb fractionation and subsequent decay over *ca.* 4.4 Ga. The Hf-W and Pb-Pb systems are mutually consistent with an end of core formation at ~80–200 Myr after Solar System formation (Fig. 5.4).

5.2.4.2. Evidence from Pd-Ag Systematics

The Pd/Ag system is in some ways similar to the Hf/W system. ^{107}Pd decays to ^{107}Ag with a half-life of 6.5 Myr, and Pd partitions more strongly into the core than Ag. As a result, early core formation results in a deficit in ^{107}Ag relative to undifferentiated material. However, the major difficulty in using Pd-Ag isotope systematics to constrain the timescales of core formation in the Earth is that Ag is moderately volatile, so that the Pd/Ag ratio of the bulk Earth is not known. Nevertheless, when combined with other independent constraints on the timing of core formation, Pd-Ag systematics are useful to constrain the timing of volatile delivery to the Earth.

The Ag isotope composition of the bulk silicate Earth (BSE) is roughly the same as that of chondritic material [*Schönbächler et al.*, 2010]. Given that on average the Earth is more depleted in volatiles than chondrites, the bulk Earth should have a more radiogenic ε_{Ag} than chondrites. Thus, relative to the expected Ag isotope composition of the bulk Earth, the BSE exhibits a ^{107}Ag deficit, which would require a rapid core formation timescale (two-stage age of less than 10 Myr). This result is inconsistent with the Hf-W constraints, which yields a two-stage age of roughly 30 Myr. This apparent discrepancy can be accounted for by invoking a change in the nature of the accreting material, initially volatile-depleted and later volatile-rich [*Schönbächler et al.*, 2010]. In this classical heterogeneous accretion model, the Ag isotope composition of the BSE is almost entirely dominated by the later accreted material, which on average had chondritic Pd/Ag and ^{107}Ag abundances. This model is consistent with numerical modeling that shows a progressive radial expansion of the feeding zone (see Section 5.1.1.1), thus delivering more distant (and thus more volatile-rich) impactors at later times. If the volatile-depleted and volatile-enriched impactors are reduced and oxidized, respectively, then this scenario is also consistent with the concentrations of siderophile elements in the BSE [*Wade and Wood*, 2005; *Rubie et al.*, 2011]. Thus, although the Pd-Ag systematics provide no additional constraints on the core formation timescale, they seem to require that the Earth accreted a mix of early volatile-depleted and later volatile-enriched material.

The mass fraction of volatile-rich material added to account for the chondritic ^{107}Ag of the BSE is significant, and cannot merely be the ~0.5 wt% "late veneer," which is thought to have been added after core formation was effectively complete (see chapter 4). A pre-late-veneer Earth that already had a significant volatile budget is also consistent with estimates based on S, Se, and Te concentrations [*Wang and Becker*, 2013].

5.2.5. Timing of Moon-forming Impact

The final large impactor to strike the Earth was probably the one that formed the Moon. This final impact thus represents the effective end of core formation on Earth. As a result, evidence from the Moon can be used to help determine the timescale of terrestrial core formation.

The bulk silicate Moon and Earth appear to have very similar ε_w, and this has been interpreted to indicate a 'late' formation of the Moon, more than ~50 Myr after the start of the Solar System [*Touboul et al.*, 2007, 2009]. The interpretation relies on the assumption that the Moon formed from terrestrial mantle material and as such had the same initial W isotope composition as the Earth's mantle at the time of the giant impact. If the Hf/W ratios of the bulk silicate Moon and bulk silicate Earth are different, then their similar W isotope compositions require formation of the Moon after extinction of ^{182}Hf, that is, after ~50 Myr. There is currently debate as to whether the Hf/W ratios of the lunar and terrestrial mantles are distinct or not [e.g., *König et al.*, 2011]. However, two observations indicate that it is unlikely that they are exactly identical. First, the Hf/W ratio of the bulk silicate Moon was established by lunar core formation and, hence, is not simply inherited from the proto-Earth's mantle. Second, the conditions of core formation in the Moon and Earth were likely different and so there is no reason to assume that the lunar and terrestrial mantles have the same Hf/W ratio. Thus, the most straightforward interpretation of the similar ^{182}W anomalies of the Earth's mantle and Moon is that the Moon is 'young.' Note that there seems to be a small difference between the ε_w of the present-day bulk silicate Earth and the Moon (see Section 5.2.2). This difference most likely reflects the different amounts of post-core formation additions of material to both bodies (the "late veneer") [*Kruijer et al.*, 2015; *Walker*, 2014]. The amount of late veneer material added to the Earth's mantle has also been used to argue for a young Moon based on dynamical arguments [*Jacobson et al.*, 2014].

An alternative approach for determining the age of the Moon is to date the formation of lunar crustal rocks, the ferroan anorthosites. These are thought to have formed as flotation cumulates on top of the lunar magma ocean, and their age should thus approximate (and certainly not precede) the age of the Moon. This is because the lunar magma ocean, which formed in the immediate aftermath of the giant impact, is expected to have cooled very rapidly, at least until the formation of the first insulating crust. Most ferroan anorthosites dated so far have ages around ~4.45 Ga [*Norman et al.*, 2003; *Nyquist et al.*, 2006]; however, a recent study reported a very precise age of 4360±3 Ma for ferroan anorthosite 60025 [*Borg et al.*, 2011]. *Boyet et al.* [2015] argued that the oldest ferroan anorthosite has a crystallization age of between ~4.50

and ~4.44 Ga, and that younger ages may reflect resetting or that not all anorthosites are flotation cumulate. Finally, chronological evidence from Mg-suite crustal rocks indicate a major period of crust formation on the Moon at ~4.4 Ga [*Carlson et al.*, 2014]. Collectively, there does not seem to be evidence for significant crust formation on the Moon prior to ~4.45 Ga, implying that either the Moon did not form earlier than ~100 Myr or that the ferroan anorthosites are not flotation cumulates of the lunar magma ocean.

Additional evidence arises from the I-Xe system. The existence of isotopically distinct Xe reservoirs within the BSE indicates differentiation prior to 100 Myr [*Mukhopadhyay*, 2012] and may suggest two episodes of outgassing at ~20–50 Myr and ~100 Myr [*Pepin and Porcelli*, 2006]. If the Moon-forming impact occurred more recently than 100 Myr, then it must not have destroyed pre-existing mantle heterogeneities, a conclusion also suggested by the survival of early-established ε_w anomalies [*Touboul et al.*, 2012].

5.2.6. Summary: When Did Core Formation Occur?

At this point, it should be obvious that for an Earth-mass body, the idea of there being a single instant of core formation is inappropriate. Instead, core formation proceeded in the same manner as planetary growth, episodically, in large, stochastic increments (Fig 5.1). Nonetheless, some bounds can be established. Based on Figure 5.4, core formation cannot have ended prior to 30 Myr, and based on the apparent timing of the Moon-forming impact, it is unlikely to have continued past 150–200 Myr. These isotopically-derived limits are consistent with typical timescales derived from dynamical N-body simulations. Indeed, the good agreement between these two wildly different ways of quantifying the Earth's accretion timescale represents a major success in geosciences.

A more precise picture of how the Earth grew is currently elusive and may remain so. The available isotopic clocks are hampered by the problem of partial equilibration (Section 5.2.3.1), and this is particularly true for the Hf-W system. Scenarios that satisfy several different isotopic systems simultaneously can help overcome this issue. For instance, combining Hf-W and U-Pb isotope systematics provides a consistent age for the Earth of 80–200 Myr if the degree of metal-silicate re-equilibration during core formation was only ~40% on average (Section 5.2.4.1). This age is also in good agreement with ages of lunar ferroan anorthosites, which should closely approximate the formation of the Moon (provided they are flotation cumulates of the lunar magma ocean) and hence the termination of the main stages of Earth's accretion.

The timescales for the accretion of the Earth indicated by these isotopic approaches are roughly consistent with the results of dynamical models. Similarly, Pd-Ag isotope systematics seem to require that the Earth accreted from volatile-poor material initially, followed by a shift to more volatile-rich material during the later stages of accretion. This observation is also consistent with some dynamical simulations showing that more volatile-rich bodies located further away from the sun were added to the Earth during the late stages of accretion [*O'Brien et al.*, 2014]. Dynamical models themselves are inevitably stochastic and are constantly under revision. Nonetheless, testing the outcomes of such models with the isotopic constraints is an important consistency test for any model and as such can provide powerful new constraints. Such combined studies are likely to become increasingly standard practice going forward [*Nimmo and Agnor*, 2006; *Kobayashi and Dauphas*, 2013; *Morishima et al.*, 2013].

5.3. SUMMARY AND CONCLUSIONS

The Earth's core formed via a series of high-energy collisions with already-differentiated objects, resulting in several distinct magma ocean epochs. The cores of these impactors probably underwent only limited emulsification and moderate (~50%) isotopic re-equilibration with the target mantle during the collision. Later impactors likely originated from more distant regions of the inner Solar System and were plausibly more volatile-rich and more oxidized than earlier impactors. Dynamical models and short-lived isotopes provide a mutually self-consistent, albeit approximate, chronology: terrestrial core formation took more than 30 Myr, but less than about 200 Myr, to complete.

Figure 5.7 is a summary sketch of how differentiation and core formation are likely to have proceeded with

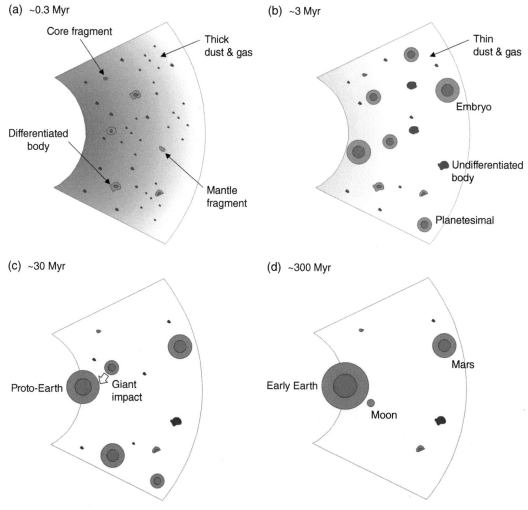

Figure 5.7 Schematic view looking down on the ecliptic, centered at about 1 AU, at different stages during the accretion of the terrestrial planets. Brown, green, and red indicate undifferentiated, silicate, and iron materials, respectively. ^{26}Al is effectively extinct by 3 Myr after CAI; at roughly the same time any remaining gas and dust is swept away by young stellar activity. Timescales are only approximate.

time. By ~0.3 Myr after Solar System formation, some bodies had grown large enough (>30 km diameter) to undergo melting and core formation via ^{26}Al decay (Fig. 5.7a). Collisional disruption will have resulted in core and mantle fragments being produced. At ~3 Myr, the nebular gas and dust were dissipating, and ^{26}Al was effectively extinct. Differentiated, Moon- to Mars-sized planetesimals and embryos had formed, while more slowly growing, smaller bodies avoided differentiating via either ^{26}Al decay or gravitational energy release (Fig. 5.7b). The growth of Mars may have effectively stalled at this point. At ~30 Myr, late-stage accretion involving massive collisions between differentiated protoplanets was taking place (Fig. 5.7c). By ~300 Myr, the Earth and Moon had finished forming and the neighborhood around 1 AU approximately resembled its current configuration (Fig. 5.7d).

Many aspects of core formation remain enigmatic. In particular, the fluid dynamics of an impactor core passing through the target mantle is very poorly understood, and yet is crucial for predicting the extent of re-equilibration (Section 5.2.3.1). Only very recently has the inefficient nature of accretion become obvious (Section 5.1.1.1). Exploration of the effect of including more realistic accretion physics on the isotopic evolution of terrestrial bodies has barely begun (*Dwyer et al.*, 2015). Partitioning behavior at high pressures is only imperfectly understood but can have a significant effect on the conditions inferred to characterize core formation (Section 5.2.3.3).

Remedying these lacunae in our knowledge will certainly improve our understanding of core formation. Ever more precise isotopic measurements and ever more sophisticated accretion simulations will also improve our understanding. Ultimately, however, the biggest advances are likely to arise from a combination of these two latter approaches. Like many problems in Earth sciences, the perspectives of both geochemists and geophysicists will be required to elucidate exactly how the core formed.

ACKNOWLEDGMENTS

We thank Dave Rubie and Qing-Zhu Yin for thoughtful comments that improved this manuscript. Partial support was provided by NASA-NNX11AK60G and an ERC Advanced Grant "ACCRETE" (contract number 290568).

REFERENCES

Abe, Y. and T. Matsui (1986), Early evolution of the Earth—accretion, atmosphere formation and thermal history, *J. Geophys. Res.*, *91*, E291–E302.

Agnor, C.B., R. M. Canup, and H. F. Levison (1999), On the character and consequences of large impacts in the late stage of terrestrial planet formation, *Icarus*, *142*, 219–237.

Albarede, F. (2009), Volatile accretion history of the terrestrial planets and dynamic implications, *Nature*, *461*, 1227–1233.

Albarede , F., C. Ballhaus, J. Blichert-Toft, C. T. Lee, B. Marty, F. Moynier, and Q.Z. Yin (2013), Asteroidal impacts and the origin of terrestrial and lunar volatiles, *Icarus*, *222*, 44–52.

Alfe, D., M. J. Gillan, and G. D. Price (2002), Ab initio chemical potentials of solid and liquid solutions and the chemistry of the Earth's core, *J. Chem. Phys.*, *116*, 7127–7136.

Araki, T. et al. (2005), Experimental investigations of geologically produced antineutrinos with KamLAND, *Nature*, *436*, 499–503.

Asphaug, E., C.B. Agnor and Q. Williams (2006), Hit-and-run planetary collisions, *Nature*, *439*, 155–160.

Bond, J. C., D. S. Lauretta, and D. P. O'Brien (2010), Making the Earth: Combining dynamics and chemistry in the solar system, *Icarus*, *205*, 321–337.

Borg, L.E., J. N. Connelly, M. Boyet, and R. W. Carlson (2011), Chronological evidence that the Moon is either young or did not have a global magma ocean, *Nature*, *477*, 70–72.

Bouhifd, M. A., L. Gautron, N. Bolfan-Casanova, V. Malavergne, T. Hammouda, D. Andrault, and A. P. Jephcoat (2007), Potassium partitioning into molten alloys at high pressure: Implications for Earth's core, *Phys. Earth Planet. Inter.*, *160*, 22–33.

Boyet, M., R. W. Carlson, L. E. Borg, and M. Horan (2015), Sm-Nd systematics of lunar ferroan anorthositic suite rocks: constraints on lunar crust formation. *Geochim. Cosmochim. Acta.*, *148*, 203–218.

Burkhardt, C., T. Kleine, N. Dauphas, and R. Wieler (2012), Nucleosynthetic tungsten isotope anomalies in acid leachates of the Murchison chondrite: implications for hafnium-tungsten chronometry *Astrophys. J. Lett.*, *753*, L6.

Canup, R. M. (2004), Simulations of a late lunar-forming impact, *Icarus*, *168*, 433–456.

Canup, R. M. (2012), Forming a Moon with an Earth-like composition via a giant impact, *Science*, *338*, 1052–1055.

Carlson, R. W., L. E. Borg, A. M. Gaffney, and M. Boyet (2014), Rb-Sr, Sm-Nd and Lu-Hf isotope systematics of the lunar Mg-suite: the age of the lunar crust and its relation to the time of Moon formation. *Phil. Trans. R. Soc. A*, *372*, 20130246.

Chambers, J. (2010), Terrestrial planet formation, in *Exoplanets*, S. Seager, Ed., pp. 297–318, Univ. Ariz. Press.

Chambers, J. E. (2013), Late-stage planetary accretion including hit-and-run collisions and fragmentation, *Icarus*, *224*, 43–56.

Corgne, A., S. Keshav, Y. W. Fei, and W. F. McDonough (2007), How much potassium is in the Earth's core? New insights from partitioning experiments, *Earth Planet. Sci. Lett.*, *256*, 567–576.

Cottrell E., M. J. Walter, and D. Walker (2009), Metal-silicate partitioning of tungsten at high pressure and temperature: Implications for equilibrium core formation in Earth, *Earth Planet. Sci. Lett.*, *281*, 275–287.

Cuk, M. and S. T. Stewart, Making the Moon from a fast-spinning Earth: A giant impact followed by resonant despinning (2012), *Science*, *338*, 1047–1052.

Dahl, T. and D. J. Stevenson (2010), Turbulent mixing of metal and silicate during planet accretion—An interpretation of the Hf-W chronometer, *Earth Planet. Sci. Lett.*, *295*, 177–186.

Dauphas, N. and A. Pourmand (2011), Hf-W-Th evidence for rapid growth of Mars and its status as a planetary embryo, *Nature, 473,* 489–491.

Dauphas, N., C. Burkhardt, P. H. Warren, and F.-Z. Teng (2014), Geochemical arguments for an Earth-like Moon-forming impactor. *Phil. Trans. R. Soc. A, 372,* 20130244.

Debaille, V., A. D. Brandon, Q. Z. Yin, and B. Jacobsen (2007), Coupled Nd-142-Nd-143 evidence for a protracted magma ocean in Mars, *Nature, 450,* 525–528.

Deguen, R., P. Olson, and P. Cardin (2011), Experiments on turbulent metal-silicate mixing in a magma ocean, *Earth Planet. Sci. Lett., 310,* 303–313.

Dwyer, C. A., F. Nimmo, and J. E. Chambers (2015), Bulk chemical and isotopic consequences of incomplete accretion during planet formation, *Icarus, 245,* 145–152.

Elkins-Tanton, L. T., B. P. Weiss, and M. T. Zuber (2011), Chondrites as samples of differentiated planetesimals, *Earth Planet. Sci. Lett., 305,* 1–10.

Fischer, R. A. and F. J. Ciesla (2014), Dynamics of the terrestrial planets from a large number of N-body simulations, *Earth Planet. Sci. Lett., 392,* 28–38.

Foley, C. N., M. Wadhwa, L. E. Borg, P. E. Janney, R. Hines, and T. L. Grove (2005), The early differentiation history of Mars from W-182-Nd-142 isotope systematic in the SNC meteorites, *Geochim. Cosmochim. Acta., 69,* 4557–4571.

Georg, R. B., A. N. Halliday, E. A. Schauble, and B. C. Reynolds (2007), Silicon in the Earth's core, *Nature, 447,* 1102–1106.

Halliday, A. N. (2004), Mixing, volatile loss and compositional change during impact-driven accretion of the Earth, *Nature, 427,* 505–509.

Huang, H. F., Y. W. Fei, L. C. Cai, F. Q. Jing, X. J. Hu, H. S. Xie, L. M. Zhang, and Z. Z. Gong (2011), Evidence for an oxygen-depleted liquid outer core of the Earth, *Nature, 479,* 513–516.

Jacobsen, S. B. (2005), The Hf-W isotopic system and the origin of the Earth and Moon, *Ann. Rev. Earth Planet. Sci., 33,* 531–570.

Jacobson, S. A., A. Morbidelli, S. N. Raymond, D. P. O'Brien, K. J. Walsh, and D. C. Rubie (2014), Highly siderophile elements in Earth's mantle as a clock for the Moon-forming impact. *Nature, 508,* 84–87.

Kleine, T. and J. F. Rudge (2011), Chronometry of meteorites and the formation of the Earth and Moon, *Elements, 7,* 41–46.

Kleine, T., C. Münker, K. Mezger, and H. Palme (2002), Rapid accretion and early core formation on asteroids and terrestrial planets from Hf-W chronometry. *Nature, 418,* 952–955.

Kleine, T., K. Mezger, H. Palme, and C. Munker (2004), The W isotope evolution of the bulk silicate Earth: constraints on the timing and mechanisms of core formation and accretion, *Earth Planet. Sci. Lett., 228,* 109–123.

Kleine, T., M. Touboul, B. Bourdon, F. Nimmo, K. Mezger, H. Palme, S. B. Jacobsen, Q. Z. Yin, and A. N. Halliday (2009), Hf-W chronology of the accretion and early evolution of asteroids and terrestrial planets, *Geochim. Cosmochim. Acta., 73,* 5150–5188.

Kleine, T., U. Hans, A. J. Irving, and B. Bourdon (2012), Chronology of the angrite parent body and implications for core formation in protoplanets, *Geochim. Cosmochim. Acta., 84,* 186–203.

Kobayashi, H. and N. Dauphas (2013), Small planetesimals in a massive disk formed Mars, *Icarus, 225,* 122–130.

Kokubo, E. and S. Ida (1998), Oligarchic growth of protoplanets, *Icarus, 131,* 171–178.

Kokubo, E. and H. Genda (2010), Formation of terrestrial planets from protoplanets under a realistic accretion condition, *Astrophys. J. Lett., 714,* 21–25.

Kruijer, T. S., M. Fischer-Goedde, T. Kleine, P. Sprung, I. Leya, and R. Wieler (2013), Neutron capture on Pt isotopes in iron meteorites and the Hf-W chronology of core formation in planetesimals, *Earth Planet. Sci. Lett., 361,* 162–172.

Kruijer, T. S., M. Touboul, M. Fischer-Gödde, K. R. Bermingham, R. J. Walker, and T. Kleine (2014a), Protracted core formation and rapid accretion of protoplanets. *Science, 344,* 1150–1154.

Kruijer, T. S., T. Kleine, M. Fischer-Gödde, C. Burkhardt, and Wieler, R. (2014b), Nucleosynthetic W isotope anomalies and Hf-W chronometry of Ca-Al-rich inclusions. *Earth Planet. Sci. Lett., 403,* 317–327.

Kruijer, T. S., T. Kleine, M. Fischer-Gödde, and P. Sprung (2015), Lunar tungsten isotopic evidence for the late veneer. *Nature, 520,* 534–537.

McCoy, T. J., D. W. Mittlefehldt, and L. Wilson (2006), Asteroid differentiation, in Meteorites and the early solar system II, Univ. Ariz. Press, pp. 733–745.

McDonough, W. F. (2003), Compositional model for the Earth's core, in *Treatise on Geochemistry,* pp. 517–586.

McKenzie, D. (1984), The generation and compaction of partially molten rock, *J. Petrology, 25,* 713–765.

Mezger, K., V. Debaille, and T. Kleine, (2013), Core formation and mantle differentiation on Mars. *Space Sci. Rev., 174,* 27–48.

Mittlefehldt, D. W., T. J. McCoy, C. A. Goodrich, and A. Kracher (1998), Non-chondritic meteorites form asteroidal bodies. In Planetary Materials (Ed. J.J. Papike). Rev. Mineral., Mineralogical Society of America, Washington DC, vol. 36, chap. 4, pp. 4–1 – 4–495.

Morard, G., J. Siebert, D. Andrault, N. Guignot, G. Garbarino, F. Guyot, and D. Antonangeli (2013), The Earth's core composition from high pressure density measurements of liquid iron alloys, *Earth Planet. Sci. Lett., 373,* 169–178.

Morbidelli, A., J. I. Lunine, D. P. O'Brien, S. N. Raymond, and K. J. Walsh (2012), Building terrestrial planets, *Ann. Rev. Earth Planet. Sci., 40,* 251–275.

Morishima, R., G. J. Golabek, and H. Samuel (2013), N-body simulations of oligarchic growth of Mars: Implications for Hf-W chronology, *Earth Planet. Sci. Lett., 366,* 6–16.

Mukhopadhyay, S. (2012), Early differentiation and volatile accretion recorded in deep-mantle neon and xenon, *Nature, 486,* 101–104.

Nimmo, F. (2015), Thermal and compositional evolution of the core, *Treatise Geophys, 9,* 201–209.

Nimmo, F. and T. Kleine (2007), How rapidly did Mars accrete? *Uncertainties in the Hf-W timing of core formation, Icarus, 191,* 497–504.

Nimmo, F. and C. B. Agnor (2006), Isotopic outcomes of N-body accretion simulations: Constraints on equilibration processes during large impacts from Hf/W observations, *Earth Planet. Sci. Lett., 243,* 26–43.

Nimmo, F., D. P. O'Brien, and T. Kleine (2010), Tungsten isotopic evolution during late-stage accretion: Constraints on Earth-Moon equilibration, *Earth Planet. Sci. Lett.*, *292*, 363–370.

Norman, M. D., L. E. Borg, L. E. Nyquist, and D. D. Bogard (2003), Chronology, geochemistry and petrology of a ferroan noritic anorthosite clast from Descartes breccias 67215: Clues to the age, origin, structure and impact history of the lunar crust, *Meteorit. Planet. Sci.*, *38*, 645–661.

Nyquist, L., D. Bogard, A. Yamaguchi, C.-Y. Shih, Y. Karouji, M. Ebihara, Y. Reese, D. Garrison, G. McKay, and H. Takeda (2006), Feldspathic clasts in Yamato-86032L: Remnants of the lunar crust with implications for its formation and impact history, *Geochim. Cosmochim. Acta.*, *70*, 5990–6015.

O'Brien, D. P., A. Morbidelli, and H.F. Levison (2006), Terrestrial planet formation with strong dynamical friction, *Icarus*, *184*, 39–58.

O'Brien, D. P., K. J. Walsh, A. Morbidelli, S. N. Raymond, and A. M. Mandell (2014), Water delivery and giant impacts in the 'Grand Tack' scenario, *Icarus*, *239*, 74–84.

O'Neill, H.S.C and H. Palme (2008), Collisional erosion and the non-chondritic composition of the terrestrial planets, *Phil. Trans. R. Soc. Lond. A*, *366*, 4205–4238.

Patterson, C. (1956), Age of meteorites and the Earth, *Geochim. Cosmochim. Acta.*, *10*, 230–237.

Pepin, R.O. and D. Porcelli (2006), Xenon isotope systematic, giant impacts, and mantle degassing on the early Earth, *Earth Planet. Sci. Lett.*, *250*, 470–485.

Poirier, J.-P. (1994), Light elements in the Earth's outer core: A critical review, *Phys. Earth Planet. Inter.*, *85*, 319–337.

Raymond, S. N., T. Quinn, and J. I. Lunine (2006), High resolution simulations of the final assembly of Earth-like planets I. *Terrestrial accretion and dynamics, Icarus*, *183*, 265–282.

Ricard, Y., O. Sramek, and F. Dubuffet (2009), A multi-phase model of runaway core-mantle segregation in planetary embryos, *Earth Planet. Sci. Lett.*, *284*, 144–150.

Ricolleau, A., Y. W. Fei, A. Corgne, J. Siebert, and J. Badro (2011), Oxygen and silicon contents of Earth's core from high pressure metal-silicate partitioning experiments, *Earth Planet. Sci. Lett.*, *310*, 409–412.

Rubie, D. C., F. Nimmo, and H. J. Melosh (2007), Formation of the Earth's core, *in Treatise Geophysics*, *9*, 43–79.

Rubie, D. C., H. H. Melosh, J. E. Reid, C. Liebske, and K. Righter (2003), Mechanisms of metal-silicate equilibration in the terrestrial magma ocean, *Earth Planet. Sci. Lett.*, *205*, 239–255.

Rubie, D.C. et al. (2011), Heterogeneous accretion, composition and core-mantle differentiation of the Earth, *Earth Planet. Sci. Lett.*, *301*, 31–42.

Rudge, J. F., T. Kleine, and B. Bourdon (2010), Broad bounds on Earth's accretion and core formation constrained by geochemical models, *Nature Geosci.*, *3*, 439–443.

Samuel, H. (2012), A re-evaluation of metal diapir breakup and equilibration in terrestrial magma oceans, *Earth Planet. Sci. Lett.*, *313*, 105–114.

Sasaki, T. and Y. Abe (2007), Rayleigh-Taylor instability after giant impacts: Imperfect equilibration of the Hf-W system and its effect on the core formation age, *Earth Planet. Space*, *59*, 1035–1045.

Schersten, A., T. Elliott, C. Hawkesworth, S. Russell, and J. Masarik (2006), Hf-W evidence for rapid differentiation of iron meteorite parent bodies, *Earth Planet. Sci. Lett.*, *241*, 530–542.

Schönbächler, M., R. W. Carlson, M. F. Horan, T. D. Mock, and E. H. Hauri (2010), Heterogeneous accretion and the moderately volatile element budget of Earth, *Science*, *328*, 884–886.

Schönberg, R., B. S. Kamber, K. D. Collerson, and O. Eugster (2002), New W-isotope evidence for rapid terrestrial accretion and very early core formation. *Geochim. Cosmochim. Acta.*, *66*, 3151–3160.

Siebert, J., J. Badro, D. Antonangeli, and F. J. Ryerson (2013), Terrestrial accretion under oxidizing conditions, *Science*, *339*, 1194–1197.

Solomatov, V. S. (2000), Fluid dynamics of a terrestrial magma ocean, in Canup, R. M. and K. Righter, Eds., *Origin of the Earth and Moon*, Univ. Arizona Press, pp. 323–338.

Sramek, O., Y. Ricard, and F. Dubuffet (2010), A multiphase model of core formation, *Geophys. J. Int.*, *181*, 198–220.

Stevenson, D. J. (1990), Fluid dynamics of core formation, in Newsom, H. E. and J. E. Jones, Eds., *Origin of the Earth*, Oxford Univ. Press, pp. 231–249.

Tang, H. and N. Dauphas (2012), Abundance, distribution and origin of Fe-60 in the solar protoplanetary disk, *Earth Planet. Sci. Lett.*, *359*, 248–263.

Tonks, W. B. and H. J. Melosh (1993), Magma ocean formation due to giant impacts, *J. Geophys. Res.*, *98*, 5319–5333.

Touboul, M., I. S. Puchtel, and R. J. Walker (2012), W-182 evidence for long-term preservation of early mantle differentiation products, *Science*, *335*, 1065–1069.

Touboul M., T. Kleine, B. Bourdon, Van Orman J. A., C. Maden, and J. Zipfel (2009), Hf-W thermochronometry II: Accretion and thermal history of the acapulcoite-lodranite parent body, *Earth Planet. Sci. Lett.*, *284*, 168–178.

Touboul, M., T. Kleine, B. Bourdon, H. Palme, and R. Wieler (2007), Late formation and prolonged differentiation of the Moon inferred from W isotopes in lunar metals, *Nature*, *450*, 1206–1209.

Touboul, M., J. Liu, J. O'Neil, I. S. Puchtel and R. J. Walker (2014), New insights into the Hadean mantle revealed by ^{182}W and highly siderophile element abundances of supracrustal rocks from the Nuvvuagittuq Greenstone Belt, Quebec, Canada. *Chem. Geol.*, *383*, 63–75.

Touboul, M., P. Sprung, S. A. Aciego, B. Bourdon, and T. Kleine (2015b), *Hf-W chronology of the eucrite parent body*, *Geochim. Cosmochim. Acta*, *156*, 106–121.

Touboul, M., I. Puchtel, and R. J. Walker (2015), Tungsten isotopic evidence for disproportional late accretion to the Earth and Moon, *Nature*, *520*, 530–533.

Tsuno, K., D. J. Frost, D. C. Rubie (2013), Simultaneous partitioning of silicon and oxygen into the Earth's core during early Earth differentiation, *Geophys. Res. Lett.*, *40*, 66–71.

Tucker, J.M. and S. Mukhopadhyay (2014), Evidence for multiple magma ocean outgassing and atmospheric loss episodes from mantle noble gases, *Earth Planet. Sci. Lett.*, *393*, 254–265.

Wade, J. and B. J. Wood (2005), Core formation and the oxidation state of the Earth, *Earth Planet. Sci. Lett.*, *236*, 78–95.

Walker, R. J. (2014), Siderophile element constraints on the origin of the Moon, Phil. *Trans. R. Soc. Lond., A, 372*, 20130258.

Walsh, K. J., A. Morbidelli, S. N. Raymond, D. P. O'Brien, and A. M. Mandell (2011), A low mass for Mars from Jupiter's early gas-driven migration, *Nature, 475*, 206–209.

Wang, Z. and H. Becker (2013), Ratios of S, Se and Te in the silicate Earth require a volatile-rich late veneer, *Nature, 499*, 328–331.

Willbold, M., T. Elliott, and S. Moorbath (2011), The tungsten isotopic composition of the Earth's mantle before the terminal bombardment, *Nature, 477*, 195–197.

Wood, B. J. and A.N. Halliday (2010), The lead isotopic age of the Earth can be explained by core formation alone. *Nature, 465*, 767–771.

Wood, B. J., J. Wade, and M. R. Kilburn (2008), Core formation and the oxidation state of the Earth: additional constraints from Nb, V and Cr partitioning. *Geochim. Cosmochim. Acta., 72*, 1415–1426.

Wood, B. J., A. N. Halliday, and M. Rehkamper (2010), Volatile accretion history of the Earth, *Nature, 457*, E6–E7.

Yang, J. and J. I. Goldstein (2006), Metallographic cooling rates of the IIIAB iron meteorites, *Geochim. Cosmochim. Acta., 70*, 3197–3215.

Yang, J., J. I. Goldstein, and E. R. D. Scott (2007), Iron meteorite evidence for early formation and catastrophic disruption of protoplanets, *Nature, 446*, 888–891.

Yin Q. Z., S. B. Jacobsen, K. Yamashita, J. Blichert-Toft, P. Telouk, and F. Albarede (2002), A short timescale for terrestrial planet formation from Hf-W chronometry of meteorites. *Nature, 418*, 949–952.

Yoshino, T., M. J. Walter, and T. Katsura (2003), Core formation in planetesimals triggered by permeable flow, *Nature, 422*, 154–157.

Zahnle, K., N. Arndt, C. S. Cockell, A. Halliday, E. Nisbet, F. Selsis, and N. H. Sleep (2007), Emergence of a habitable planet, *Space Sci. Rev., 129*, 35–78.

Zhang, J. J., N. Dauphas, A. M. Davis, I. Leya, and A. Fedkin (2012), The proto-Earth as a significant source of lunar material, *Nature Geosci., 5*, 251–255.

6

An Experimental Geochemistry Perspective on Earth's Core Formation

Julien Siebert[1] and Anat Shahar[2]

ABSTRACT

In this chapter, we present a review of experimental data in geochemistry used to place constraints on Earth's accretion and core formation. Siderophile element abundances combined with partitioning experiments potentially give insights about pressure, temperature, and oxygen fugacity conditions during core formation. The interplay between siderophile partitioning and light elements in the core can help depict accurate models of accretion and core formation eventually. Additionally, resolvable metal-silicate isotopic fractionations in core formation experiments have been evidenced over the past few years. This new experimental tool merging the fields of experimental petrology and isotope geochemistry represents a promising approach, providing new independent constraints on the nature of light elements in the core and the conditions of Earth's core formation.

6.1. INTRODUCTION

Differentiation is the process with which a planetary body separates into layers based on the physical and chemical characteristics of the body. Based on the temperature, pressure (or size of body), oxygen fugacity, and impact history, the body will separate into layers and the final product will be a unique planet, moon, or asteroid. In this chapter we focus on the differentiation of a silicate mantle and a metallic core. This process occurred on the terrestrial planets, on the moon, and on many asteroids, because these objects were hot and big enough (above 10-km radius) to melt the metal so it could percolate through the silicate [*Taylor*, 1992; *Walter et al.*, 2004; *Greenwood et al.*, 2005]. The heat

comes from energy released from impacts together with heat from the decay of radiogenic isotopes [*Chambers*, 2005]. A large fraction of the iron and siderophile elements sink to the center to form a core, while the silicates and lithophile elements form a mantle. Exactly how this process occurs is unclear; however, a simplified model suggests equilibrium segregation between metal and silicate in a magma ocean setting. There are many ways to study differentiation including calculating models [e.g., *Elkins-Tanton*, 2012; *Moskovitz and Gaidos*, 2012], measuring radiogenic isotopic ratios [e.g., *Borg et al.*, 2011; *Caro et al.*, 2006; *Kleine et al.*, 2005], analyzing meteorites [e.g., *Armytage et al.*, 2011; *Weyer et al.*, 2005; *Poitrasson et al.*, 2005; *Moynier et al.*, 2011], experimentally determining and analyzing siderophile element abundances [e.g., *Righter et al.*, 1997; *Righter*, 2011; *Chabot et al.*, 2003; *Wade and Wood*, 2005; *Rubie et al.*, 2011; *Siebert et al.*, 2011], and most recently using stable isotopes as tracers of the conditions during differentiation [e.g., *Shahar et al.*, 2009; *Moynier et al.*,

[1]*Institut de Physique du Globe de Paris, Université Paris Diderot, Paris, France*

[2]*Geophysical Laboratory, Carnegie Institution for Science, Washington, DC, USA*

The Early Earth: Accretion and Differentiation, Geophysical Monograph 212, First Edition.
Edited by James Badro and Michael Walter.
© 2015 American Geophysical Union. Published 2015 by John Wiley & Sons, Inc.

2011; *Hin et al.*, 2014]. This review examines how recent experimental data in geochemistry may be used to progress in our understanding of core formation and differentiate between the various and highly debated scenarios of core mantle-equilibration. We will show how studying the partitioning of siderophile elements and stable isotopes between metal and silicate can be used to investigate the scenario of accretion and the nature of Earth's building blocks as well as the nature and budget of light elements in the core.

6.2. THE GEOCHEMISTRY OF SIDEROPHILE ELEMENTS

6.2.1. Introduction to Siderophile Elements Geochemistry

The primary chemical evidence for core formation in the early Earth is provided by the elements that are more soluble in metal than in silicate phases, the so-called siderophile elements. The bulk silicate Earth composition (BSE, that is, the mantle before any crust was formed) inferred from multiple studies on primitive mantle rocks (e.g., massif peridotite, peridotite xenoliths) is depleted in siderophile elements with respect to chondrites, the most primitive undifferentiated meteorites, taken as a canonical reference for the bulk Earth composition (Figure 6.1). This is undoubtedly assumed to reflect the extraction of siderophile elements into the core as the degree of depletion of these elements is correlated to their respective affinity for Fe-rich metal and as these elements are depleted with respect to lithophile elements of same volatility (Figure 6.1). Accordingly, the present distribution of siderophile elements in the BSE was likely governed by metal-silicate equilibrium and holds important constraints about the possible mechanisms of core formation and accretion.

The degree of affinity for Fe-rich metal of an element M is described by the metal-silicate partition coefficient, which is expressed as:

$$D_M^{met-sil} = \frac{x_M^{met}}{x_M^{sil}} \qquad (6.1)$$

where x_M^{met} and x_M^{sil} are the concentrations of M (in wt.% or at.%) in metal and silicate, respectively. Among the large class of siderophile elements, one can distinguish between slightly siderophile elements (SSE) having metal-silicate partitioning values below or close to 1, moderately siderophile elements (MSE) with $D^{met-sil}$ values between 10 and 10^4 and highly siderophile elements (HSE) with $D^{met-sil}$ values greater than 10^4 [*Walter et al.*, 2000]. This classification is established experimentally at 1 bar and relatively low temperature.

Partition coefficients are a function of pressure and temperature. This feature is particularly notable for siderophile elements as their abundances in the BSE are different than those predicted from 1 bar partitioning values, in some cases by a few orders of magnitude (Figure 6.1). A widely accepted explanation for this inconsistency is to consider that core-mantle equilibration took place at high pressure and high temperature (HP-HT); understanding the P-T conditions capable of producing the pattern of siderophile elements in the present mantle can accordingly reveal the conditions of Earth's core formation [e.g., *Li and Agee*, 1996; *Chabot et al.*, 2005; *Mann et al.*, 2009; *Siebert et al.*, 2013]. The process of core-mantle separation requires at least some degree of partial melting allowing the transport of liquid metal through either solid silicate (crystals) or fully molten silicate (magma ocean). The main sources of energy to produce melting are impacts, gravitational energy due to core segregation, and decay of short-lived isotopes. The geochemistry of siderophile elements in the mantle holds information to determine whether core formation occurred by percolation from a solid matrix of silicates or through metal sinking in a magma ocean because such mechanisms can affect the conditions and degree of metal-silicate equilibration. Moderately siderophile elements (MSE) such as Ni, Co, or W have been extensively studied because their partitioning values are highly P-T sensitive. Consequently, these elements may be useful for discriminating between core segregation mechanisms and can more generally provide information about the thermal state of the proto-Earth (e.g., the extent of mantle melting).

Partitioning of an element is also controlled by another critical parameter, the oxygen fugacity (fO_2). The fO_2 of the present-day core-mantle system is described by the proportions of reduced iron (core) and oxidized ferrous iron (mantle) and is defined relatively to the Iron-Wüstite buffer (IW, Fe-FeO). At the end of core formation, core-mantle equilibration needs to produce a mantle with its current FeO content (i.e., 8 wt.%) that is to say at IW-2 (i.e., 2 log units below the IW buffer). Beside this imposed constraint, oxygen fugacity can change during core formation, and this constitutes the main basis for heterogeneous accretion models [e.g., *O'Neill*, 1991] in which composition of accreting materials and subsequent fO_2 evolve during accretion and core formation of the Earth. The effect of fO_2 on the partitioning of an element *M* is related to its valence state *n* in the silicate phase following metal-silicate equilibrium:

$$MO_{n/2}^{silicate} = M^{metal} + \frac{n}{4}O_2 \qquad (6.2)$$

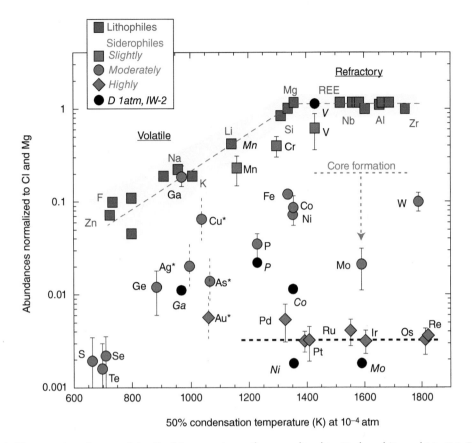

50% condensation temperature (K) at 10⁻⁴ atm

Figure 6.1 Element abundances of the Earth's present mantle normalized to Cl chondrite and Mg [*McDonough*, 2003; *Palme and O'Neil*, 2003; *Becker et al.*, 2006]. Lithophile elements (blue square symbols) follow a depletion trend from volatile behavior, which is considered to be function of the elements 50% condensation temperatures from the solar nebula at 10^{-4} atm [*Lodders*, 2003]. The Siderophile elements (red symbols) plot below this planetary volatility trend (blue shaded area), which indicates that they are depleted in the mantle. Error bars include errors on Cl and mantle abundances from different works (*elements with large error bars on their estimated depletions. *i.e.* error exceeding 50 % on mantle abundance). The remaining planetary complement of these elements is in the core. Black dots show the depletion of some selected siderophile elements if metal-silicate equilibration occurred at 1 atm, 1600 K and IW-2 (calculated from results in *Kegler et al.*, 2008; *Capobianco et al.*, 1999; *Schmitt et al.*, 1989; *Siebert et al.*, 2011). The imprint of core formation on the geochemistry of the present mantle is used to constrain the P-T-fO_2 conditions of core-mantle equilibration.

Thus, reducing conditions (i.e., low FeO content) increase the siderophile behavior of most elements (except anions such as O or S) by driving the equilibrium (2) toward the right hand-side. Over the past three decades, heterogeneous accretion models of all kinds [e.g., *Wänke*, 1981; *O'Neill*, 1991; *Rubie et al.*, 2011] have been promoted to explain the larger depletions in BSE of slightly siderophile elements (SSE, such as V and Cr) or elements normally regarded as lithophile elements such as Nb or Si with respect to their low pressure and temperature partitioning behaviors (Figure 6.1). Therefore, understanding the pattern of SSEs in the BSE, linked to the redox state during core-mantle differentiation, holds important constraints on the nature of the building blocks of the Earth, the nature and budget of light elements in the core (Si, O, S, C), and the composition of the bulk Earth.

As for the MSEs, highly siderophile elements (HSE, Au, and PGE) are overabundant in the BSE with respect to their low P-T partitioning values [see *Walker*, 2009]. Moreover, a key role of HSEs in understanding Earth accretion and differentiation is linked to their near-chondritic relative abundances (Figure 6.1), though their partitioning values at atmospheric pressure are greatly different from one to another, sometimes by a few orders of magnitude. The widely accepted explanation for this feature is to invoke the replenishment of the mantle in HSEs late in accretion, after core formation was complete, by addition of a small amount (~1 wt.%) of oxidized material of chondritic composition. This so-called "late-veneer" or "late-accretion" hypothesis would represent sufficient material to establish the chondritic abundances of HSEs in the BSE and not sufficient material to modify the abundances of MSEs and SSEs established from core-mantle equilibrium. The alternative hypothesis proposes that the pattern of HSEs in the BSE simply reflects core-mantle equilibration because partitioning values of this group of elements could strongly decrease at pressure and temperature conditions relevant for core formation. Thus, the study of HSEs metal-silicate partitioning is crucial to solving this issue and placing quantitative constraints on the required amount of "late veneer," an important parameter because the presence of volatile elements (e.g., H, C, N) on Earth is often associated to this late accreting complement [e.g., *Drake and Righter*, 2002; *Albarède et al.*, 2009; *Rubie et al.*, 2011].

The exchange mechanism of a siderophile element between metal and silicate can be described as a redox reaction with iron [*Wade and Wood*, 2005] following:

$$MO_{n/2}^{silicate} + \frac{n}{2}Fe^{metal} = \frac{n}{2}FeO^{silicate} + M^{metal} \tag{6.3}$$

Where $\log K = -\dfrac{\Delta G^{\circ}_{P,T(3)}}{RT}$ relates the equilibrium constant K to the free energy of reaction (6.3) and

$$K = \underbrace{\frac{\left(x_M^{metal}\right)\left(x_{FeO}^{silicate}\right)^{n/2}}{\left(x_{MO_{n/2}}^{silicate}\right)\left(x_{Fe}^{metal}\right)^{n/2}}}_{D_M/(D_{Fe})^{n/2}} \cdot \frac{\left(\gamma_M^{metal}\right)\left(\gamma_{FeO}^{silicate}\right)^{n/2}}{\left(\gamma_{Fe}^{metal}\right)^{n/2}\left(\gamma_{MO_{n/2}}^{silicate}\right)}$$ where x is the

molar fraction and γ the activity coefficient of the chemical species in metal or silicate. The ratio of oxide activity coefficients describes the effect of silicate melt composition on partitioning. Ideally the activities of individual oxides should be considered in describing the influence of melt composition on partitioning. However, estimating activities in silicate melts is difficult given the scarcity of data. Therefore, compositional variation of silicate melt is often expressed as a first order approximation by the melt structural parameter nbo/t [*Mysen et al.*, 1982], the molar ratio of non-bridging oxygens to tetrahedrally coordinated cations. It has been shown that the relative influence of melt composition on trace element partitioning is a function of cation oxidation state [e.g., *Walter and Thibault*, 1995; *Jana and Walker*, 1997; *Siebert et al.*, 2011]. Low valence cations (Cr2+, Mn2+, Co2+, Ni2+,...) are essentially independent of the nbo/t parameter. However, high valence cations (e.g., P5+, Nb5+, Mo4+, W4+,...) are strongly dependent of the nbo/t parameter with their affinity for silicate melts increasing in highly depolymerised melts (high nbo/t). This is because high valence cations may form polyhedral oxyanionic units with non-bridging oxygens and require more oxygens than divalent cations to be stabilized in the silicate melt [e.g., *Ryerson*, 1985]. Basic silicate melts with high nbo/t

(nbo/t=2,8 for a peridotite primitive mantle) ratio thus favor the solubility of high valence cations. Then, the partitioning coefficient D_M of a siderophile element can be related to pressure, temperature, oxygen fugacity (ΔIW), and nbo/t using the following expression:

$$\log D_M = a + \frac{b}{T} + \frac{c \cdot P}{T} + d \cdot nbo/t - \frac{n}{2}\Delta IW - \log\frac{\left(\gamma_M^{metal}\right)}{\left(\gamma_{Fe}^{metal}\right)^{\frac{n}{2}}} \tag{6.4}$$

Where the *a, b, c* terms result from the expansion of the free energy term $\left(\Delta G^{\circ} = \Delta H^{\circ} - T\Delta S^{\circ} + P\Delta V^{\circ}\right)$ and are related to $\Delta S^{\circ}/R$, $\Delta H^{\circ}/R$, and $\Delta V^{\circ}/R$, respectively. Equation (6.4) displays the influence of another important parameter on partitioning, the composition of metal expressed by the activity coefficients ratio of solutes in metal. The Earth's core contains light elements responsible for the observed ~10% density deficit relative to pure Fe or Fe-Ni alloy [*Birch*, 1964]. Addition of these elements can affect the activities of the other metallic components and modify the partitioning behavior of siderophile elements. Because the separate influence of potential light elements for entering core composition (i.e., S, Si, O, C, H) on partitioning can greatly differ from one to another, this effect constitutes a powerful tool for the study of Earth's core composition [e.g., *Corgne et al.*, 2009].

Experimental petrology has played a fundamental role in the modern understanding of terrestrial accretion and core-mantle differentiation. Over the last three decades, a substantial effort has been dedicated to performing large volume press (piston-cylinder press, multi-anvil press) experiments at the highest possible P-T conditions. The common approach to place constraints on the conditions of core formation from siderophile element geochemistry is to systematically measure their partitioning behaviors at various P-T-fO_2-X conditions. Then, the determination of the regression constants *a, b, c* of equation (6.4) using experimental results allows subsequent modeling of siderophile partitioning and evaluation of the required conditions for core-mantle equilibration to account for the observed depletions of siderophile in the BSE. However, *a, b, c* terms may not be constant over the possible large P-T range for core mantle-equilibration. For instance, transitions in the structure of liquids or changes in coordination have been reported to occur around 5 GPa and 35 GPa for metal and silicate liquids, respectively [*Sanloup et al.*, 2011, 2014]. The associated volume changes for reaction (6.3) prevents safe extrapolations of partitioning values below 5 GPa to higher pressure, as parameter *c* varies. Similarly, valence changes of siderophile elements in silicate melt can occur over the large range of fO_2

conditions (i.e., from IW-6 to IW) relevant to heterogeneous accretion models [*Cartier et al.*, 2014]. Experiments covering the full range of P-T-fO_2-X conditions for core-mantle equilibration are required to derive reliable accretion and core formation models.

In the next sections, we briefly review the current understanding of the specific pattern of siderophile element from HP-HT partitioning experiments; discuss resulting controversial hypotheses for Earth's accretion and core formation; and their implications for the composition of the core.

6.2.2. The Overabundance of Moderately Siderophile Elements: Hadean Bathymetry

The so-called "excess siderophile problem" for moderately siderophile elements clearly illustrates that mantle concentrations of these elements cannot be explained by equilibrium at low pressure and low temperature (Figure 6.1; Table 6.1). In other words, the typical signature of MSEs needs to be inherited from substantial re-equilibration of pre-existing cores from accreting planetesimals and proto-planets. Among moderately siderophile elements (e.g., P, Ni, Co, Ge, Cu, Mo, W...), those that are unaffected or poorly affected by volatility have received the most attention as their mantle depletions are imputable to core formation only and as their bulk Earth contents are well constrained. Siderophile elements showing little variations in concentration of mantle samples are also valuable for estimating their core-mantle required partitioning values with small uncertainties (Table 6.1). In this framework, a great deal of effort has been dedicated to the study of Ni and Co partitioning with the emergence of large volume press experiments (mostly multi-anvil press) at HP-HT since the late 1990s [e.g., *Thibault and Walter*, 1995; *Li and Agee*, 1996, 2001; *Righter et al.*, 1997; *Gessmann and Rubie*, 2000; *Chabot et al.*, 2005; *Wade and Wood*, 2005; *Kegler et al.*, 2008]. The general consensus emerging from these works is that partitioning of Ni and Co decreases significantly with increasing pressure and temperature. This has led to the popular idea that metal-silicate equilibration at the base of a deep magma ocean could account for the abundances of MSE at least. The P-T dependences of Ni and Co partitioning have been used to develop constraints on the depth of the Magma Ocean (MO) and extent of mantle melting during core formation. However, the extrapolation of low pressure results to more extreme conditions following Equation (6.4) by these studies have yielded disparate estimates for the base of the magma ocean, ranging from 25 GPa to 60 GPa and 2000 °C to over 4000 °C [*Chabot et al.*, 2005; *Corgne et al.*, 2009; *Gessmann and Rubie*, 2000; *Li and Agee*, 1996; *Li and Agee*, 2001; *Righter et al.*, 1997;

Table 6.1 Apparent Core-Mantle Partition Coefficients for Siderophile Elements

Element	D core/mantle	Lithophile element[a]	References[e]
Ag	10–40	K (T$_c$~1010 K)	(1), (2), (3)
As	40–100	Li (T$_c$~1150 K)	(1), (2)
Au	350–550	-	(2), (4), (5), (6), (7)
Co	23–26	-	(1), (2), (3), (4)
Cr	2–4	-	(1), (2), (3), (4)
Cu	4–10	-	(1), (2), (3), (4)
Ga	0–0.8	Na (T$_c$~950 K)	(1), (2)
Ge	9–17	Rb (T$_c$~950 K)	(1), (2)
Ir	650–830	RLE	(2), (5), (6), (7)
Mn	0–2	-	(1), (2), (3), (4)
Mo	90–150	RLE	(2), (5)
Nb	0–1	RLE	(1), (2)
Ni	24–28	- [b]	(1), (2), (3), (4)
Os	650–830	RLE	(2), (5), (6), (7)
P	20–50	-	(1), (2), (3), (4)
Pd	400–560	RLE	(2), (5), (6), (7)
Pt	670–830	RLE	(2), (5), (6), (7)
Re	600–800	RLE	(2), (5), (6), (7)
Rh	600–800	RLE	(2), (5), (6), (7)
Ru	550–700	RLE	(2), (5), (6), (7)
S	70–140	Zn	(1), (2), (3)
Se	60–160	Zn	(1), (2), (3)
Ta	0–0.4	RLE	(1), (2)
Te	60–170	Zn	(1), (2), (3)
V	1.3–2.5	RLE [d] (T$_c$ >1400 K)	(1), (2)
W	20–55	RLE	(8)
Zn	0–8	F (T$_c$~730 K)[c]	(1), (2)

[a] Lithophile element used to determine the bulk Earth abundance of a siderophile element with similar volatility trend.
[b] Bulk Earth abundance was taken directly from references.
[c] Reference temperature at which half the mass of a specific element condensates into a mineral from a cooling solar nebula under a given oxygen partial pressure. Condensation temperatures and volatility order are from *Wasson* [1985], *Allègre et al.* [2001], and *Lodders* [2003].
[d] Refractory lithophile elements (e.g., Al, Ti, Ca, Sc, REE).
[e] References for siderophile and lithophile element abundances in both upper mantle rocks and chondrites CI used to calculate the ranges of apparent core-mantle partition coefficients. (1) *McDonough and Sun* [1995]; (2) *Palme and O'Neill* [2003]; (3) *Allègre et al.* [2001]; (4) *Allègre et al.* [1995]; (5) *McDonough* [2003]; (6) *Becker et al.* [2006]; (7) *Fischer-Gödde et al.* [2011]; (8) *Wade and Wood* [2012].

Righter, 2011; *Siebert et al.*, 2011; *Wade and Wood*, 2005] Figure 6.2 displays selected results from these works. Moreover, a non-linear pressure effect on partitioning reported by *Kegler et al.* [2008] has questioned the

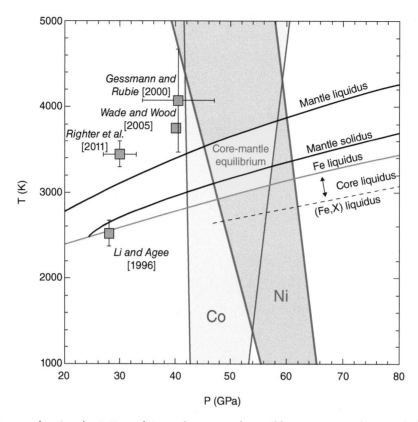

Figure 6.2 Diagram showing the P–T conditions of core-mantle equilibration required to match the present-day mantle concentrations of Ni and Co. The individual solutions for Ni (red) and Co (blue) are obtained from recent diamond cell experiments at direct P–T conditions of predicted core-mantle equilibration [*Siebert et al.*, 2012]. The solution domain is further constrained to lie within the mantle solidus–liquidus interval (averaged after *Fiquet et al.* [2010] and *Andrault et al.* [2011]). Square symbols are the P–T solutions proposed from large volume press experiments [*Li and Agee*, 1996; *Gessmann and Rubie*, 2000; *Righter et al.*, 2011; *Wade and Wood*, 2005] and display solutions below the required pressure. The liquidus of pure iron [*Anzellini et al.*, 2013] and the liquidus of a core containing about 10 wt.% of light elements X [e.g., S, Si; *Morard et al.*, 2011] are reported and show that core forming material is also molten at these conditions.

possibility of producing Ni and Co BSE signatures by core-mantle equilibration in a deep magma ocean. Such discrepancies have highlighted the need for additional direct measurements of partitioning at very high P-T. The laser heated diamond anvil cell technique is the only static pressure device capable of producing these conditions (above 25 GPa and 3000 °C). A major breakthrough in this field has been the recent development of the Laser Heated Diamond Anvil Cell (LH-DAC) as a new tool for the study of experimental petrology [*Bouhifd and Jephcoat*, 2003, 2011; *Siebert et al.*, 2012, 2013]. Sample recovery after quench, and suitable analytical instrument at small spatial scale, have strongly limited the use of such HP-HT devices for measuring metal-silicate partitioning behaviors. It is only recently that the first partitioning data at such extreme conditions have been reported for refractory siderophile elements. These data have confirmed the deep magma ocean hypothesis for core-mantle equilibration at pressure between

45–60 GPa (Figure 6.2) [*Bouhifd and Jephcoat*, 2011; *Siebert et al.*, 2012]. In such a model, metal from accreting planetesimals is supposed to break up and become emulsified as it travels toward the bottom of the MO. Liquid metal in the form of small droplets, with stable size of a few cm in diameter, equilibrates with molten silicate at HP-HT until it reaches the bottom of the MO before quick segregation in the solid lower mantle without further re-equilibration. (see *Rubie et al.*, 2007, for further descriptions of the physics of core formation and references therein). Accordingly, the temperature of equilibration in this model should lie between the solidus and liquidus of peridotite, as the liquidus of metal is always lower than the peridotite solidus (Figure 6.2). Thus, a temperature of equilibration between 3100 and 3800 K is required to account for Ni and Co abundances between 45 and 60 GPa (Figure 6.2). Above 60 GPa, Ni and Co partitioning becomes too low [*Siebert et al.*, 2012]. This provides an upper bound for the depth of core formation corresponding to a magma ocean of around 1500 km depth, and argues against core formation scenarios in a fully molten silicate mantle.

Single P-T conditions (single stage event) is an unrealistic physical representation of core formation but could simply represent the average of a wide range of conditions for the magma ocean generated from a series of various accretion events including late giant impacts. An alternative model, the continuous core formation model, physically more plausible than single-stage core formation, has been proposed by *Wade and Wood* [2005] (see Figure 6.3).

In this model, the core grows continuously and constantly from 0 to full planetary mass over the course of accretion. P-T conditions of core-mantle equilibration at the base of the magma ocean (MO), still constrained by the peridotite solidus-liquidus, increase with increasing depth of the MO as the Earth grows. Applying continuous core formation to account for Ni and Co leads to a depth of the MO around 50–55 GPa [*Siebert et al.*, 2012], similar to those derived from single-stage core formation model.

Eventually, the objective of partitioning experimental studies is to find a unique (single-stage model) or progressive (continuous core formation model) set of P-T conditions that can account for abundances of other moderately siderophile elements (e.g., Mo, W, Ge, Ga, As...). Some recent studies [e.g., *Corgne et al.*, 2008; *Cottrell et al.*, 2009; *Righter et al.*, 2010; *Siebert et al.*, 2011; *Wade and Wood*, 2012] have determined the regression constants a, b, c for a large spectrum of MSE from large volume press experiments. Figure 6.2a displays the final partitioning for MSE during a continuous core formation process using a compilation of recent experimental results. (See caption of Figure 6.2 for further details of the modeling.) The results are consistent with the generally stated conclusion that equilibration in a deep magma ocean decreases the siderophile behavior of MSEs and can account for their observed concentrations. However, the temperature defined by the Ge and Ga solutions are at least few hundreds of kelvin above and below the peridotite liquidus, respectively. The effect of other parameters affecting partitioning, such as fO_2

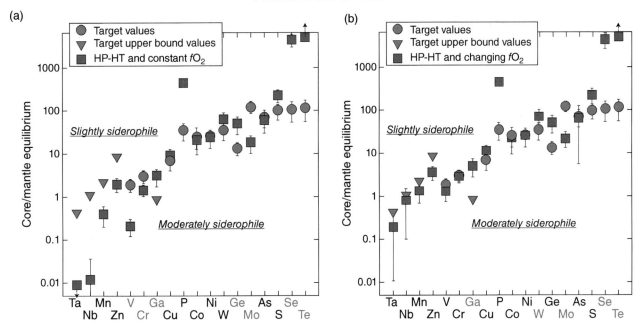

Continuous core formation models

Figure 6.3 Summary of final core-mantle partitioning results for slightly and moderately siderophile elements with a continuous core formation model. The results are calculated from the combination of recent experimental works [*Corgne et al.*, 2008; *Rose-Weston*, 2008; *Mann et al.*, 2009; *Righter et al.*, 2010; *Siebert et al.*, 2011, 2012, 2013; *Boujibar et al.*, 2014; *Wade and Wood*, 2012]. Filled red dots and triangles give respectively the required target values and upper limits for observed core-mantle partitioning in the Earth following Table 6.1. Blue squares give final results at the end of the accretion process with a final core-mantle equilibration at 50 GPa and T of mantle liquidus averaged after *Fiquet et al.* [2010] and *Andrault et al.* [2011]. Errors bars give 1σ uncertainties. (a) Shows results for a continuous core formation model with constant oxygen fugacity during accretion (IW-2.3 corresponding to 8 wt. % FeO in the silicate Earth). (b) Shows results for a continuous core formation model with increasing oxygen fugacity conditions during accretion (from 0.8 wt. % FeO to 8 wt. % FeO in the silicate Earth).

or metallic composition, needs to be explored thoroughly for MSEs and integrated into core formation models. For instance, *Wade et al.* [2012] showed that sulfur has a positive effect on the partitioning of Mo. A core containing 2 wt% sulfur [*Allègre et al.*, 1995; *Dreibus and Palme*, 1996] could account for Mo concentrations in the mantle but only through heterogeneous accretion with addition of S during the 10–20% of accretion. Then, the abundance of Mo in the BSE is likely to hold information on the identity of light elements in the core and the origin and timing of accretion of moderately volatile elements on Earth.

The next section of this chapter shows how another category of siderophile elements, the slightly siderophile elements (SSE), has been used to provide additional constraints on the redox state during core formation and accretion of the Earth and is implications for the composition of its core.

6.2.3. Slightly Siderophile Elements: A Probe for Understanding the Redox of Earth's Accretion and Core Formation

Contrary to MSE, slightly siderophile elements (e.g., V, Cr, Mn, Nb) are more depleted in the mantle than expected from their low P-T partitioning values (Figure 6.1). The P-T conditions required to produce the observed depletions for SSE like V and Cr (at the present-day fO_2) require temperatures that greatly exceed that of the mantle liquidus [*Wade and Wood*, 2005; *Corgne et al.*, 2008, 2009; *Wood et al.*, 2008; *Siebert et al.*, 2011]. Such conditions are physically inconsistent with the magma ocean hypothesis, where the temperature at the base of a magma ocean necessarily lies between the mantle solidus and liquidus, hence creating a rheological boundary that enables the metal to pond and equilibrate with the silicate.

To overcome this issue, recent models of core formation invoke early accretion of highly reduced materials with an FeO-poor silicate component. These initially low fO_2 conditions enhance the siderophile behavior of the SSE at high mean pressures and temperatures of silicate-core equilibration. A subsequent gradual oxidation of the mantle over the course of core formation is required to account for MSE abundances and to eventually reach the present-day terrestrial redox around IW-2.3 (i.e., 8 wt% FeO in silicate). The increase in FeO content of the silicate mantle has been proposed to occur either through a change in the redox state of the accreting material [i.e., heterogenous accretion [*Rubie et al.*, 2011] or through self-oxidation processes involving Si, Fe, and O with a single homogenous accreting composition [*Javoy et al.*, 2010; *Wade and Wood*, 2005; *Wood et al.*, 2006; *Corgne et al.*, 2008; *Wood et al.*, 2008; *Tuff et al.*, 2011]. For instance, reduction of Si

to the core acts as a powerful oxygen "pump," releasing FeO to the mantle through the reaction:

$$SiO_2^{silicate} + 2Fe^{metal} = 2FeO^{silicate} + Si^{metal} \quad (6.5)$$

Disproportionation of ferrous iron into ferric iron plus metal in perovskite also represents an additional mechanism to raise the iron content of the magma ocean [*Frost et al.*, 2004; *Wade and Wood*, 2005]. A common point however is that Earth needs to accrete largely from highly-reduced material, such as enstatite chondrites. Figure 6.2b displays the final partitioning results obtained for changing redox conditions from roughly IW-4 (<1 wt% FeO) to IW-2 (~8 wt% FeO) during a continuous core formation. Still, this model does not seem capable of producing the abundance of Mo, W, or Ga in the mantle. Determining under what conditions gallium becomes almost lithophile (Figure 6.1) would certainly provide important constraints on core formation.

Such redox conditions have strong implication for the light-element content of the core, where silicon is likely to be the main light element entering the core in large amounts under highly reduced conditions [e.g., *Siebert et al.*, 2004; *Wood et al.*, 2008; *Rubie et al.*, 2011; *Ricolleau et al.*, 2011; *Tuff et al.*, 2011]. However, that model relies on extensive extrapolation of high pressure and high temperature partitioning data of SSE. Using extremely high pressure and high temperature diamond anvil cell partitioning experiments, *Siebert et al.* [2013] showed recently that the observed depletions of SSE (V, Cr) could also be produced by core formation under relatively oxidizing conditions. These results demonstrate not only the influence of increased pressure and temperature, but also that of the associated increase in oxygen solubility in the metallic phase, on siderophile element partitioning [*Corgne et al.*, 2009]. Presence of oxygen in metal increases the siderophility of V and Cr and enables core formation model in a deep magma ocean at constant fO_2 (Figure 6.4) and even under more oxidizing conditions. Thus, accretion under fO_2 such as that of the present-day mantle (~IW-2) and perhaps as oxidized as carbonaceous or ordinary chondrites (~IW-1) may be relevant with respect to SSE composition of the BSE. Transfer of oxygen from the mantle to the core provides a mechanism to reduce the initial magma ocean redox state to that of the present-day mantle [*Rubie et al.*, 2004; *Asahara et al.*, 2007; *Siebert et al.*, 2013].

Under such conditions, simultaneous presence of Si and O as dominant light elements for the core is predicted, with approximately between 2 and 5 wt% each depending on the initial redox conditions [*Bouhifd and Jephcoat*, 2011; *Siebert et al.*, 2013; *Tsuno et al.*, 2013]. Such composition is consistent with both siderophile trace elements concentrations in the mantle and geophysical constraints on the composition of the core [*Antonangeli et al.*, 2010; *Badro et al.*, 2007].

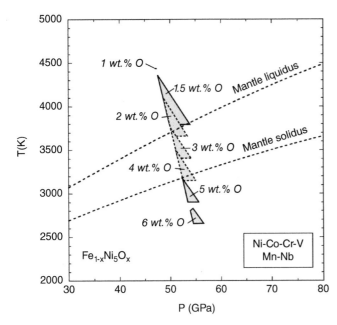

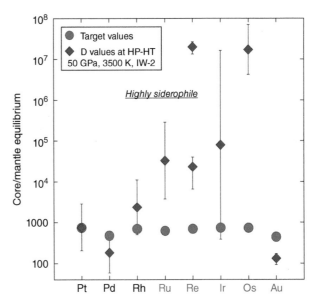

Figure 6.4 Diagram showing the P-T conditions of core-mantle equilibration required to match the present-day mantle concentrations of Co-Cr-Mn-Nb-Ni-V during a single-stage core formation at IW-2.3 (the fO_2 is fixed by the present FeO content of the mantle) as a function of oxygen contents in the core. The solutions are calculated using regression models for each considered siderophile element [*Wood et al.*, 2008; *Tuff et al.*, 2011; *Siebert et al.*, 2012, 2013]. The solution domain is further constrained to lie within the mantle solidus-liquidus interval and overlap this required P-T conditions for a core containing between roughly 1.5 and 5 wt.% of oxygen.

Figure 6.5 Summary of final partitioning results for highly siderophile elements (Au and PGEs, Platinum Group Elements) for core-mantle equilibration at the bottom of a deep magma ocean. The results are calculated from the combination of recent experimental works [*Cottrell and Walker*, 2006; *Righter et al.*, 2008; *Brenan and McDonough*, 2009; *Mann et al.*, 2012; *Bennett and Brennan*, 2013; *Bennett et al.*, 2014]. Filled red dots give respectively the target values observed for required core-mantle partitioning in the Earth following Table 6.1. Blue diamonds give partitioning results for core-mantle equilibration at 50 GPa, 3600 K and IW-2.3. For Re, diagram displays the two very different partitioning obtained using either parameterization from *Mann et al.* [2012] (lowest value) or *Bennett et al.* [2013].

6.2.4. The Late Veneer Hypothesis

Up to now, the apparent overabundance of highly siderophile elements (HSE) in the mantle of the Earth is likely best explained by late accretion of a chondritic component. Such a small proportion of late-accreting material, estimated generally between 0.4 and 1% of the mass of the Earth [e.g., *Walker*, 2009; *Brennan and McDonough*, 2009] would have negligible or small effect on bulk Earth composition of other elements. Still, metal-silicate equilibration in a magma ocean cannot be ruled out due to the lack of experiments at the direct P-T conditions relevant for core formation. Extrapolations of the current partitioning dataset to these conditions lead to very large uncertainties on the predicted partitioning values of HSE (Figure 6.5), allowing the option of core-mantle equilibrium to account for the observed signatures of most HSE in the mantle (Figure 6.5). The solubility of HSE in silicate liquid has been mostly studied (see *Ertel et al.*, 2008, for review and references therein) at 1 atm and moderate temperatures. However, some recent experimental studies brought significant constraints on

their partitioning at HP-HT up to 18 GPa and 2773 K [e.g., *Cottrell and Walker*, 2006; *Ertel et al.*, 2006; *Righter et al.*, 2008; *Brennan and McDonough*, 2009; *Mann et al.*, 2012; *Bennett et al.*, 2013, 2014]. Most of the HSEs partitioning show a significant decrease of partitioning with increasing temperature (e.g., Au, Pt, Pd, Re), and a potential influence of pressure is more controversial and difficult to assess over the small pressure range covered by most of the studies. *Mann et al.* [2012] report a small effect of pressure (above 6 GPa) for most HSEs. Another source of large discrepancy, for predicting partitioning at the conditions of core formation, relies on the discrepancies between the considered valence state of the HSEs in silicate melt under reducing conditions (<IW) [e.g., *Mann et al.*, 2012; *Bennett et al.*, 2014]. Although the conclusions of *Righter et al.* (2008) showed that an equilibration scenario at HP-HT could account for Pd in the BSE, other works [*Brennan and McDonough*, 2009; *Mann et al.*, 2012; *Bennett et al.*, 2014] based on the partitioning of multiple HSEs showed that the large disparity between the partitioning behaviors of HSEs at HP-HT would

strongly fractionate some elements (e.g., Au, Pd, Pt) from others (e.g., Os, Ir, Re) (Figure 6.5). Theses results are in contradiction to the observed near-chondritic proportions of HSEs in the BSE. Hybrid models, including both the influence of metal-silicate equilibration (>3000 K) and addition of late-accreting material, appear relevant to account for the budget of HSEs in the Earth's mantle [*Mann et al.*, 2012; *Bennett et al.*, 2014]. Finally, the partitioning of some moderately siderophile and volatile elements such as S, Se, and Te argues for the late-veneer scenario. The large disparity between the partitioning of these elements argues against their similar depletions in the BSE set solely by core-mantle equilibrium at the base of a deep magma ocean (Figure 6.2a, b) [*Rose-Weston et al.*, 2008]. The addition of 0.5 wt% of meteorite component such as ordinary chondrites [*Rose-Weston et al.*, 2008] or carbonaceous chondrites CM [*Wang and Becker*, 2013] is debated to account for S, Se, and Te concentrations in the BSE and may have supplied a large fraction (up to 100%) of highly volatile elements on Earth such as hydrogen and carbon. However, the process of volatile accretion remains highly controversial as *Boujibar et al.* [2014] showed that sulfur abundance in the mantle could be produced by core-mantle equilibration alone, and as sulfur isotope constraint requires that sulfur must have been present to a considerable extent before late accretion event [*Labidi et al.*, 2014].

6.3. CONSTRAINTS ON CORE FORMATION FROM STABLE ISOTOPES

6.3.1. Introduction to Stable Isotope Geochemistry

The principle of using stable isotopes to probe the bulk chemical composition of planets lies with the combination of isotope fractionation and sequestration of elements in unseen reservoirs like the core. Isotope fractionation will exist between phases with distinct bonding environments (e.g., Earth's core and mantle), and separation of elements between reservoirs manifests this fractionation. Theoretical calculations predict that as temperature increases and mass increases the equilibrium isotopic fractionation factor becomes negligible (or at least unresolvable). However, with the advent of the Multiple Collector Inductively Coupled Plasma Mass Spectrometer (MC-ICPMS), a larger portion of the periodic table has become available, and better precision has allowed new research to be done. Suddenly with this new technique it has become apparent that stable isotopes of even the "heavy" elements could have resolvable fractionations.

Equilibrium isotope fractionation is driven by the effects of atomic mass on bond vibrational energy. The relationships are easily understood using a simple molecule as an example. All molecules have a zero-point

vibrational energy (ZPE) = $1/2\ h\nu$, where ν is the vibrational frequency and h is Planck's constant. The vibrational frequency of a particular mode for a molecule can be approximated using Hooke's Law, $\nu = 1/2\pi\sqrt{(k/\mu)}$, where k is the force constant and μ is the reduced mass of the molecule (e.g., $\mu = m_a m_b/(m_a+m_b)$) where m_a and m_b are the atomic weights of two atoms in a diatomic molecule. When a light isotope in a molecule is substituted by a heavy isotope, the potential energy curve does not change shape to a first approximation (hence the force constant does not change,), but the vibrational frequency does change. When a more massive isotope is substituted, the reduced mass of the molecule increases, which decreases the vibrational frequency and the energy. Therefore, equilibrium-stable isotope fractionations are quantum-mechanical effects that depend on the zero point energy of the molecule being investigated. The same principle applies to crystalline materials.

In general, the most important factor that determines the magnitude of the isotopic fractionations is differences in bond strength; stiffer bonds concentrate the heavy isotope. Bond strength (stiffness) determines vibrational frequency, and vibrational frequency determines internal energy. Increasing the pressure in a given phase stiffens the bonds and increasing the temperature weakens the bonds. *Urey* [1947] and *Bigeleisen and Mayer* [1947] calculated equilibrium constants for isotopic exchange reactions and found that at high temperature, the equilibrium constant becomes proportional to the inverse square of temperature (i.e., isotope fractionation decreases proportional to $1 / T^2$). Thus, at the high temperatures involved in core-mantle studies, the fractionation predicted for the heavy isotopes (isotope fractionation also decreases as $\sim 1/Mass^2$) is relatively small.

However, it is thought that pressure has the opposite effect on isotope fractionation. *Polyakov and Kharlashina* [1994] outlined the three most important variables for the pressure effect on equilibrium isotope fractionation, namely, the Grüneisen parameter, the derivative of the β-factor with respect to temperature, and the isothermal bulk modulus. The effect of pressure was also found to be extremely temperature dependent, and it is the interplay between these two variables that will determine if pressure will play a role in the fractionation found in a system of choice. In general however, pressure will decrease volume and therefore increase vibrational frequencies for all materials relevant to planetary interiors. To a first order, since a pressure-induced shift in frequency for mode i is $\Delta v_i = \Delta V^{-\gamma}$ where v_i is frequency, V is volume, and γ is the Grüneisen parameter, and because fractionation depends on the differences in the squares of frequencies [e.g., $(v_{i,2}+\Delta v_i)^2 - (v_{i,1}+\Delta v_i)^2$ where subscripts 2 and 1 refer to two minerals], pressure will generally amplify fractionation effects.

Joy and Libby [1960] first calculated the effect of pressure on isotope fractionation. They suggested that oxygen isotope fractionation might be pressure dependent at low temperatures. However, in 1961, *Hoering* (the first to examine the pressure effect experimentally) did not observe a pressure effect on oxygen isotope partitioning between water and bicarbonate at 43.5°C and 0.1 and 400 MPa. Then in 1975 *Clayton et al.* found no pressure effect on the calcite-water fractionation at 500°C 0.1–2GPa and at 700°C 50–100 MPa. The Chicago group continued studying the pressure effect on fractionation by studying quartz, wollastonite, and albite at a range of pressures and temperatures (up to 2 GPa) and found no oxygen isotope fractionation with pressure [*Matsuhisa et al.*, 1979; *Matthews et al.*, 1983]. Due to these initial studies, the effect of pressure on isotope fractionation has been assumed to be negligible for all elements (except for hydrogen) and pressures. However, in 1994, a study again predicted that pressure should have an effect on isotopic fractionation [*Polyakov and Kharlashina*, 1994], and this has been confirmed for hydrogen isotopes [*Horita et al.*, 1999].

6.3.2. Chondrite vs. Earth

The most direct method to determine if there is an isotope fractionation during core formation is to analyze chondritic meteorites, which are thought to represent the bulk Earth, and compare the isotope ratios to that of a typical mantle rock from Earth. If there is a difference between these samples then it can be concluded that the fractionation occurred during core formation assuming no present day physical or chemical processes affected the mantle rocks. An added benefit of this method is that the composition of the core can be determined as well. During the differentiation process, the iron molten metal will alloy with other elements on its route to the center of the planetary body. Which elements it bonds with will be a function of the conditions attending the core formation. For example, at high temperature under reducing conditions, Si is likely to alloy with iron [*Ricolleau et al.*, 2011], however, in more oxidizing conditions sulfur is likely to enter the metal [*Sanloup and Fei*, 2004]. As discussed above, on Earth, seismic data exist that show the density difference between pure iron and the inferred density from the velocities of the seismic waves within the Earth's interior. This discrepancy implies that there are elements other than Fe (and Ni since Ni does not change the density of Fe much) within the core of the Earth. The main candidates considered are carbon, sulfur, oxygen, silicon, and hydrogen. Determining which elements reside in the Earth's interior requires indirect methods such as high pressure and temperature experimental studies. The nature of the light element in an Fe-Ni core is directly related to the mode of formation, for example, whether differentiation occurred at high pressure or low pressure, the redox conditions present, or the composition of the melt. For example, *Georg et al.* [2007] discovered that the silicon isotopic ratio of Earth was different than chondrites and hypothesized that this difference was probably due to Si hidden in the Earth's core. This was the first time that stable isotopes were used not only to shed light on the conditions of core formation but also to identify the composition of the light elements in the core.

After the *Georg et al.* study was published several other groups also measured the Si isotope ratios in chondrites and rocks thought to represent the BSE, and varying results were found [*Fitoussi et al.*, 2009; *Savage et al.*, 2013; *Armytage et al.*, 2011; *Chakrabarti and Jacobsen*, 2010]. Today most will agree that there is a difference in $^{29}Si/^{28}Si$ between chondrites and BSE and that it is approximately 0.14 to 0.20‰. The isotope ratio differences between chondrites and Earth have also been reported for Mg [*Young et al.*, 2009; *Wiechert and Halliday*, 2007; *Teng et al.*, 2010], Cr [*Moynier et al.*, 2011], S [*Labidi et al.*, 2014], Ca [*Simon and DePaolo*, 2010], Mo [*Burkhard et al.*, 2014], V [*Nielsen et al.*, 2014] and Fe [*Weyer et al.*, 2005; *Poitrasson et al.*, 2005; *Craddock et al.*, 2013]. Although most of the differences from chondrite are extremely small (Figure 6.6), that small difference can still be used to explain large-scale processes on Earth and other planetary bodies. Once it became clear that the differences found between chondrite isotope ratios and BSE *could* be due to core formation processes, a more thorough understanding of that process with respect to stable isotopes was needed. This is where experiments were noticeably missing.

6.3.3. Experiments to Date

Determining whether there is an equilibrium fractionation between metal and silicate is central to understanding the isotope signatures found within different meteorites and planetary bodies. However, the field of combing high pressure and temperature experiments with non-traditional stable isotope geochemistry is still quite new. This technique is not as common as analyzing natural rocks, however, it can provide crucial information. Although there are many problems with conducting these sorts of experiments, if done correctly, the information gained can elucidate the temperature, pressure, fO_2, and compositional conditions during the differentiation process. Experiments so far have been published on the Fe, Si, Mo, and Ni isotope systematics for an equilibrium metal-silicate fractionation.

Most of the experimental studies published on this topic have focused on iron [*Williams et al.*, 2012; *Shahar et al.*, 2008, 2014; *Poitrasson et al.*, 2009; *Hin et al.*, 2012], though some also focusing on silicon [*Shahar et al.*, 2009,

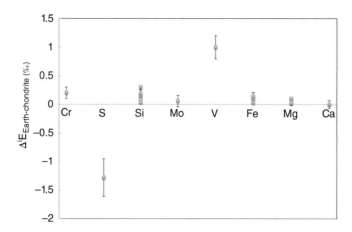

Figure 6.6 Diagram showing the isotopic ratio difference between Earth and chondrite as analyzed in the laboratory by the studies cited in the text. Each element has a different isotopic ratio plotted: $^{53/52}$Cr, $^{34/32}$S, $^{30/28}$Si, $^{98/95}$Mo, $^{51/50}$V, $^{56/54}$Fe, $^{25/24}$Mg, $^{44/40}$Ca.

Table 6.2 Experimentally Determined Fractionation Factors to Date

Isotopic System	Phase 1	Phase 2	Fractionation	Reference
Fe	Magnetite	Fayalite	Δ^{57}Fe = 0.30 ± 0.024 ×10^6 / T^2	*Shahar et al.* [2008]
Fe	Silicate	Metal	No Fractionation	*Poitrasson et al.* [2009]
Fe	Silicate	Metal	No Fractionation	*Hin et al.* [2012]
Fe	Silicate	Metal	>0.45‰/amu at 1850 °C	*Williams et al.* [2012]
Fe	Metal	Silicate	+0.12 ± 0.04‰ at 1650 °C	*Shahar et al.* [2014]
Si	Silicate	Metal	Δ^{30}Si = 7.45 ± 0.41 ×10^6 / T^2	*Shahar et al.* [2009, 2011]
Si	Metal	Silicate	Δ^{30}Si = −4.42 ± 0.05 ×10^6 / T^2	*Hin et al.* [2013]
Si	Silicate	Metal	Δ^{30}Si between 0.7 – 1.6‰ at 1400–1600 °C	*Kempl et al.* [2013]
Mo	Metal	Silicate	Δ^{98}Mo = −4.80 ± 0.55 ×10^6 / T^2	*Hin et al.* [2014]
Ni	Metal	Talc	Δ^{62}Ni = 0.25 ± 0:02 × 10^6 / T^2	*Lazar et al.* [2012]

2011; *Kempl et al.*, 2013; *Hin et al.*, 2013], Mo [*Hin et al.*, 2014] and Ni [*Lazar et al.*, 2012] (Table 6.2). For Fe isotopes, the four experimental studies performed to date that focus on iron metal-silicate fractionation at high temperature are split. Two papers cite that no fractionation between liquid metal and liquid silicate at high temperature is present [*Poitrasson et al.*, 2009; *Hin et al.*, 2012], one argues that it is small [*Shahar et al.*, 2014], and one argues that the fractionation during perovskite disproportionation is quite large [*Williams et al.*, 2012]. However, it is very difficult to compare experiments as small differences in the experimental set-up or composition of the run products can cause large differences in the resulting fractionation. It is imperative that a systematic approach be taken so that the true equilibrium values are found. Many experiments are required to obtain one fractionation factor, because there are several kinetic factors that could play a role in the fractionation (such as interaction with capsule material or Soret diffusion as in *Richter et al.*, 2009).

One way to show that equilibrium has been attained is through the use of an isotopic 'spike' and is termed the three-isotope exchange method. This method was pioneered at the University of Chicago in the 1970s [*Matsuhisa et al.*, 1978; *Matthews et al.*, 1983] for oxygen isotopes between a mineral and an aqueous phase. *Shahar et al.* [2008] then adapted the method to apply to mineral-mineral fractionation factors. The principle of the three-isotope exchange method is to replace the terrestrial fractionation line (TFL) having a zero intercept with a secondary fractionation line (SFL) that has a non-zero intercept by use of a spike. The secondary fractionation line will have the same slope as the TFL but will be displaced from it in proportion to the amount of spike. At isotopic equilibrium all phases containing the element of interest must lie on the SFL. Figure 6.7 depicts the typical trend to equilibrium expected in the experiments. In the context of three Fe isotopes, a condition for equilibrium between two phases *i* and *j* is $\alpha_{i-j}^{56/54} = \left(\alpha_{i-j}^{57/54}\right)^{\gamma}$ where

$\alpha^{56/54}_{i-j} = {}^{56}R_i / {}^{56}R_j$, ${}^{56}R_i = {}^{56}Fe/{}^{54}Fe$, for phase i, and the exponent γ relating the three isotopes at equilibrium is

$$\gamma = \frac{\dfrac{1}{m_{56}} - \dfrac{1}{m_{54}}}{\dfrac{1}{m_{57}} - \dfrac{1}{m_{54}}}$$

where m is the mass of subscripted Fe isotope. In this case $\gamma = 0.67795$. The exponent γ is effectively the slope of a fractionation line on a plot of $\delta^{56}Fe$ vs. $\delta^{57}Fe$ [e.g., *Young et al.*, 2002].

In silicate/metal differentiation experiments the sample thoroughly mixes before equilibrating. When the isotopes all mix, all the phases lie on the bulk value (star in Figure 6.7). Following mixing there is an 'unmixing' among the isotopes of Fe along the secondary fractionation line (SFL) as the system equilibrates isotopically as well as chemically. In this case, equilibrium is reached when the two phases of interest lie on the same fractionation line. This is critically important, and the most rigorous way to prove equilibrium in these experiments. The utility of the three-isotope method is in the ability to trace the isotopes as the experiments progress. Without an isotope tracer it would be more difficult to prove that isotopic equilibrium has been reached or to know

whether the different components had exchanged isotopes at all.

For iron isotopes, the utility of the three-isotope method is not only in the ability to trace the isotopes as the experiments progress but also to make sure that contamination or loss of Fe has not occurred. Iron is very difficult to contain within a capsule, as it is highly reactive. By spiking the experiments there is better control on mass balance and it can be determined if the experiment was truly a closed system. Proving that the isotopes have come to equilibrium is the most crucial part of conducting the high temperature experiments. If the isotopes are mixed but have not equilibrated yet then the fractionation will be determined as zero. The majority of the iron in the experiment is in the metallic phase, so any contamination from the metal to the silicate (which cannot be seen by the eye) will cause the fractionation factor to seem as though it is near zero. However, with the three-isotope exchange method this can be easily seen.

Once a study has determined an experimental fractionation factor between metal and silicate, it can be used to determine variables such as the composition of the light element in the core [e.g., *Shahar et al.*, 2009], the oxygen fugacity conditions during accretion [*Georg and Shahar*, 2014], the building blocks of the planets [*Fitoussi and Bourdon*, 2012], and the temperature of formation of meteorites [*Ziegler et al.*, 2010].

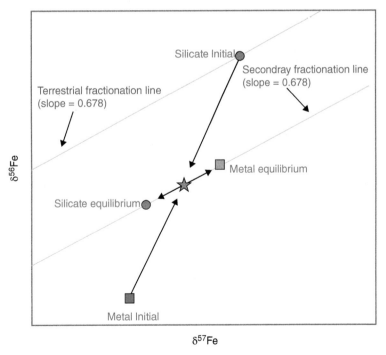

Figure 6.7 A schematic of the three-isotope exchange method shown for iron isotopes. The silicate (green circles) starts on the terrestrial fractionation line and moves to lighter values, while the metal (red squares) is spiked off the terrestrial fractionation line and moves towards heavier values. The star represents the bulk value of the system. Equilibrium is reached when both minerals reach the secondary fractionation line.

Determining if there is a pressure effect on isotope fractionation is a large feat and one that should be approached simply and carefully. *Polyakov and co-workers* [2005, 2007, 2009] have pioneered the use of NRIXS (Nuclear Resonant Inelastic X-ray Scattering) synchrotron data to obtain vibrational properties of minerals from which isotopic fractionation factors can be calculated. In NRIXS an x-ray beam produced by a synchrotron is used as a photon source to excite the ^{57}Fe nucleus in the target material. NRIXS data can be used to derive reduced partition function ratios (β-factors) from which equilibrium isotopic fractionation factors can be calculated: $\delta_A - \delta_B = 1000*(\ln \beta_A - \ln \beta_B)$, where A and B are two different phases of interest. This technique has been widely used by the mineral physics community to investigate seismic velocities and phonon density of states of high-pressure minerals [*Sturhahn et al.*, 1995] but is relatively new to the isotope geochemistry community.

Dauphas et al. [2012] derived a force constant approach to calculate the equilibrium fractionation factors from the NRIXS spectrum. They also found that the high-pressure results used by *Polyakov* [2009] were inaccurate due to the truncation of the high-energy tail during the data acquisition. The authors point out that existing NRIXS data at high pressure cannot be used for determining isotope fractionation factors because the data are noisy and not of high enough quality. They suggest that longer acquisition times are necessary to determine the partial density of states (PDOS) of the phases at high pressure with particular emphasis on the high-energy tail. Although the technique is still new to geochemistry, it has enormous potential to provide equilibrium fractionation factors that do not suffer from the usual experimental difficulties. As shown in *Murphy et al.* [2013] there is a pronounced effect on the ^{57}Fe β-factor as pressure increases.

Once a fractionation is found it is not straightforward to determine its cause. For Fe isotopes this is particularly evident. Hints that there should be Fe isotope fractionation between metal that becomes the core and silicate is found in comparisons of the Fe isotope ratios of iron meteorites, representing core material of some body, and chondrites, representing the isotopic composition prior to differentiation [*Poitrasson et al.*, 2005]. Thus, there seems to be a discrepancy between the experimental work, theoretical calculations, and natural record. Many studies have shown that, in both low and high temperature environments, the oxidation state (Fe^{3+}, Fe^{2+}) exerts strong control on the magnitude of the iron isotope fractionation [*Williams et al.*, 2004; *Shahar et al.*, 2008]. Theoretical and experimental [*Hill et al.*, 2009] results at low temperature have shown that speciation (bond partner and coordination in a solution) also regulates mass fractionation of Fe. In high-temperature solids or melts, the ligand to which Fe is bound also should affect the

iron isotopic fractionation [*Schauble et al.*, 2004]. Teasing out which of these mechanisms along with others plays a role in the fractionation found can only be accomplished with more experiments aimed at understanding each mechanism on its own.

An example of this is illustrated in a recent study [*Shahar et al.*, 2014] in which the authors focused on the effect of the Fe isotope fractionation between metal and silicate as a function of the sulfur content in the liquid Fe alloy. That work showed that high-temperature equilibrium iron isotopic fractionation between metal and silicate increases when sulfur is incorporated into the metal portion of the system. The similarity in Fe content of the glasses and metals in all the experiments suggests that all experiments have nearly the same fO_2 as demonstrated by a simple calculation comparing the mole fraction of FeO in the silicate to the mole fraction of Fe in the metal. Although fO_2 has been argued to be the leading mechanism for iron isotopic fractionation at high temperature, in this case, the addition of a light element into the metal seems to be causing the fractionation. The nature of the light element in an Fe-Ni core is directly related to the mode of formation, for example, the pressure of differentiation and the redox conditions present. It is therefore possible that the light elements (e.g., S, C, O, H, Si) in the core of a differentiated body control the iron isotope fractionation during differentiation, however, many more experiments need to be conducted before that is known.

6.4. WHAT DOES THIS ALL MEAN? WHERE DO WE GO FROM HERE?

The extensive dataset existing for the partitioning of slightly and moderately siderophile elements strongly indicate that metal-silicate equilibration must have occurred at high pressures and high temperatures (45–60 GPa, 3000–4000 K), and therefore support core formation that involve a deep magma ocean. Such conclusion supports substantial re-equilibration of impactor cores of smaller bodies [*Rudge et al.*, 2010]. There are a number of issues that can be addressed to allow a distinction between the different hypotheses for accretion and core formation and that can increase the precision and accuracy of these models. For instance, (1) an assessment of activity coefficients of siderophile elements in the metallic phase is required to investigate the effect of light elements on partitioning [e.g., *Tuff et al.*, 2011; *Siebert et al.*, 2013]; (2) modeling of core formation needs to include notably a parameter of partial re-equilibration of metal as well as mass balance calculation to evaluate reliable compositions of equilibrating core and mantle at HP-HT [*Rudge et al.*, 2010; *Rubie et al.*, 2011]; (3) Potential changes of valence state of some siderophile elements in the silicate melt (e.g., W, Nb) with the large redox variation during core formation needs to be investigated [*Wade et al.*, 2013]; and (4) most importantly, obtaining partitioning data covering the full range of P-T conditions of core formation for a large number of siderophile elements (using LH-DAC) is key for providing accurate regressions of partitioning that would include potential non-linear pressure effect notably [*Bouhifd and Jephcoat*, 2011; *Siebert et al.*, 2012, 2013]. Regression coefficients used to model the partitioning of siderophile elements are currently too imprecise to allow a statistically conclusive discrimination between core formation models [*Walter and Cottrell*, 2014]. Advances in modeling of core formation require better isolation of the independent variables that affect partition coefficients through systematic experimental works.

The field of high pressure and temperature stable isotope geochemistry has come a long way in the last ten years though there have only been a handful of studies to date. There is great potential to use the stable isotope ratios found in meteorites and Earth rocks as tracers of the conditions in which the Earth formed. It has become clear that there are resolvable equilibrium fractionations even at high pressures and temperatures, however, we are still very far from the conditions in which core formation occurred. A systematic and thorough experimental program is warranted in which the effects of pressure, temperature, oxygen fugacity, and composition are investigated for each isotopic system. This would allow extrapolation to the conditions required for understanding the Earth's differentiation or other planetary bodies.

ACKNOWLEDGMENTS

Careful reviews by M. A. Bouhifd and an anonymous reviewer helped improve the manuscript. We thank M. J. Walter and J. Badro for editorial handling. JS acknowledges financial support from the PNP research program from INSU, the French National Research Agency (ANR project VolTerre, grant no. ANR-14-CE33-0017-01) and the UnivEarthS Labex program at Sorbonne Paris Cité (ANR-10-LABX-0023 and ANR-11-IDEX-0005 02). AS acknowledges support from NSF EAR 1321858.

REFERENCES

Albarède, F. (2009), Volatile accretion history of the terrestrial planets and dynamic implications, *Nature*, *461*, 1227–1233.

Allègre, C. J., G. Manhes, and E. Lewin (2001), Chemical composition of the Earth and the volatility control on planetary genetics. *Earth Planet. Sci. Lett.*, *185*, 49–69.

Allègre, C. J., J. P. Poirier, E. Humler, and A. W. Hofmann (1995), The chemical composition of the Earth, *Earth Planet. Sci. Lett.*, *134*, 515–526.

Andrault, D., N. Bolfan-Casanova, G. Lo Nigro, M.A. Bouhifd, G. Garbarino, and M. Mezouar (2011), Solidus and liquidus profiles of chondritic mantle: Implication for melting of the Earth across its history, *Earth Planet. Sci. Lett.*, *304*, 251–259.

Antonangeli, D., J. Siebert, J. Badro, D. L. Farber, G. Fiquet, G. Morard, and F. J. Ryerson (2010), Composition of the Earth's inner core from high-pressure sound velocity measurements in Fe-Ni-Si alloys, *Earth Planet. Sci. Lett.*, *295*, 292–296.

Anzellini, S., A. Dewaele, M. Mezouar, P. Loubeyre, and G. Morard (2013), Melting of Iron at Earth's Inner Core Boundary Based on Fast X-ray Diffraction, *Science*, *340*, 464–466.

Armytage, R. M. G., R. B. Georg, P. S. Savage, H. M. Williams, and A. N. Halliday (2011), Silicon isotopes in meteorites and planetary core formation, *Geochimica et Cosmochimica Acta.*, *75*, 3662–3676.

Asahara, Y., D. J. Frost, and D. C. Rubie (2007), Partitioning of FeO between magnesiowüstite and liquid iron at high pressures and temperatures: Implications for the composition of the Earth's outer core, *Earth Planet. Sci. Lett.*, *257*, 435–449.

Badro, J., G. Fiquet, F. Guyot, E. Gregoryanz, F. Occelli, D. Antonangeli, and M. D'Astuto (2007), Effect of light elements on the sound velocities in solid iron: Implications for the composition of Earth's core, *Earth Planet. Sc. Lett.*, *254*, 233–238.

Becker, H., M. F. Horan, R. J. Walker, S. Gao, J.-P Lorand, and R. L. Rudnick (2006), Highly siderophile element composition of the Earth's primitive upper mantle: constraints from new data on peridotite massifs and xenoliths, *Geochim. Cosmochim. Acta.*, *70*, 4528–4550.

Bennett, N. R. and J. M. Brenan (2013), Controls on the solubility of Rhenium in silicate melt: Implications for the osmium isotopic composition of the mantle, *Earth Planet Sci. Lett.*, *361*, 320–332.

Bennett, N. R., J. M. Brenan, and K. T. Koga (2014), The solubility of platinum in silicate melt under reducing conditions: Results from experiments without metal inclusions, *Geochim. Cosmochim. Acta.*, *133*, 422–442.

Bigeleisen, J. and M. G. Mayer (1947), Calculation of equilibrium constants for isotopic exchange reactions, *J. Chem. Phys.*, *15*, 261.

Birch, F. (1964), Density and composition of mantle and core, *Journal of geophysical research*, *69*, 4377–4388

Borg, L. E., J. N. Connelly, M. Boyet, and R. W. Carlson (2011), Chronological evidence that the Moon is either young or did not have a global magma ocean, *Nature*, *1–4*, doi:10.1038/nature10328.

Bouhifd, M. A. and A. P. Jephcoat (2003), The effect of pressure on partitioning of Ni and Co between silicate and iron-rich metal liquids: a diamond anvil cell study, *Earth Planet. Sci. Lett.*, *209*, 245–255.

Bouhifd, M. A. and A. P. Jephcoat (2011), Convergence of Ni and Co metal-silicate partition coefficients in the deep magma-ocean and coupled silicon-oxygen solubility in iron melts at high pressures, *Earth Planet. Sci. Lett.*, *307*, 341–348.

Boujibar, A., D. Andrault, M. A. Bouhifd, N. Bolfan-Casanova, J. L. Devidal, and N. Trcera, (2014), Metal–silicate partitioning of sulphur, new experimental and thermodynamic constraints on planetary accretion, *Earth Planet. Sci. Lett.*, *391*, 42–54.

Brenan, J. M. and W. F. McDonough (2009), Core formation and metal-silicate fractionation of osmium and iridium from gold, *Nature Geosciences*, *2*, 798–801.

Burkhardt, C., R. C. Hin, T. Kleine, and B. Bourdon (2014), Evidence for Mo isotope fractionation in the solar nebula and during planetary differentiation, *Earth and Planetary Science Letters*, *391*, 201–211.

Capobianco, C. J., M. J. Drake, and J. De'Aro (1999), Siderophile geochemistry of Ga, Ge, and Sn: Cationic oxidation states in silicate melts and the effect of composition in iron-nickel alloys, *Geochim. Cosmochim. Acta.*, *63*, 2667–2677.

Caro, G., B. Bourdon, J. Birck, and S. Moorbath (2006), High-precision 142Nd/144Nd measurements in terrestrial rocks: Constraints on the early differentiation of the Earth's mantle, *Geochimica et Cosmochimica Acta.*, *70*, 164–191.

Cartier, C., T. Hammouda, M. Boyet, M. A. Bouhifd, and J. L. Devidal (2014), Redox control of the fractionation of niobium and tantalum during planetary accretion and core formation. *Nature Geoscience*, doi:10.1038/ngeo2195.

Chabot, N. L. and C. B. Agee (2003), Core formation in the Earth and Moon: new experimental constraints from V, Cr, and Mn, *Geochim. Cosmochim. Acta.*, *67*, 2077–2091.

Chabot, N. L., D. S. Draper, and C. B. Agee (2005), Conditions of Earth's core formation: constraints from nickel and cobalt partitioning, *Geochim. Cosmochim. Acta.*, *69*, 2141–2151.

Chabot, N. L., W. F. McDonough, J. H. Jones, S. A. Saslow, R. D. Ash, D. S. Draper, and C. B. Agee (2011), Partitioning behavior at 9 GPa in the Fe–S system and implications for planetary evolution, *Earth and Planetary Science Letters*, *305*, 425–434.

Chakrabarti, R. and S. B. Jacobsen (2010), Silicon isotopes in the inner Solar System: Implications for core formation, solar nebular processes and partial melting, *Geochimica et Cosmochimica Acta.*, *74*, 6921–6933.

Chambers, J. E. (2003), Planet formation. *Treatise on Geochemistry*, 1–17.

Clayton, R. N., J. R. Goldsmith, K. J. Karel, T. K. Mayeda, and R. C. Newton (1975), Limits on the effect of pressure on isotopic fractionation, *Geochimica et Cosmochimica Acta.*, *39*, 1197–1201.

Corgne, A., S. Keshav, B. J. Wood, W. F. McDonough, and Y. Fei (2008), Metal-silicate partitioning and constraints on core composition and oxygen fugacity during Earth accretion, *Geochim. Cosmochim. Acta.*, *72*, 574–589.

Corgne, A., J. Siebert, and J. Badro (2009), Oxygen as a light element: A solution to single-stage core formation, *Earth Planet. Sci. Lett.*, *288*, 108–114.

Cottrell, E. and D. Walker (2006), Constraints on core formation from Pt partitioning in mafic silicate liquids at high temperatures, *Geochim. Cosmochim. Acta.*, *70*, 1565–1580.

Cottrell, E., M. J. Walter, and D. Walker (2009), Metal-silicate partitioning of tungsten at high pressure and temperature: Implications for equilibrium core formation in Earth, *Earth Planet. Sci. Lett.*, *281*, 275–287

Craddock, P. R., J. M. Warren, and N. Dauphas (2013), Abyssal peridotites reveal the near-chondritic Fe isotopic composition of the Earth, *Earth and Planetary Science Letters*, *365*, 63–76.

Dauphas, N., M. Roskosz, E. E. Alp, D. C. Golden, C. K. Sio, F. L. H. Tissot, M. Hu, J. Zhao, L. Gao, and R. V. Morris (2012), A general moment NRIXS approach to the determination of equilibrium Fe isotopic fractionation factors: application to goethite and jarosite, *Geochimica et Cosmochimica Acta.*, *94*, 254–275.

Drake, M. J. and K. Righter (2002), Determining the composition of the Earth, *Nature*, *416*, 39–44.

Dreibus, G. and H. Palme (1996), Cosmochemical constraints on the sulfur content in the Earth's core, *Geochim. Cosmochim. Acta.*, *60*, 1125–1130.

Elkins-Tanton, L. T. (2012), Magma Oceans in the Inner Solar System. *Annu. Rev. Earth Planet. Sci.*, *40*, 113–139.

Ertel, W., M. J. Walter, M. J. Drake, and P. J. Sylvester (2006), Experimental study of platinum solubility in silicate melt to 14 GPa and 2273 K: Implications for accretion and core formation in Earth, *Geochim. Cosmochim. Acta.*, *70*, 2591–2602.

Fiquet, G., A. L. Auzende, J. Siebert, A. Corgne, H. Bureau, H. Ozawa, and G. Garbarino (2010), Melting of peridotite to 140 gigapascals, *Science*, *329*, 516–518.

Frost, D. J., C. Liebske, F. Langenhorst, C. A. McCammon, R. G. Tronnes, and D. C. Rubie (2004), Experimental evidence for the existence of iron-rich metal in the Earth's lower mantle, *Nature*, *428*, 409–412.

Fischer-Gödde, M., H. Becker, and F. Wombacher (2010), Rhodium, gold and other highly siderophile element abundances in chondritic meteorites, *Geochim. Cosmochim. Acta.*, *74*, 356–379.

Fischer-Gödde, M., H. Becker, and F. Wombacher (2011), Rhodium, gold and other highly siderophile elements in orogenic peridotites and peridotite xenoliths, *Chem. Geol.*, *280*, 365–383.

Fitoussi, C., B. Bourdon, T. Kleine, F. Oberli, and B. C. Reynolds (2009), Si isotope systematics of meteorites and terrestrial peridotites: implications for Mg/Si fractionation in the solar nebula and for Si in the Earth's core, *Earth and Planetary Science Letters*, *287*, 77–85.

Georg, R. B., A. N. Halliday, E. A. Schauble, and B. C. Reynolds (2007). Silicon in the Earth's Core, *Science*, *447*, 1102–1006.

Georg, R. B. and A. Shahar (2014), Constraining conditions of core formation and the building blocks of Earth from the extended silicon system, Submitted to American Mineralogist.

Gessmann, C. K. and D. C. Rubie (2000), The origin of the depletions of V, Cr and Mn in the mantles of the Earth and Moon, *Earth Planet. Sci. Lett.*, *184*, 95–107.

Greenwood, R. C., I. A. Franchi, A. Jambon, and P. C. Buchanan (2005), Widespread magma oceans on asteroidal bodies in the early solar system, *Nature*, *435*, 916–918.

Hill, P. S., E. A. Schauble, A. Shahar, E. Tonui, and E. D. Young (2009), Experimental studies of equilibrium iron isotope fractionation in ferric aquo-chloro complexes, *Geochimica et Cosmochimica Acta.*, *73*, 2366–2381.

Hin, R. C., M. W. Schmidt, and B. Bourdon (2012), Experimental evidence for the absence of iron isotope fractionation between metal and silicate liquids at 1 GPa and 1250–1300 °C and its cosmochemical consequences, *Geochimica et Cosmochimica Acta.*, *93*, 164–181.

Hin, R. C., C. Fitoussi, M. W. Schmidt, and B. Bourdon (2014), Experimental determination of the Si isotope fractionation factor between liquid metal and liquid silicate, *Earth and Planetary Science Letters*, *387*, 55–66.

Hin, R. C., C. Burkhardt, M. W. Schmidt, B. Bourdon, and T. Kleine (2013), Experimental evidence for Mo isotope fractionation between metal and silicate liquids, *Earth and Planetary Science Letters*, *379*, 38–48.

Hoering, T. C. (1961), The physical chemistry of isotopic substances: The effect of physical changes on isotope fractionation, Carnegie I. *Wash.*, *60*, 201–204.

Horita, J., T. Driesner, and D. R. Cole (1999), Pressure effect on hydrogen isotope fractionation between brucite and water at elevated temperatures, *Science*, *286*, 1545–1547.

Jana, D. and D. Walker (1997), The influence of silicate melt composition on distribution of siderophile elements among metal and silicate liquids, *Earth Planet. Sci. Lett.*, *150*, 463–472.

Javoy, M., E. Kaminski, F. Guyot, D. Andrault, C. Sanloup, M. Moreira, S. Labrosse, A. Jambon, P. Agrinier, A. Davaille, and C. Jaupart (2010), The chemical composition of the Earth: Enstatite chondrite models, *Earth and Planetary Science Letters*, *293*, 259–268.

Joy, H. W. and W. F. Libby (1960), Size effects among isotopic molecules, *J. Chem. Phys.*, *33*, 1276.

Kegler, P., A. Holzheid, D. J. Frost, D. C. Rubie, R. Dohmen, and H. Palme (2008), New Ni and Co metal-silicate partitioning data and their relevance for an early terrestrial magma ocean, *Earth Planet. Sci. Lett.*, *268*, 28–40.

Kempl, J., P. Z. Vroon, E. Zinngrebe, and W. van Westrenen (2013), Si isotope fractionation between Si-poor metal and silicate melt at pressure-temperature conditions relevant to metal segregation in small planetary bodies, *Earth and Planetary Science Letters*, *368*, 61–68.

Kleine, T., H. Palme, K. Mezger, and A. N. Halliday (2005), Hf-W chronometry of lunar metals and the age and early differentiation of the Moon, *Science*, *310*, 1671–1674.

Labidi, J., P. Cartigny, and M. Moreira (2014), Non-chondritic sulphur isotope composition of the terrestrial mantle, *Nature*, *501*, 208–211.

Lazar, C., E. D. Young, and C. E. Manning (2012), Experimental determination of equilibrium nickel isotope fractionation between metal and silicate from 500 °C to 950 °C, *Geochimica et Cosmochimica Acta.*, *86*, 276–295.

Li, J. and C. B. Agee (1996), Geochemistry of mantle-core differentiation at high pressure, *Nature*, *381*, 686–689.

Li, J. and C. B. Agee (2001), The effect of pressure, temperature, oxygen fugacity and composition on the partitioning of nickel and cobalt between liquid Fe-Ni-S alloy and liquid silicate: Implications for the Earth's core formation, *Geochim. Cosmochim. Acta.*, *65*, 1821–1832.

Lodders, K. (2003), Solar system abundances and condensation temperatures of the elements, *Astrophys. J.*, *591*, 1220–1247.

Mann, U., D. J. Frost, and D. C. Rubie (2009), Evidence for high-pressure core-mantle differentiation from the metal-silicate partitioning of lithophile and weakly-siderophile elements, *Geochim. Cosmochim. Acta.*, *73*, 7360–7386.

Matsuhisa, Y., J. R. Goldsmith, and R. N. Clayton (1979), Oxygen isotope fractionation in the system quartz-albite-anorthite-water, *Geochimica et Cosmochimica Acta.*, *43*, 1131–1140.

Matthews, A., J. R. Goldsmith, and R. N. Clayton (1983), Oxygen isotope fractionations involving pyroxenes: The calibration of mineral-pair geothermometers, *Geochimica et Cosmochimica Acta.*, *47*, 631–644.

McDonough, W. F. (2003), Compositional model for the Earth's core. In *Treatise on Geochemistry*, vol. 2 (Ed. R. W. Carlson), Elsevier-Pergamon, Oxford. pp. 547–568.

McDonough W. F. and S. S. Sun (1995), The composition of the Earth, *Chem. Geol.*, *120*, 223–253.

Morard, G., D. Andrault, N. Guignot, J. Siebert, G. Garbarino, and D. Antonangeli (2011), Melting of Fe-Ni-Si and Fe-Ni-S alloys at megabar pressure: implications for the core-mantle boundary temperature, *Phys. Chem. Minerals*, *38*, 767–776.

Moskovitz, N. and E. Gaidos (2011), Differentiation of planetesimals and the thermal consequences of melt migration, *Meteoritics & Planetary Science*, *46*, 903–918.

Moynier, F., Q. Z. Yin, and E. Schauble (2011), Isotopic Evidence of Cr Partitioning into Earth's Core, *Science*, *331*, 1417–1420.

Murphy, C. A., J. M. Jackson, and W. Sturhahn (2013), Experimental constraints on the thermodynamics and sound velocities of hcp-Fe to core pressures, *J. Geophys. Res.*, *118*, 1999–2016.

Mysen, B. O., D. Virgo, and F. A. Seifert (1982), The structure of silicate melts: Implications for chemical and physical properties of natural magma, *Rev. Geophys. Space Phys.*, *20*, 353–383.

Nielsen, S. G., J. Prytulak, B. J. Wood, and A. N. Halliday (2014), Vanadium isotopic difference between the silicate Earth and meteorites, *Earth and Planetary Science Letters*, *389*, 167–175.

O'Neill, H. S. C. (1991), The origin of the Moon and the early history of the Earth—A chemical model. Part 2: The Earth, *Geochim. Cosmochim. Acta.*, *55*, 1159–1172.

Palme, H. and H. S. C. O'Neill (2003), Cosmochemical estimates of mantle composition, In *Treatise on Geochemistry*, vol. *2* (Ed. R. W. Carlson), Elsevier-Pergamon, Oxford. pp. 1–38.

Poitrasson, F., S. Levasseur, and N. Teutsch (2005), Significance of iron isotope mineral fractionation in pallasites and iron meteorites for the core-mantle differentiation of terrestrial planets, *Earth and Planetary Science Letters*, *234*, 151–164.

Poitrasson, F., M. Roskosz, and A. Corgne (2009), No iron isotope fractionation between molten alloys and silicate melt to 2000C and 7.7 GPa: Experimental evidence and implications for planetary differentiation and accretion, *Earth and Planetary Science Letters*, *278*, 376–385.

Polyakov, V. B. (2009), Equilibrium iron isotope fractionation at core-mantle boundary conditions, *Science*, *323*, 912–914.

Polyakov, V. B. and N. N. Kharlashina (1994), Effect of pressure on equilibrium isotopic fractionation, *Geochimica et Cosmochimica Acta.*, *21*, 4739–4750.

Polyakov, V. B., S. D. Mineev, R. Clayton, G. Hu, and K. S. Mineev (2005), Determination of tin equilibrium isotope fractionation factors from synchrotron radiation experiments, *Geochimica et Cosmochimica Acta.*, *69*, 5531–5536.

Polyakov, V. B., R. N. Clayton, J. Horita, and S. D. Mineev (2007), Equilibrium iron isotope fractionation factors of minerals: Reevaluation from the data of nuclear inelastic resonant X-ray scattering and Mössbauer spectroscopy, *Geochimica et Cosmochimica Acta.*, *71*, 3833–3846.

Ricolleau, A., Y. Fei, A. Corgne, J. Siebert, and J. Badro (2011), Oxygen and silicon contents of Earth's core from high pressure metal-silicate partitioning experiments, *Earth and Planetary Science Letters*, *310*, 1–13.

Richter, F., N. Dauphas, and F. Teng (2009), Non-traditional fractionation of non-traditional isotopes: Evaporation, chemical diffusion and Soret diffusion, *Chemical Geology*, *258*, 92–103.

Righter K., M. J. Drake, and G. Yaxley (1997), Prediction of siderophile element metal-silicate partition coefficients to 20 GPa and 2800°C: the effects of pressure, temperature, oxygen fugacity and silicate and metallic melt compositions, *Phys. Earth Planet. Inter.*, *100*, 115–134.

Righter K., M. Humayun, and L. Danielson (2008), Partitioning of palladium at high pressures and temperatures during core formation, *Nat. Geosci.*, *1*, 321–323

Righter, K., C. King, L. R. Danielson, K. Pando, and C.T. Lee (2011), Experimental determination of the metal/silicate partition coefficient of Germanium: Implications for core and mantle differentiation, *Earth Planet. Sci. Lett.*, *304*, 379–388.

Righter, K. (2011), Prediction of metal-silicate partition coefficients for siderophile elements: An update and assessment of PT conditions for metal-silicate equilibrium during accretion of the Earth, *Earth Planet. Sci. Lett.*, *304*, 158–167.

Rose-Weston, L., J. M. Brenan, Y. Fei, R. A. Secco, and D. J. Frost (2009), Effect of pressure, temperature, and oxygen fugacity on the metal-silicate partitioning of Te, Se, and S: Implications for earth differentiation, *Geochim. Cosmochim. Acta.*, *73*, 4598–4615.

Rubie, D. C., C. K. Gessmann, and D. J. Frost (2004), Partitioning of oxygen during core formation on the Earth and Mars, *Nature*, *429*, 58–61.

Rubie, D. C., F. Nimmo, and H.J. Melosh (2007), Formation of Earth's core, In: Stevenson, D. (Ed.), *Treatise on Geophysics*: Volume 9>—Evolution of the Earth, Elsevier.

Rubie, D. C., D. J. Frost, U. Mann, Y. Asahara, F. Nimmo, K. Tsuno, P. Kegler, A. Holzheid, and H. Palme (2011), Heterogeneous accretion, composition and core-mantle differentiation of the Earth, *Earth and Planetary Science Letters*, *301*, 31–42.

Rudge, J. F., T. Kleine, and B. Bourdon (2010), Broad bounds on Earth's accretion and core formation constrained by geochemical models, *Nat. Geosci.*, *3*, 439–443.

Ryerson, F. J. (1985), Oxide solution mechanisms in silicate melts—systematic variations in the activity-coefficient of SiO_2, *Geochim. Cosmochim. Acta.*, *49*, 637–649.

Sanloup, C., W. van Westrenen, R. Dasgupta, H. Maynard-Casely, and J. P. Perrillat (2011), Compressibility change in iron-rich melt and implications for core formation models, *Earth Planet. Sci. Lett.*, *306*, 118–122.

Sanloup, C., J. W. E. Drewitt, Z. Konopkova, P. Dalladay-Simpson, D. M. Morton, N. Rai, W. van Westrenen, and W. Morgenroth (2013), Structural change in molten basalt at deep mantle conditions, *Nature*, *503*, 104–107.

Savage, P. S. and F. Moynier (2013), Silicon isotopic variation in enstatite meteorites: Clues to their origin and Earth-forming material, *Earth and Planetary Science Letters*, *361*, 487–496.

Schauble, E.A. (2004), Applying stable isotope fractionation theory to new systems, In C. M. Johnson, B. Beard, and F. Albarède, Eds. Geochemistry of non-traditional stable isotopes. *MSA Reviews in Mineralogy & Geochemistry, v. 55*, Ch. 3, 65–111.

Schmitt, W., H. Palme, and H. Wänke (1989), Experimental determination of metal/silicate partition coefficients for P, Co, Ni, Cu, Ga, Ge, Mo, and W and some implications for the early evolution of the Earth, *Geochim. Cosmochim. Acta.*, *53*, 173–185.

Shahar, A., E. D. Young, and C. E. Manning (2008), Equilibrium high-temperature Fe isotope fractionation between fayalite and magnetite: An experimental calibration, *Earth and Planetary Science Letters*, *268*, 330–338.

Shahar, A., K. Ziegler, E. D. Young, A. Ricolleau, E. A. Schauble, and Y. Fei (2009), Experimentally determined Si isotope fractionation between silicate and metal and implications for Earth's core formation, *Earth Planet. Sc. Lett.*, *288*, 228–234.

Shahar, A., V. J. Hillgren, E. D. Young, Y. Fei, C. A. Macris, and L. Deng (2011), High-Temperature Si isotope fractionation between iron metal and silicate, *Geochimica et Cosmochimica Acta.*, *75*, 7688–7697.

Shahar, A., J. Mesa-Garcia, L. A. Kaufman, V. J. Hillgren, M.F. Horan, and T.D. Mock (2014), Sulfur-controlled iron isotope fractionation experiments of core formation in planetary bodies, *Geochim. Cosmochim. Acta.*, in press.

Siebert, J., A. Corgne, and F. J. Ryerson (2011), Systematics of metal-silicate partitioning for many siderophile elements applied to Earth's core formation, *Geochim. Cosmochim. Acta.*, *75*, 1451–1489.

Siebert, J., V. Malavergne, F. Guyot, R. Combes, and I. Martinez (2004), The Behaviour of sulphur in metal-silicate core segregation experiments under reducing conditions, *Physics of the Earth and Planetary Interiors*, *143–144*, 433–443.

Siebert, J., J. Badro, D. Antonangeli, and F. J. Ryerson (2012), Metal-silicate partitioning of Ni and Co in a deep magma ocean, *Earth Planet. Sc. Lett.*, *321–322*, 189–197.

Siebert, J., J. Badro, D. Antonangeli, and F. J. Ryerson (2013), Terrestrial accretion under oxidizing conditions, *Science*, *339*, 1194–1197.

Simon, J. I. and D. J. DePaolo (2010), Stable calcium isotopic composition of meteorites and rocky planets, *Earth and Planetary Science Letters*, *289*, 457–466.

Sturhahn, W., T. Toellner, E.E. Alp, X. Zhang, M. Ando, Y. Yoda, S. Kikuta, M. Seto, C.W. Kimball, and B. Dabrowski (1995), Phonon density of states measured by inelastic nuclear resonant scattering, *Physical Review Letters*, *74*, 3832–3835.

Taylor, G. J. (1992), Core formation in asteroids, *J. Geophys. Res.*, *97*, 14717–14726.

Teng, F.-Z., W.-Y Li, S. Ke, B. Marty, N. Dauphas, S. Huang, F.-Y. Wu, and A. Pourmand (2010), Magnesium isotopic composition of the Earth and chondrites, *Geochimica et Cosmochimica Acta.*, *74*, 4150–4166.

Thibault, Y. and M. J. Walter (1995), The influence of pressure and temperature on the metal-silicate partition coefficients of nickel and cobalt in a model C1 chondrite and implications for metal segregation in a deep magma ocean, *Geochim. Cosmochim. Acta.*, *59*, 991–1002.

Tsuno, K., D. J. Frost, and D. C. Rubie (2013), Simultaneous partitioning of silicon and oxygen into the Earth's core during early Earth differentiation, *Geophys. Res. Lett.*, *40*, 1–12.

Tuff, J., B. J. Wood, and J. Wade (2011), The effect of Si on metal-silicate partitioning of siderophile elements and implications for the conditions of core formation, *Geochimica et Cosmochimica Acta.*, *75*, 673–690.

Urey, H. C. (1947), The thermodynamic properties of isotopic substances, *Journal of the Chemical Society* (Resumed), 562–581.

Wade, J. and B. J. Wood (2005), Core formation and the oxidation state of the Earth, *Earth Planet. Sci. Lett.*, *236*, 78–95.

Wade, J., B. J. Wood, and J. Tuff (2012), Metal-silicate partitioning of Mo and W at high pressures and temperatures: Evidence for late accretion of sulphur to the Earth, *Geochim. Cosmochim. Acta.*, *85*, 58–74, 2012.

Wade, J., B. J. Wood, and C. A. Norris (2013), The oxidation state of tungsten in silicate melt at high pressures and temperatures, *Chemical Geology*, *335*, 189–193.

Walker, R. J. (2009), Highly siderophile elements in the Earth, Moon and Mars: update and implications for planetary accretion and differentiation, *Chemie der Erde– Geochemistry*, *69*, 101–125.

Walter, M. J. and Y. Thibault (1995), Partitioning of tungsten and molybdenum between metallic liquid and silicate melt, *Science*, *270*, 1186–1189.

Walter, M. J. and E. Cottrell (2013), Assesing uncertainty in geochemical models for core formation in Earth, *Earth and Planet. Sci. Lett.*, *365*, 165–176.

Walter, M. J., H. E. Newsom, W. Ertel, and A. Holzheid (2000), Siderophile elements in the Earth and Moon. Metal/silicate partitioning and implication for core formation. In *Origin of the Earth and Moon* (Eds. R. M. Canup and K. Righter), University of Arizona Press, Tucson. pp. 265–289.

Walter, M. J. (2004), Early Earth differentiation, *Earth Planet. Sci. Lett.*, *225*, 253–269.

Wang, Z. C. and H. Becker (2013), Ratios of S, Se and Te in the silicate Earth require a volatile-rich late veneer, Nature, *499*, 328–331.

Wänke, H. (1981), Constitution of terrestrial planets, Phil. Trans. Roy. Soc. London, *A303*, 287–302.

Wasson, J. T. (1985), *Meteorites: Their Record of the Early Solar- System History*, Freeman, New York.

Weyer, S., A. D. Anbar, G. P. Brey, C. Münker, K. Mezger, and A. B. Woodland (2005), Iron isotope fractionation during planetary differentiation, *Earth and Planetary Science Letters*, *240*, 251–264.

Wiechert, U. and A. Halliday (2007), Non-chondritic magnesium and the origins of the inner terrestrial planets, *Earth and Planetary Science Letters*, *256*, 360–371.

Williams, H. M., B. J. Wood, J. Wade, D. J. Frost, and J. Tuff (2012) Isotopic evidence for internal oxidation of the Earth's mantle during accretion, *Earth and Planetary Science Letters*, *321–322*, 54–63.

Williams, H. M., C. A. McCammon, A. H. Peslier, A. N. Halliday, N. Teutsch, S. Levasseur, and J. P. Burg (2004), Iron isotope fractionation and the oxygen fugacity of the mantle, *Science*, *304*, 1656–1659.

Wood, B.J., M. J. Walter, and J. Wade (2006), Accretion of the Earth and segregation of its core, *Nature*, *441*, 825–833.

Wood, B. J., J. Wade, and M. R. Kilburn (2008), Core formation and the oxidation state of the Earth: Additional constraints from Nb, V and Cr partitioning, *Geochim. Cosmochim. Acta.*, *72*, 1415–1426.

Young, E. D., E. Tonui, C. E. Manning, E. Schauble, and C. A. Macris (2009), Spinel-olivine magnesium isotope thermometry in the mantle and implications for the Mg isotopic composition of Earth, *Earth and Planetary Science Letters*, *288*, 524–533.

7

Fractional Melting and Freezing in the Deep Mantle and Implications for the Formation of a Basal Magma Ocean

Stéphane Labrosse[1], John W. Hernlund[2], and Kei Hirose[2]

ABSTRACT

Processes that operated in the early Earth have largely been erased or overprinted by subsequent evolution. However, some traces may persist in the deep Earth as imaged by seismology. Large-scale features with reduced seismic velocities are simply explained as variations of composition, and small-scale ultra-low velocity zones are explained by the presence of Fe-rich material that may be partially molten. Both can originate from fractional crystallization of an originally thick basal magma ocean (BMO). Many questions are raised by this scenario regarding properties of melts with various compositions, in particular the partition coefficients between melt and crystals of various elements and their relative densities. After reviewing recent progress on both the structure of the lower mantle and the mineral physics associated with partial melting/freezing of silicates at high pressure, we discuss several ways in which a BMO can be produced.

We argue that, in most cases, independently of whether a melt of composition similar to that of the bulk mantle is more or less dense than crystals in equilibrium, the compositional evolution of both the magma and the solid should lead to the formation of a dense, Fe-rich BMO whose subsequent slow evolution would explain some features of the present lower mantle.

7.1. INTRODUCTION

Most scenarios for the formation and early evolution of the Earth favor a hot start as a result of several processes: giant impacts thought to have punctuated the latest stages of the formation of the Earth; core formation and the associated release of gravitational energy [*Flasar and Birch*, 1973; *Ricard et al.*, 2009]; radioactive heating by short-lived isotopes in the planetesimals and embryos that accreted to form the Earth [*Šrámek et al.*, 2012]. The

giant impact thought to have resulted in the formation of the moon likely melted a large fraction of the silicate Earth [e.g., *Canup*, 2012; *Ćuk and Stewart*, 2012; *Nakajima and Stevenson*, 2014]. The ensuing crystallization of the magma ocean is thought to happen in a few tens of millions of years because of the enormous radiative surface heat flow, even in presence of a steam atmosphere [*Abe*, 1997; *Sleep*, 2000].

Conventional models of magma ocean evolution [*Abe*, 1997; *Solomatov*, 2000] postulate that crystallization occurred from the bottom up, because the increase of the liquidus with pressure is steeper than the isentropic temperature gradient in silicate liquid [*Zhang and Herzberg*, 1994], the temperature that is thought to prevail in the vigorously stirred convecting magma ocean. However, recent high-pressure measurements of the melting curves

[1] *Laboratoire de géologie de Lyon, ENS de Lyon, Université Lyon-1, CNRS, Lyon, France*
[2] *Earth-Life Science Institute, Tokyo Institute of Technology, Meguro, Tokyo, Japan*

The Early Earth: Accretion and Differentiation, Geophysical Monograph 212, First Edition.
Edited by James Badro and Michael Walter.

of mantle minerals suggest that this may not be true all the way to the pressure of the core-mantle boundary. Independently of where crystallization starts from a fully molten state, the resulting evolution depends on the comparative density of melt and crystals. For an univariant freezing process, crystals are usually denser than melt (although the case of water-ice shows this is not always true), but the increase of pressure makes the volume difference decrease because melts are generally more compressible than solids. For the complex chemistry of mantle minerals, liquid and solids in equilibrium have different compositions, and in particular the liquids generally have a larger Fe/Mg ratio than solids. The diminished volume difference and enhanced mass of melt owing to Fe-enrichment can cause melt to become more dense than solids at pressures of the deepest mantle. Recent developments in mineral physics regarding the melting temperatures, partitioning of Fe, and comparative densities of liquids and solids in the pressure range of the lower mantle will occupy a large part of this review.

The possibility of liquids to be denser than solids opens a wealth of different scenarios for the evolution of magma oceans. *Labrosse et al.* [2007] proposed that the mantle crystallized from an unspecified mid-depth both upward and downward, with two magma oceans co-existing in the early Earth, the surface (usual) magma ocean, and one at the bottom of the mantle, the basal magma ocean (BMO). They showed that the crystallization of the BMO occurs on a timescale that is comparable to the age of the Earth. The reason for this is threefold. Firstly, the full crystallization of the basal magma ocean requires the temperature to decrease to the lowest point in the phase diagram, which, for complex compositions like the mantle, could mean a total cooling of order 1000K. Secondly, because of the low viscosity of the magma ocean, only a minute super-isentropic temperature gradient is permitted across the BMO [*Ulvrová et al.*, 2012], and the cooling of the BMO requires the same cooling of the core. Finally, typical values of the heat flow that solid-state convection in the mantle can sustain at its base is of the order 10TW, and the thermal inertia of the core makes it cool down by 1000K over the age of the Earth at that rate. One major implication of such a slow evolution of the BMO is that traces of its early stage may still be seen in the present day mantle and the traces were a large part of the motivation for the original BMO model. Arguments in favor of this basal magma ocean will first be reviewed along two main lines of thought: the observation of the present state of the lower mantle and the long-term evolution of the core-mantle system.

Since the existence of the basal magma ocean is mostly supported by backward extrapolation of present state observations, it leaves the question of its formation open. This chapter concludes with a discussion of several possibilities for its formation process with particular attention paid to the constraints needed from mineral physics as to which of these is most likely and the implication it would have on the early Earth processes.

7.2. STRUCTURE OF THE LOWER MANTLE AS IMAGED BY SEISMOLOGY

Unlike the Moon, Earth's surface does not strictly exhibit any well-preserved rocks of Hadean age, a characteristic that is sometimes considered to define this period of Earth's history. The virtual absence of Hadean domains in the crust suggests significant mass exchange between the surface and interior of the early Earth, consistent with the occurrence of plate tectonics, or at least mobile lid mantle convection, in the early Earth [e.g., *Hopkins et al.*, 2008; *Condie and Pease*, 2008; *Turner et al.*, 2014]. For example, it has been proposed that fractional crystallization of a completely molten mantle could produce dense iron-rich material near the surface, leading to a Rayleigh-Taylor instability that buries this material above the CMB [*Hess and Parmentier*, 1995; *Elkins-Tanton et al.*, 2005; *Elkins-Tanton*, 2008]. Numerous geochemical arguments have been used to suggest that some early crust or other material has been sequestered in the deep mantle as a "hidden reservoir" formed within the first ~100 Myrs of Earth's history [e.g., *Boyet and Carlson*, 2005; *Tolstikhin and Hofmann*, 2005; *Mukhopadhyay*, 2012]. Removal of a portion of Earth's early crust and lithosphere into the deeper mantle suggests the possibility that geologic structures from the Hadean may be found in the Earth's deep interior. Thus, seismological examination of Earth's deep interior may constitute an important element of Hadean structural geology, and it is important to understand how major features could be inherited from processes operating in the early Earth.

7.2.1. Observations

The Earth's lowermost mantle is characterized by a variety of seismically heterogeneous features at scales ranging from <1 km up to ~1,000 km. These features exhibit seismic shear S and compressional P wave speed (V_S and V_P, respectively) variations ranging from ~–50% to ~+5% relative to average mantle. Some similar features exhibit different characteristics in different locations, rendering a simple interpretation and/or single causative mechanism implausible. However, the same degree of complexity is also present in the shallow mantle and crust, thus the geological precedent is not inconsistent with observations of the mantle's other major boundary layer. Here we give a brief overview of structures in the deep mantle, followed by the ways in which they can be interpreted in light of high CMB heat flow, large secular core cooling, and fractional crystallization of an early dense basal magma ocean. A more exhaustive survey of deep mantle structure and its relationship to Earth's dynamical evolution can be found in recent reviews [*Lay et al.*, 2004; *Garnero and McNamara*, 2008; *Lay and Garnero*, 2011; *Hernlund and McNamara*, 2015]. At the largest scales the deep mantle can be roughly divided into two provinces on the basis of normal modes and long-period seismic wave speed variations: (1) regions exhibiting moderate to fast seismic wave speeds, and (2) regions exhibiting very slow to moderately slow seismic wave speeds [*Dziewonski et al.*, 1977]. The faster regions underlie past subduction zones, and are interpreted to host ponded subducted lithosphere that is cooler than the surrounding mantle [*Ricard et al.*, 1993]. The slower regions underlie the Pacific and Africa, and are usually called "large low shear velocity provinces" (LLSVP), although in the past they have also been called "super-plumes," "mega-plumes," and "high bulk modulus structures" [*Hernlund and McNamara*, 2015]. The distinction between the faster and slower regions is most apparent in V_S tomography models but is less striking in V_P models. This is readily apparent when both V_S and V_P models are plotted on the same color scale (Figure 7.1), although they can also be delineated using more objective criteria. For example, *Hernlund and Houser* [2008] showed that distributions of S velocity at a given depth become nearly bimodal in the lowermost ~700 km of the mantle, a feature that is present in all S models but is not present in P models. These anomalously slow patches can be separated from the normal variations, and clearly distinguish the anomalously slow V_S regions. Additionally, *Lekic et al.* [2012] performed a cluster analysis of radial profiles of V_S models and found that they naturally segregate into the two dominant (faster and slower) clusters, with a boundary corresponding to the edges of the LLSVP.

In addition to the large-scale decoupling in the strength of S and P seismic wave speed, there is additional evidence that the edges of the LLSVP represent the boundaries of compositionally distinct domains, such as reflections and/or large lateral gradients in seismic wave speed near the edges [e.g., *Luo et al.*, 2001; *Wen*, 2001; *Wang and Wen*, 2004; *To et al.*, 2005; *Ford et al.*, 2006; *Sun et al.*, 2007]. When coupled to modes that have slight sensitivity to density, the LLSVP also appear to exhibit a density variation that is decoupled from the velocity variation [*Ishii and Tromp*, 2004]. Combined geodynamical and seismic velocity inversions also suggest that these features exhibit an anomalous buoyancy that does not scale proportionally to V_S variations [*Simmons et al.*, 2006]. Many hotspots are geographically correlated with the edges of LLSVP [*Thorne et al.*, 2004]. When geographically shifted to the same reference frame by plate motion history models, the loci of large igneous provinces (LIP) erupted over the past several hundred million years are also clustered around the edges of the LLSVPs [*Wen*, 2006; *Burke et al.*, 2008], although there still exist uncertainties in paleo-longitude, and it is not a unique interpretation.

Some concerns persist regarding the robustness of S and P model differences, owing to the relatively better coverage of S waves traversing the deep mantle. These concerns were especially important before 2002, when P coverage was, by itself, relatively poor and P tomography models produced by different researchers exhibited significant differences [*Becker and Boschi*, 2002]. Data coverage in P models has improved significantly in subsequent years, and resolution tests of more recent models confirm that differences in S and P distributions cannot be simply

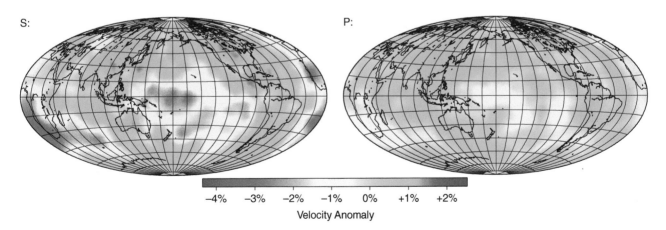

Figure 7.1 Maps of S (left) and P (right) wave velocity anomalies at 2,800 km depth. Identical color scales have been used for both images to emphasize the relative variations. The small amplitude of variations in P wave velocity relative to the larger variation in S suggests that the bulk modulus increases relative to the shear modulus. This is sometimes referred to as the "anti-correlation in bulk and shear wave speeds."

explained as an artifact of relatively poor coverage in *P* models [*Hernlund and Houser*, 2008]. Addition of a new *P*-diffracted phase that has good coverage at the base of the mantle to tomographic models also supports the main finding that *P* velocity variations are less dramatic than *S* velocity [*Manners*, 2008]. It has also been proposed that long-period *P* waves passing through the slow regions could be subject to wavefront healing, an effect that would distort the seismic waveforms in predictable ways and could cause *P* models to under-resolve the amplitude of *P* wave speed variations if relative travel times were obtained by aligning peaks of seismic waveforms [*Schuberth et al.*, 2012]. However, the predicted waveform distortion has not been demonstrated in actual *P* traces, and differences in *P* and *S* models persist even when travel times are determined by means other than aligning peaks [e.g., *Houser et al.*, 2008].

In addition to the ~1,000km scale LLSVP, there also exist anomalous thin (~10 km) layers at the CMB exhibiting *P* wave decrease of ~10% and *S* wave speed decreases in the range 10–40% [*Garnero and Helmberger*, 1996]. These extreme regions extend laterally as patches of order ~100–1000km wide and are called ultra-low velocity zones (ULVZ). In some locations ULVZ have sharp upper boundaries, producing *S* to *P* conversions that suggest a density increase of ~10% relative to average mantle [*Reasoner and Revenaugh*, 2000; *Rost and Revenaugh*, 2003; *Rost et al.*, 2005, 2006]. The scale and anomalous properties of ULVZ are analogous to those of the Earth's crust (but with opposite density change), and it could be called the "crust of the CMB."

7.2.2. Interpretations

The existence of large-scale compositional domains induced by intermittent layers of material more dense than the mantle by ~2% offers the simplest mechanism to explain seismic observations of LLSVP [e.g., *Tackley*, 2002]. The separation in *S* velocities between faster and slower regions in the deep mantle is also enhanced in the lowermost several hundred km of the mantle by the occurrence of $(Mg,Fe)SiO_3$-post-perovskite, which becomes stable at cooler temperatures and is expected to exhibit a V_S increase of 1–2% while having little effect on V_P [*Wookey et al.*, 2005], consistent with seismic observations of discontinuities arising from a post-perovskite transition [*Hutko et al.*, 2008]. If the CMB itself is too hot to stabilize post-perovskite, then this phase could be entirely confined to laterally limited lens-like structures that traverse only the cooler regions of the Earth's deep mantle, outside the LLSVP [*Hernlund et al.*, 2005; *Nakagawa and Tackley*, 2005; *Monnereau and Yuen*, 2007]. Although a post-perovskite lens-like structure has been suggested inside the Pacific LLSVP [*Lay et al.*,

2006], perhaps stabilized at higher temperatures by an iron-rich composition [*Mao et al.*, 2004], more recent work demonstrates that the structures are more complicated than previously thought [*Thorne et al.*, 2013] and therefore unlikely to be explained by post-perovskite.

Although thermo-chemical piles are a favored mechanism for explaining LLSVP, numerical models of mantle convection including dense layers produces "tent-shaped" structures that are inconsistent with the observed steep-sided morphology and focusing of plumes at the edges [*Tackley*, 1998; *McNamara and Zhong*, 2004]. However, steep sides can be produced in thermo-chemical convection experiments in a "dome-like" regime [*Davaille*, 1999a, 1999b; *Le Bars and Davaille*, 2002; *Gonnermann et al.*, 2004]. Focusing of plume activity at the edges of compositionally distinct piles in addition to the steep sides are reproduced in mantle convection models if a distinct bulk modulus consistent with seismological inferences is employed [*Tan et al.*, 2011; *Samuel and Bercovici*, 2006]. In these models, the density anomaly associated with LLSVP-material varies with depth owing to a lower compressibility relative to the surrounding mantle.

For geochemical reasons it has been suggested that LLSVP could form by segregation of MORB from its harzburgitic residue following subduction of lithospheric slabs into the deep mantle, an effect that has been induced in numerical mantle convection models by imposing a significant buoyancy difference between these two compositions [e.g., *Christensen and Hofmann*, 1994; *Coltice and Ricard*, 1999; *Xie and Tackley*, 2004; *Nakagawa and Tackley*, 2004; *Brandenburg and van Keken*, 2007]. However, these models typically include a MORB crust that is up to one order of magnitude thicker than actual oceanic crust, and the degree of segregation between MORB and harzburgite residue decreases significantly when more realistic thicknesses are employed [*Li and McNamara*, 2013]. On the other hand, segregation of thin MORB crust is enhanced in the presence of pre-existing "primitive" thermo-chemical piles, by ingestion of basalt veneers into the tops of the piles, perhaps providing a mechanism for a variety of geochemical heterogeneity that can be stored over long-time scales in thermo-chemical piles [*Li et al.*, 2014].

Compositional variations compatible with LLSVP have been explored in mineral physics models of the deep mantle, and suggest that an increased bulk modulus and decreased shear modulus are most easily explained by a coupled enrichment in iron and silicon, such as an increase in $FeSiO_3$ in bridgmanite [$(Mg,Fe)SiO_3$- perovskite] [*Kiefer et al.*, 2002; *Stixrude and Lithgow-Bertelloni*, 2005; *Cobden et al.*, 2009]. Mantle convection modeling demonstrates that a dense chemical pile will also be anomalously hot [e.g., *Tackley*, 2002], thus the effects of both temperature and composition need to be untangled.

The combined influence of temperature and composition consistent with dynamical and mineral physics models has recently been considered by *Deschamps et al.* [2012], who find that a coupled enrichment in Fe by ~3% and $(Mg,Fe)SiO_3$-perovskite by ~20% yields the best over all agreement with seismological constraints, in particular compared to the recycled MORB hypothesis. Another important finding is that MORB crust may increase the *S* velocity of thermo-chemical piles in a way that is difficult to reconcile with seismological observations. The increase in *S*-velocity of MORB arises from the increased modal fraction of $CaSiO_3$-perovskite expected in a MORB-based mineral assemblage at conditions of the deep mantle, and therefore relies on the robustness of elasticity models for this particular phase [*Stixrude and Lithgow-Bertelloni*, 2005].

The large seismic velocity changes exhibited by ULVZs are suggestive of the presence of up to ~10% partial melt [*Williams and Garnero*, 1996; *Berryman*, 2000; *Lay et al.*, 2004; *Hier-Majumder*, 2008]. However, a density increase of ~10% cannot plausibly be produced by 10% partial melting of ordinary mantle rock, thus ULVZs are likely compositionally distinct from the overlying mantle. It is usually thought that an iron-oxide rich composition is needed to satisfy all of the constraints, whether or not they are additionally partially molten [*Wicks et al.*, 2010]. It is also possible that partial melting may be facilitated in ULVZs owing to their compositionally distinct state, or that they may be partially molten in some regions and melt-free in others [*Nomura et al.*, 2011]. Although the tendency of melts to segregate through solids raises concerns about the long-term stability of partial melt inside ULVZs [*Hernlund and Tackley*, 2007], it has been shown that even a dense melt can be sustained in ULVZs by viscous stirring-induced pressure gradients sufficient to support/sustain topography of ULVZs [*Hernlund and Jellinek*, 2010].

It is also useful to consider the relationship between ULVZs and LLSVPs. The existence of chemically distinct large-scale LLSVPs predicts that steady horizontal mantle flow across the CMB converges at the edges of LLSVPs, in such a way that ULVZs will tend to cluster and thicken near the edges of LLSVPs [*Hernlund and Tackley*, 2007], analogous to the way continental crust is thickened in convergent tectonic regions near Earth's surface. This is consistent with the locations of several large ULVZ provinces underneath hot spots [*Williams et al.*, 1998; *Helmberger et al.*, 1998; *Cottaar and Romanowicz*, 2012] and the hot spot-LLSVP-edge correlation mentioned previously [*Thorne et al.*, 2004; *Burke et al.*, 2008]. Mantle convection models containing both LLSVPs and ULVZs as distinct compositional anomalies confirm this behavior, although ULVZs may sometimes be found in the middle of larger chemical piles as a transient feature, such as

following the merger of two smaller piles into a larger one [*McNamara et al.*, 2010]. As discussed in greater detail below, ULVZs and LLSVPs have also been proposed to represent fractionated dense melt at the base of the mantle, suggesting that their origins are intimately linked with one another [*Labrosse et al.*, 2007].

In summary of this section, we argue that the deep Earth offers a view on processes happening in the early times of the Earth. The important improvements of imaging techniques in the last decades uncover a complexity that is akin to that known in the shallow silicate Earth, which should not be surprising. In particular, several independent lines of evidence call for variations of composition in the lower mantle. Each of these observations, taken independently from the others, might be explained without invoking composition effects but considering them together, it becomes difficult to avoid it entirely. ULVZs themselves require compositional variations to explain their high density and patchy nature and even if they can be explained by Fe-rich solids [*Mao et al.*, 2006; *Wicks et al.*, 2010], their melting temperature is likely low enough that they would in fact be partially molten at the conditions prevailing now at the CMB or in the recent past. Moreover, a good way to create the composition of solids that have been proposed to explain ULVZs is by fractional crystallization. Alternatively, it has been proposed to be formed by chemical interaction between the core and the lower mantle, but this has been suggested to result in depletion of the lower mantle in FeO and its dissolution in the core [*Ozawa et al.*, 2009; *Frost et al.*, 2010]. This would likely not produce the Fe-rich solid composition required to explain ULVZs. The compositional variations explaining the LLSVPs have been termed crypto-continents by *Stacey* [1991], and the analogy can be pursued by calling ULVZs crypto-crust (Fig. 7.2). The origin of the lower mantle structure is rooted in its long-term evolution. As discussed above, if the density of subducted MORBs at high pressure is large enough, the crustal material could accumulate at the base of the mantle, participating in the formation of LLSVPs. Some of those minerals could also remelt by coming in contact with the hot core [*Andrault et al.*, 2014] and join ULVZs. Alternatively, both LLSVPs and ULVZs could result from the slow fractional crystallization of a magma, which gives the basal magma ocean model [*Labrosse et al.*, 2007]. The main reason to support this latter option comes from the presence at the base of the mantle of partially molten regions or highly fusible solids that form ULVZs and the necessary important cooling of the core as evidenced by the existence of a geomagnetic field for at least 3.5 Gyr. But before showing how it works, we need to review the recent results on the melting and freezing behavior in the deep mantle.

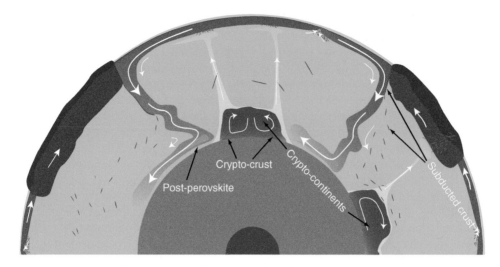

Figure 7.2 Cartoon model of the present mantle. The large low shear velocity provinces can be seen as the lower mantle equivalent of the continental lithosphere (crypto-continents) and ULVZs as the CMB equivalent of crust (crypto-crust). In any case, their scale and seismological contrast are very similar to their shallow counterparts.

7.3. LOWER MANTLE MELTING PHASE DIAGRAMS AND PARTIAL MELT COMPOSITIONS

7.3.1. Melting Phase Relations

Melting phase relations of lower mantle materials have been examined by multi-anvil experiments only up to 33 GPa [*Ito et al.*, 2004]. *Trønnes and Frost* [2002] reported a detailed phase diagram for pyrolitic mantle above solidus temperatures to 24.5 GPa, demonstrating that (Mg,Fe)O ferropericlase is the liquidus (the first crystallizing) phase [*Hirose and Fei*, 2002] followed by $MgSiO_3$-rich and $CaSiO_3$ perovskite (Ca-perovskite virtually disappears on the solidus). They also showed that the interval between the solidus and the liquidus temperatures is narrow, less than 200 K in the uppermost lower mantle. The melting experiments performed by *Ito et al.* [2004] showed that the liquidus phase changes from ferropericlase to bridgmanite above 31 GPa, suggesting that partial melt continuously extracted from peridotitic rock becomes progressively Mg-rich and Si-poor with increasing depth in the lower mantle. More recently, *Liebske and Frost* [2012] studied melting phase relations in the MgO–$MgSiO_3$ binary system, demonstrating that the eutectic composition becomes more MgO-rich with increasing pressure from 16 to 26 GPa. On the basis of thermodynamical models using the melting curves of MgO and $MgSiO_3$ they predicted that the eutectic composition becomes further MgO-enriched with Mg/Si = 1.5 (in molar abundance) at 60 GPa, but is almost constant at higher pressures to 135 GPa (Figure 7.3).

Melting phase relations and partial melt compositions have also been explored under deep lower mantle conditions

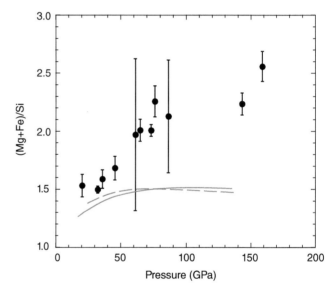

Figure 7.3 Change in (Mg+Fe)/Si molar ratio of melt formed by 46 –77% partial melting of a natural pyrolitic mantle [*Tateno et al.*, 2014]. The eutectic melt compositions in the MgO-$MgSiO_3$ binary system are shown by the solid curve (thermodynamical modeling by *Liebske and Frost* [2012]) and broken curve (first-principles cal- culations by *de Koker et al.* [2013]).

in recent diamond-anvil cell (DAC) studies. *Fiquet et al.* [2010] carried out melting experiments on peridotite to 140 GPa, in order to determine melting phase relations as well as solidus and liquidus temperatures by X-ray diffraction measurements. They found that bridgmanite is the liquidus phase between 36 GPa and 135 GPa (i.e., CMB pressure). The crystallization sequence changed from (1) ferropericlase → bridgmanite → Ca-perovsite to

(2) bridgmanite → ferropericlase → Ca-perovsite above ~36 GPa, and further to (3) bridgmanite → Ca-perovsite → ferropericlase above ~60 GPa. Another study by *Tateno et al.* [2014] also measured melting phase relations in a natural peridotite (pyrolite) composition to 179 GPa using laser-heated DAC experiments combined with chemical and textural characterizations of recovered samples under the electron microprobe. The crystallization sequence they found is in agreement with the results by *Fiquet et al.* [2010]; the changes from (1) to (2) occurred above 34 GPa, and to (3) above ~70 GPa. The melting experiments on $(Mg_{0.89}Fe_{0.11})_2SiO_4$ performed by *Nomura et al.* [2011] to 159 GPa also demonstrated that the liquidus phase changed from ferropericlase to bridgmanite at 36 GPa, and bridgmanite (or post-perovskite) remained the first phase to crystallize at CMB pressure.

7.3.2. Partial Melt Composition

Chemical compositions of melts formed by partial melting (or fractional solidification) of lower-mantle materials were systematically studied by both *Nomura et al.* [2011] and *Tateno et al.* [2014]. The latter work on a natural pyrolitic mantle composition reported that (Mg+Fe)/Si molar ratio of partial melt increased progressively from 1.4 at 34 GPa to 1.6–1.7 at > 100 GPa (since these partial melts coexisted with bridgmanite, the chemical composition of incipient melt formed on the solidus of the $(Mg, Fe)O − (Mg, Fe)SiO_3$ system should be more (Mg, Fe)O-rich) (Figure 7.3). Such (Mg+Fe)/Si ratios are higher than those obtained by thermodynamical predictions of eutectic composition in the $MgO − MgSiO_3$ system by *Liebske and Frost* [2012] and the first-principles calculations by *de Koker et al.* [2013], because their composition lacks Fe.

The experimental results by *Nomura et al.* [2011] indicate that bridgmanite crystallizes from melts with a wide range of (Mg+Fe)/Si ratios; this is true even for Si-poor melt with (Mg+Fe)/Si ~2.5 at CMB conditions. It suggests that residual melts after partial crystallization in a basal magma ocean [*Labrosse et al.*, 2007] become SiO_2-poor and FeO-rich progressively, considering that iron is partitioned preferentially into melt [see Section 7.4 below]. *Tateno et al.* [2014] demonstrated that the melts formed by 46 to 77 wt% partial melting of a pyrolitic mantle material are ultrabasic in composition and become more depleted in SiO_2 and more enriched in FeO with increasing pressure.

7.3.3. Melting Temperature

7.3.3.1. Solidus Temperature of the Lower Mantle
The solidus temperature of the lower mantle is much less than that of single-phase bridgmanite or ferropericlase

[e.g., *de Koker et al.*, 2013] because it is a multi-component system and occurs as eutectic-like melting of three constituent phases. Melting curves of peridotitic and "chondritic mantle" materials (low Mg/Si ratio for the latter) have been determined by recent DAC experiments to the bottom of the mantle [*Fiquet et al.*, 2010; *Andrault et al.*, 2011; *Nomura et al.*, 2014]. *Fiquet et al.* [2010] reported the solidus temperature of peridotite to be 4180 ± 150K at the base of the mantle, and *Andrault et al.* [2011] found that partial melting of a "chondritic mantle" occurred above 4150 ± 150K at the CMB. Note that the "chondritic mantle" as defined by *Andrault et al.* [2011] (this is in fact a chondritic composition from which volatiles and most of the metal would have been removed) has a higher Fe/Mg ratio than peridotites as well as a lower (Mg + Fe)/Si. See *Thomas et al.* [2012] for a comparison of the different compositions. On the other hand, more recent experiments by *Nomura et al.* [2014] suggested that the solidus temperature of pyrolite is as low as 3570 ± 200K at CMB pressure (Fig. 7.4).

The lower solidus temperature obtained by *Nomura et al.* [2014] is probably in part because of the presence of 400 ppm H_2O in their sample (which could be a reasonable estimate for the water concentration of the lower mantle according to some geochemical arguments [*Dixon et al.*, 2002; *Marty*, 2012]). The discrepancy might also be partly attributed to the difference in criteria for the onset of melting in these DAC experiments.

Fiquet et al. [2010] used multiple melting criteria: (1) diffuse X-ray scattering, (2) disappearance of one

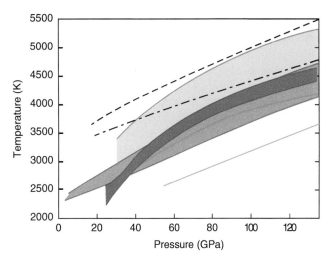

Figure 7.4 Melting (liquidus and solidus) curves of lower mantle materials compared with liquid isentropes [*Thomas and Asimow*, 2013]. Green with broken curve, peridotite mantle composition [*Fiquet et al.*, 2010]; red with dotted-broken curve, "chondritic" mantle composition [*Andrault et al.*, 2011]; purple, Fe-corrected MgO- MgSiO3 [*Liebske and Frost*, 2012]; blue, solidus curve of pyrolite [*Nomura et al.*, 2014].

of constituent solid phases, (3) rapid recrystallization, and (4) abrupt change in the relationship between laser output power and sample temperature. The diffuse X-ray scattering is characteristic of liquid and thus robust evidence for melting. It has been frequently used to determine the melting temperature of metals such as tantalum and iron [*Dewaele et al.*, 2010; *Anzellini et al.*, 2013]. It is, however, difficult to observe it from silicate melt, in particular when melt fraction is small, because it is composed of elements with relatively low atomic numbers [*Andrault et al.*, 2011]. Also, the melting (disappearance) of one of solid phases overestimates the solidus temperature. Indeed, Ca-perovskite is known to melt almost on the solidus below 30 GPa [e.g., *Trønnes and Frost*, 2002] but is present above solidus temperatures at higher pressures. Ferropericlase melts first above 60 GPa prior to Ca-perovskite as mentioned above [*Fiquet et al.*, 2010; *Tateno et al.*, 2014], possibly well above the solidus. The two-dimensional X-ray diffraction images often show the disappearance of a certain grain and its recrystallization in a short time scale, suggesting the presence of partial melt. However, a certain amount of melt fraction may be necessary for such rapid recrystallization to occur. The change in the relationship between laser-power output and sample temperature has traditionally been used to detect melting, because solid and liquid absorb laser power differently [*Zerr and Boehler*, 1993; *Shen and Lazor*, 1995]. A certain amount of melt is again likely necessary for a sample to change its laser-absorption property. *Andrault et al.* [2011] also used rapid recrystallization as a sign of solidus temperature. The most recent work by *Nomura et al.* [2014] performed textural characterization using three-dimensional X-ray computed tomography (CT) technique. The CT images provided the distribution of FeO in a sample; the enrichment in iron at the hottest part of the sample indicated the presence of partial melt [*Nomura et al.*, 2011; *Andrault et al.*, 2012, see Section 7.4]. As a result, Nomura and colleagues detected as low as 3 vol.% partial melt formed near the solidus temperature.

Liebske and Frost [2012] estimated the solidus (eutectic-like) temperature of Fe-bearing MgO–MgSiO$_3$ system to be 4400 K at 135 GPa by a combination of multi-anvil experiments up to 26 GPa and thermodynamical modeling for extrapolation (Figure 7.4). It is likely overestimated since neither the presence of CaSiO$_3$ perovskite nor the effects of other impurity elements such as aluminum, sodium, and water were considered.

7.3.3.2. Liquidus Temperature of the Lower Mantle

Experimental determination of liquidus temperature at lower mantle conditions has been challenging and certainly more difficult than determining solidus temperature. Indirect evidence such as the end of the

temperature-power plateau was previously used as the sign of liquidus temperature [*Andrault et al.*, 2011]. *Fiquet et al.* [2010] determined the liquidus temperature of peridotite to be about 5400 K at the CMB pressure, which is higher by 1200 K from the solidus (Figure 7.4). It could be overestimated because of the large temperature gradient always present in laser-heated DAC samples and the solid phase that could remain in the region that has the lowest temperature. *Andrault et al.* [2011] also reported the liquidus of a chondritic mantle to be 4725 K at the CMB, higher by 575 K than the solidus temperature.

The thermodynamical modeling by *Liebske and Frost* [2012] provided the liquidus temperature of bulk silicate Earth in the Fe-bearing MgO–MgSiO$_3$ system to be 4650 K at 135 GPa. It is only 250 K higher than the eutectic temperature, possibly because the bulk silicate Earth is close in composition to the eutectic point in the MgO–MgSiO$_3$ binary system. They also suggested that the chondritic mantle possesses lower Mg/Si ratio and thus has higher liquidus temperature. It is noted, however, that the eutectic composition predicted by *Liebske and Frost* [2012] (Mg/Si = 1.5 in molar ratio) is not consistent with the recent DAC experiments on the Fe-bearing system. *Nomura et al.* [2011] have demonstrated that melt with (Mg+Fe)/Si >2 coexisted with bridgmanite above 60 GPa, suggesting that an incipient melt should have even more Mg-rich composition.

The pressure-temperature slope of the liquidus curve with respect to that of liquid isentrope determines the depth from which crystallization would have started in a deep well-mixed magma ocean. First principles calculations by *Stixrude et al.* [2009] showed that the melting curve of MgSiO$_3$ perovskite has a gentle temperature/pressure slope in the middle to deep lower mantle, leading to the intersection with the liquid MgSiO$_3$ isentrope in the mid-lower mantle. More recently, *Thomas et al.* [2012] and *Thomas and Asimow* [2013] obtained the equations of state of liquid Mg$_2$SiO$_4$ and Fe$_2$SiO$_4$ from shock-wave experiments and calculated the isentropes for liquid peridotite and chondritic mantle composition. They found that the liquid isentrope for peridotite is tangent to its liquidus curve determined by *Fiquet et al.* [2010] at 85 GPa (Figure 7.4), suggesting that a deep magma ocean would have commenced crystallization at around 2000 km depth. On the other hand, the liquidus curve obtained for a chondritic mantle by *Andrault et al.* [2011] has a steeper pressure/temperature slope than its own liquid isentrope, indicating that crystallization for this composition would begin at the bottom of the mantle (Figure 7.4).

The depth for the onset of crystallization as well as melt-solid density relationship is crucial for the chemical evolution of a deep magma ocean. If solidification started in the mid-lower mantle and the crystals (the liquidus phase, bridgmanite) are relatively dense, then bridgmanite would have sunk and melted again at deeper levels, leading to a complex chemical evolution.

7.4. PARTITIONING OF FE AND DENSITY OF MELT

Since liquids are more compressible than solids, it has been speculated that silicate melt possibly becomes denser than solid in the deep lower mantle [*Stolper et al.*, 1981; *Agee*, 1998]. *Stixrude et al.* [2009] calculated that MgSiO₃ liquid remains buoyant with respect to bridgmanite with identical composition throughout the lower mantle. They argued, however, that a density crossover between melt and bridgmanite should occur at lower mantle pressures if the effect of iron partitioning is taken into account.

Nomura et al. [2011] examined iron partitioning between melt and coexisting bridgmanite to 159 GPa in $(Mg_{0.89}Fe_{0.11})_2SiO_4$ bulk composition. They found that the Fe-Mg distribution coefficient, $K_D = ([Fe_{Pv}]/[Mg_{Pv}])/([Fe_{melt}]/[Mg_{melt}])$, substantially decreased from 0.25 at 36 GPa to 0.06–0.08 above 76 GPa, corresponding to the deep lower mantle. A reduction in the K_D values means iron enrichment in melt, possibly leading to a density crossover between melt and coexisting bridgmanite above ~75 GPa corresponding to about 1800 km depth. More recently, *Tateno et al.* [2014] measured the Fe–Mg distribution coefficient between bridgmanite and melt in a natural pyrolitic mantle bulk composition. The results obtained by *Tateno and others* are in good agreement with those by *Nomura et al.* [2011], demonstrating that partial melt becomes strongly Fe-enriched in the deep lower mantle. This suggested the possibility that melt is denser and thus segregates downward to form a melt pool just above the CMB.

On the other hand, *Andrault et al.* [2012] measured the partition coefficient, $D_{Fe} = [Fe^{Pv}]/[Fe^{melt}]$, in a chondritic mantle bulk composition. The D_{Fe} (not K_D) values obtained by *Andrault and others* are 0.5–0.6 over a wide pressure range of the lower mantle, much higher than 0.1–0.2 obtained by *Nomura et al.* [2011] and *Tateno et al.* [2014]. *Andrault et al.* observed Fe-enrichment in partial melt with increasing pressure, but it was not as strong as that observed by *Nomura et al.* [2011] and *Tateno et al.* [2014]. *Andrault and others* argued that partial melt remains buoyant even at the CMB pressure and therefore migrates upward.

The discrepancy may be due to the difference in the method of chemical analysis rather than the difference in bulk composition. Both *Nomura et al.* [2011] and *Tateno et al.* [2014] prepared a cross section of a laser-heated hot spot in a sample and obtained the chemical compositions of quenched partial melt and neighboring bridgmanite under a field-emission-type electron microprobe (FE-EPMA) with high spatial resolution. On the other hand, *Andrault et al.* [2012] measured Fe contents of melts and bridgmanite based on X-ray fluorescence (XRF) analyses. In their analysis, the incident X-ray beam passed through the whole sample. Therefore, the XRF signals for partial melt may have come from both quenched melt pocket and surrounding Fe-poor bridgmanite owing to the three-dimensional distribution of quenched melt and bridgmanite. At the same time, the signals for bridgmanite also could have been contaminated by those from an unmelted part of the sample that was not depleted in Fe. Indeed, the microprobe data reported by *Andrault et al.* [2012] are in much better agreement with the results of *Nomura et al.* [2011] and *Tateno et al.* [2014].

The partitioning of iron between bridgmanite and melt primarily controls the buoyancy of partial melt in the lower mantle, but the Mg/Si ratio also plays a role. Lower mantle partial melts become more enriched in iron (high Fe/Mg ratio) and depleted in silicon at the solidus (for a given degree of partial melting) with increasing pressure (Figure 7.3), both of which increase their density. A sink/float relationship between bridgmanite and melt has been demonstrated as functions of Mg/(Mg+Fe) and SiO₂ content in melt [*Funamori and Sato*, 2010; *Thomas et al.*, 2012; *de Koker et al.*, 2013]. All of these models suggest that the first bridgmanite crystal formed from liquid pyrolite should have sunk, regardless of the following: (1) the iron partitioning between melt and bridgmanite and (2) the depth where crystallization started (see section 7.3.3.2 above). Nevertheless, residual melts formed after a certain degree of crystallization of bridgmanite became enriched in FeO and impoverished in SiO₂, which eventually results in a density crossover between melt and crystallizing solid. Indeed, after 50 –60% solidification of a fully molten magma ocean (liquid pyrolite) resulting melts obtained by *Tateno et al.* [2014] are denser than coexisting bridgmanite, at least above 88 GPa (corresponding to 2000-km depth), according to the models shown by *Funamori and Sato* [2010] and *Thomas et al.* [2012].

7.5. IMPLICATIONS FROM THE THERMAL EVOLUTION OF THE CORE

As discussed above, seismological observations of the present day mantle are best explained by the presence of localized pockets of partially molten material. Alternatively, a mixture of very Fe-rich magnesiowüstite and normal mantle may also explain the observations [*Wicks et al.*, 2010]. For either interpretation, this raises a question: how did the Earth evolve to this state? Some clues on this question can be obtained by considering the thermal evolution of the core and therefore past CMB temperatures.

The thermal evolution of the core can be studied by writing down its energy balance equation, which allows one to relate the CMB heat flow to different energy sources, in particular latent heat of inner core freezing, sensible heat, and compositional energy due to gradual enrichment of the outer core in light elements expelled at the inner core boundary [e.g., *Gubbins et al.*, 1979; *Braginsky and Roberts*, 1995; *Lister and Buffett*, 1995; *Labrosse et al.*, 1997; *Nimmo et al.*, 2004]. Solving this energy equation requires knowledge of the CMB heat flow at each time, which conversely can be constrained by using the entropy balance of the core [e.g., *Gubbins et al.*, 1979; *Braginsky and Roberts*, 1995; *Lister and Buffett*, 1995; *Labrosse*, 2003; *Gubbins et al.*, 2004; *Nimmo et al.*, 2004; *Labrosse*, 2015] and considering the requirement that the geomagnetic field must have existed for at least 3.5 Gyr [*Usui et al.*, 2009]. The energy and entropy balances of the core can be combined to obtain an efficiency equation in which the contribution of all energy sources maintaining the total dissipation are identified. Details of these equations can be found in the literature cited above and need not be replicated here. It is, however, useful to explain how core evolution calculations are performed and in particular how the energy and efficiency equations are parameterized. Two situations need to be considered: the present one with an inner core growing by crystallization and the period before the birth of an inner core. For the present situation, the thermal structure of the core can be parameterized by the inner core radius r_{IC}, and the energy balance and efficiency equation can both be written as function of r_{IC} and its growth rate:

$$Q_{CMB} = F\left(r_{IC}\right)\frac{dr_{IC}}{dt} + Q_R(t) \qquad (7.1)$$

$$\Phi + \Sigma = \eta_F F\left(r_{IC}\right)\frac{dr_{IC}}{dt} + \eta_R Q_R(t). \qquad (7.2)$$

Equation (7.1) states that the heat flow across the CMB, Q_{CMB}, is balanced by the sum of all energy terms that are linked to the inner core growth (secular cooling, latent heat, compositional energy), which are lumped into the function F, and the radiogenic heating Q_R, which depends explicitly on time and on the concentration in heat producing elements, K being commonly discussed (more on that below). Each of these energy terms contributes to maintain the total dissipation in the core with a different efficiency factor, η_X for term X, and this is reflected by equation (7.2).

The total dissipation contains three contributions from Ohmic, viscous, and thermal diffusion processes. The first two, related to motion in the core, are combined in Φ, while dissipation from thermal diffusion exists even without motion and is noted $\Sigma \equiv T_\Phi \int_V k\left(\nabla T / T\right)^2 dV$, with k the thermal conductivity, T the temperature, T_Φ

the effective temperature at which dissipation occurs and the integral is over the whole core volume V. The value of Φ, which is thought to be dominated by Ohmic dissipation [*Braginsky and Roberts*, 1995], cannot be measured directly because only a fraction of the magnetic field is known, the large-scale poloidal field at the CMB. Its estimates are based on scalings of results obtained in dynamo calculations [*Christensen and Aubert*, 2006; *Aubert et al.*, 2009] and in particular *Christensen* [2010] proposes a total value in the very wide range 1.1–15 TW. The lowest value gives the lowest bound on the CMB heat flow required to maintain the present core state. To this "useful" dissipation we must add the sink that dissipation associated to thermal diffusion represents. If the core is convecting, temperature should be very close to an isentropic profile, which can be computed accurately. This value is proportional to the thermal conductivity of core material, which has been upwardly revised in several recent studies [*Sha and Cohen*, 2011; *Pozzo et al.*, 2012; *de Koker et al.*, 2013; *Gomi et al.*, 2013]. The minimum value for the thermal conductivity of the core is now thought to be 85W/m/K at the top of the core and 130W/m/K at the ICB. With such high values of the thermal conductivity, $\Sigma > 2.5$TW. Figure 7.5 [from *Labrosse*, 2015] shows the different contribution to both energy and entropy balances as function of the present value of Φ. As already stated, the total ohmic dissipation in the core is poorly known but it must be strictly positive for any motion and dynamo action to occur. Taking $\Phi = 0$TW as an obvious minimum bound already brings a useful constraint, as seen on Figure 7.5, since maintaining only the dissipation associated with the isentropic temperature gradient requires a CMB heat flow around 7TW if no radioactivity is included. Including 200 ppm of K, producing about 1TW of heat, pushes the minimum CMB heat flow upward, because of the low efficiency of radiogenic heating [*Labrosse*, 2003; *Labrosse*, 2015].

This minimum value of Q_{CMB} is the one required to have a positive Ohmic dissipation in an isentropic core, but note that it is lower than the isentropic heat flow, which is about 13TW [*Pozzo et al.*, 2012; *Labrosse*, 2015]. In this case, a stably stratified layer is expected to develop at the top of the core, and its thickness is too large to go unnoticed. The increase of conductivity with depth is such that even for a CMB heat flow exactly isentropic, a thick stably stratified layer is expected [*Gomi et al.*, 2013; *Labrosse*, 2015]. This means that the isentropic CMB heat flow should in fact be considered as a minimum, and according to Figure 7.5 this means that the total dissipation is larger than 2.5TW.

With such a large value of the CMB heat flow, the age of the inner core is less than 700 Myr [*Labrosse*, 2015], and the thermodynamics of the core for the period preceding the existence of the inner core is what matters

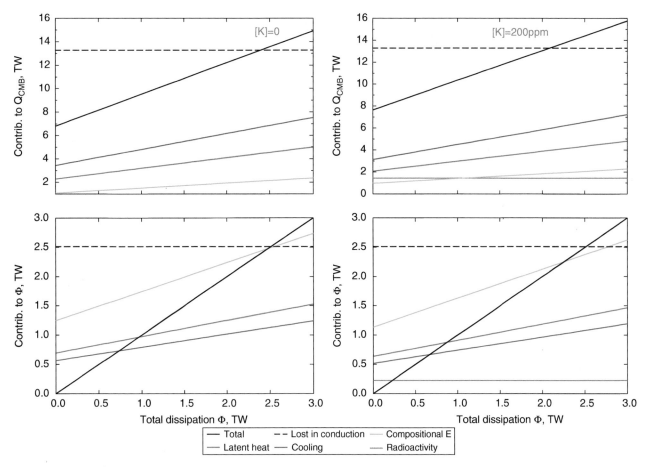

Figure 7.5 Present time contributions to the core energy balance (top) and dissipation (bottom) without radioactivity (left) or with 200ppm of K (right).

to constrain the early Earth. Two equations similar to equations (7.1) and (7.2) can be written with the rate of change of the central temperature playing the role of the inner core growth rate. The situation is simpler because only secular cooling and radioactivity can provide the energy required for the geodynamo, and the heat flow at the CMB must be larger than the isentropic value for thermal convection to occur. Figures 7.5 and 7.6 from *Labrosse* [2015] show the different contributions to the energy and entropy balances as function of Φ, with or without radioactivity. Again, the increase of conductivity with depth implies that a CMB heat flow just equal to the isentropic value is not sufficient to maintain a positive dissipation, and a heat flow in excess of 15 TW is necessary to explain the existence of the magnetic field for more than 3.5 Gyr [*Usui et al.*, 2009].

The constraint from the thermodynamics of the core on the heat flow at the CMB can be used to compute the evolution of the temperature of the core. Some trade-offs exist with radiogenic heating: the same CMB heat flow can be accommodated with a smaller cooling rate if the core contains radioactivity. Historically, potassium

has been proposed to enter the core and provide some heat to drive the geodynamo. However, experiments on the solubility of K in Fe [*Rama Murthy et al.*, 2003] show that less than 130 ppm can be expected in the Earth's core, and only if S is the main light alloying component. In addition, geochemical arguments exist against a significant amount of K in the core [*McDonough*, 2003]. The higher bound of 130 ppm of K would produce 0.8 TW at present, not enough to avoid strong cooling of the core with a CMB heat flow of order 15 TW [*Hirose et al.*, 2013].

Using the minimum constraint on the CMB heat flow, an isentropic value, *Gomi et al.* [2013] computed that for a present CMB temperature of 3750 K, its value 4.5 Gyr ago must have been at least 4500 K, well above the solidus of any reasonable bulk mantle composition (Section 7.3.3.1) and possibly above the liquidus of the Fe-rich silicate compositions invoked to explain the ULVZs. This means that, even if ULVZs are not partial melt, it is impossible to avoid the largely molten state of the lowermost mantle in the past in order to explain the maintenance of the geodynamo for the last 3.5 Gyrs.

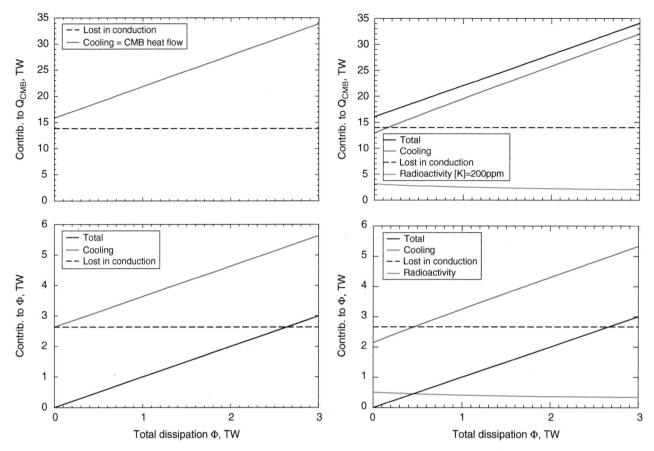

Figure 7.6 Contributions to the core energy balance (top) and total dissipation (bottom) as a function of the wanted total dissipation for the period just preceding the onset of the inner core crystallization, without radioactivity (left) or with 200 ppm of K (right).

7.6. FORMATION SCENARIOS

Given the potential importance of partial melt in the present lowermost mantle and the long-term thermal evolution of the deep Earth suggesting substantial basal melting of the mantle, we need to discuss the ways by which a BMO could have been formed. Time backward calculations from the present [*Labrosse et al.*, 2007] typically result in an initial BMO thickness of order 1000 km but tell us nothing about the way it formed in the early times. This section discusses a few aspects of this problem and proposes several options depending on the properties of melt and crystals.

We will start by assuming that giant impacts and core formation processes provide enough energy to melt a large fraction of the silicate Earth. Depending on the way these processes happen, the silicate Earth could be entirely molten and then crystallize, or only locally, which could still produce a global magma ocean by isostatic adjustment [*Tonks and Melosh*, 1993]. Following a cartoon scenario proposed by Stevenson [*Stevenson*, 1981, 1990], the depth of the magma ocean has often been estimated based on the

composition of mantle minerals in elements such as cobalt and nickel [e.g., *Li and Agee*, 1996; *Siebert et al.*, 2012] or molybdenum [*Burkemper et al.*, 2012] with pressures in the range 30 to 60 GPa and the corresponding liquidus temperature. However, such inferences on the depth of the magma ocean are model-dependent. Multiple events of magma ocean are expected and the resulting composition of the mantle in siderophile elements integrates all the processes happening at varying temperatures and pressures.

A giant impact could also produce a local melt pond that could trigger runaway core formation [*Ricard et al.*, 2009], which itself could melt a larger fraction of the silicate Earth. Where and how melting and freezing happens depends on the processes of Earth formation and differentiation and on the phase diagram of mantle minerals. The question of whether crystallization proceeds at equilibrium or fractionally was stressed in many papers [see *Solomatov*, 2007, for a recent review], in particular because evidence was lacking for differentiation of the Earth's mantle as drastic as that encountered on the Moon, for which the concept of a magma ocean was initially forged. However, recent developments both in geochemistry and geophysics argue for an initial differentiation of the mantle and we will assume that some fractional melting/crystallization did in fact happen, even though some of its results may have been erased following processing by mantle convection. The relative densities of melts and crystals then controls what happens when melting and freezing proceeds. Some of the possibilities are sketched here starting with global processes and then proceeding with local ones.

7.6.1. Freezing of a Global Magma Ocean

Considering first a global magma ocean, the position where crystallization starts can be determined by comparing the liquidus curve and the isentropic temperature profile, which is assumed to apply in the vigorously convecting magma ocean. Both the liquidus and the isentrope depend on composition, and crystallization could happen first at the bottom of the mantle or at some intermediate depth [*Thomas and Asimow*, 2013; *Thomas et al.*, 2012]. The densities of melt and crystals in equilibrium also depend on the composition of the magma and, at any given pressure, only an unlikely coincidence can make crystals neutrally buoyant. It means that, unless dense crystals start to form at the bottom of the magma ocean, the first crystals will always tend to move with respect to the coexisting magma, which would inevitably put them out of equilibrium and prone to remelting. Such dynamics has been proposed for the cores of small planets [e.g., *Stewart et al.*, 2007], and the resulting dynamics is not well understood yet.

Starting with the most classical scenario: bottom-up crystallization of dense solids, which according to *Thomas et al.* [2012], is the case with the "chondrite" liquidus of *Andrault et al.* [2011]. Even in this situation, the first crystals formed will likely be entrained by turbulent flow in the magma ocean and remelt. A slurry region would be formed in which crystals continuously form at the bottom and get mixed into the magma ocean. The thickness of the slurry layer, the volume fraction of crystals, and their typical sizes will grow as cooling of the magma ocean proceeds and eventually, when the critical fraction of crystals for rheological transition is reached [e.g., *Abe*, 1993], the high viscosity of the solid will limit its short-term deformation and inhibit entrainment, and a compacting layer of solid will stabilize at the bottom of the magma ocean. Assuming some fractional crystallization occurs, the upward progression of the freezing front induces a compositional evolution of both liquid and solid (Fig. 7.7). As discussed above (Section 7.4), Fe is partitioned preferentially in the liquid. The concentration of Fe in the residual liquid therefore increases as does that of the solid that crystallizes from it. In the cartoon of Figure 7.7, the liquid is depicted with a uniform color (i.e., composition) but the release of Fe-rich liquid at the bottom of the magma ocean would tend to slow down convection. Whether thermal convection driven by cooling at the surface is able to maintain the composition of the magma ocean in a well-mixed state depends on the balance between the buoyancy flux provided by cooling at the surface and that induced by adding dense material at the bottom. This balance can easily be worked

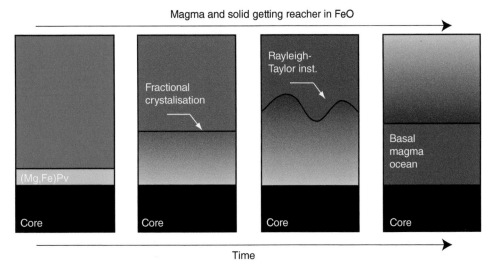

Figure 7.7 Formation of a basal magma ocean following the fractional crystallization of a global magma ocean from the bottom up. Fractional crystallization causes the composition of both liquid and solid to evolve, notably by increasing the Fe content (denoted by darker hues). The resulting solid mantle is unstably stratified and subject to a Rayleigh-Taylor instability. The large gravitational energy transformed into heat can lead to melting of the more fusible components, and produce an Fe-rich dense melt that sinks to the bottom.

out quantitatively if the phase diagram is precisely known. This is not critical for the present discussion, but it would clearly increase the extent of fractional crystallization. In any case, the gradual enrichment of the solid in FeO, inherited from the gradual enrichment of the magma, creates a gravitationally unstable situation in the solid that would eventually lead to overturn [*Hess and Parmentier*, 1995; *Elkins-Tanton et al.*, 2005]. In this process, gravitational energy is transformed into heat by viscous friction, and since all the solid is close to its melting point, a large part can remelt and the liquid hence formed should be rich in Fe and denser than the solid. This should then collect at the bottom of the mantle and form a basal magma ocean. This scenario clearly needs to be explored more quantitatively once a good parameterization of the phase diagram of the mantle is known and the numerical tools required to treat it in a self-consistent way have been developed. One drawback of this scenario is that the magma ocean at the surface is likely to strongly degas, unless thermal convection is not efficient enough to keep it well mixed or a thick and a stable crust is formed at the surface of the magma ocean, and the BMO thusly formed may not be able to explain noble gas geochemistry [*Coltice et al.*, 2011].

Consider now the case where crystallization starts somewhere in mid-mantle, which according to *Thomas et al.* [2012] is the case for the peridotite liquidus obtained by *Fiquet et al.* [2010]. Consider first the case of dense crystals, which is *a priori* the least favorable to the formation of a BMO and seems to apply to the peridotite composition, even though they are depleted in Fe compared to the magma [*Thomas et al.*, 2012]. Crystals would tend to fall in the magma ocean into a region where the temperature is still larger than the liquidus. Since these crystals are depleted in Fe compared to the initial liquid concentration, their descent and remelting drives a net upward transfer of FeO (Fig. 7.8). Each new crystal formed at the top of the slurry layer solidifies in a liquid that is richer in Fe than the one previously crystallized and the Fe content of the solid increases with time as well. In the slurry zone, the temperature should not be too far from equilibrium (although it depends on the crystal size and descent velocity), which makes the temperature increase faster with depth than the initial isentrope. The snow is therefore responsible for a downward heat transfer. Temperature, composition and crystal fraction all contribute to gravitational destabilization of the slurry layer, which should be convecting. The resulting profiles are the result of the competition between the crystallization process that creates the vertical variations and convection that tends to erase it. On the other hand, the development of this slurry layer creates stable composition and temperature profiles in the adjacent layers. Convection driven from cooling at

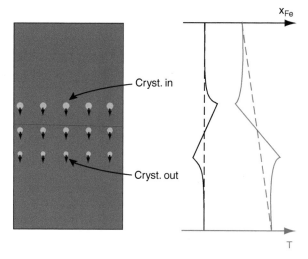

Figure 7.8 Sketch of the thermal and compositional structure expected in a region of the mid-magma ocean when crystallization of dense solid creates a slurry zone. Dashed lines represent the initial profiles of Fe content, x_{Fe} (black), and temperature, T (red), while the solid lines represent the profiles in the liquid after a finite thickness of the slurry layer has been created.

the surface may counteract this stable stratification in the shallow layer, but the deep one should stop convecting altogether.

The continuous cooling of the magma ocean should make the slurry layer thicken with time, with the size of crystals increasing as well. The whole system can evolve like this until the boundaries of the slurry layer reach the magma ocean boundary or the large fraction of crystals leads to a rheological transition. The resulting solid would have a concentration in Fe increasing with height and a temperature close to its Fe-dependent solidus, therefore steeper than the isentropic temperature gradient. This is both thermally and compositionally unstable and should lead to an overturn as in the case studied by *Hess and Parmentier* [1995]. This is again likely to produce a large amount of Fe-rich dense melt that could form the BMO. If, on the other hand, positively buoyant crystals form in the mid-mantle, the situation is reversed from that of Figure 7.8 with a stably stratified slurry bounded above and below by fully liquid layers that are convecting from both heat and solute inputs. And in this case, the lower layer becomes progressively enriched in Fe while the upper one gets depleted. This situation naturally leads to the formation of a surficial magma ocean (SMO) and a BMO. At some point, when the amount of solid in the slurry is around 60%, the rheology strengthens to that of a solid as it becomes a mush. Its subsequent freezing, mostly toward the surface at short timescales, accompanied by viscous compaction, is similar to the classical bottom-up crystallization scenario. It would then undergo the same type of Rayleigh-Taylor instability mentioned above and a secondary melting event. The melt then produced can join the preexisting BMO if it is dense enough.

7.6.2. Fractional Melting of the Mantle from Below

Whether the formation of the Earth results in a completely molten silicate part depends on how the energy from impacts and core formation is distributed. If core formation proceeds in a way that most of the gravitational energy partitions into the core, the high core temperature would make the bottom of the mantle melt. We expect fractional melting to make a Fe-rich melt, which could be more dense than the solid and create a BMO. Otherwise, if it is less dense than the solid, it would rise through the solid mantle, most likely by porous flow through the compacting solid matrix or by hydraulic fracturation. Since the mantle temperature decreases with height, the silicate melt would freeze as it moves up, thereby transferring heat and FeO toward the surface. Over time, this creates an unstable stratification in Fe content of the solid mantle, which again leads to overturn (Fig. 7.9). We are brought back to the situation described in the previous section where the overturn would bring down Fe-rich silicate, thereby transforming gravitational energy into heat and melting the Fe-rich component, now producing a melt denser than the solid. The advantage of this scenario over the one described in the preceding section is that it does not require a completely molten mantle and therefore avoids degassing of the magma ocean. This is one way to transfer both incompatible elements (Fe, He, U, etc.) and heat upward in the mantle (by porous flow) and then downward to be collected in the BMO by Rayleigh-Taylor instability.

7.6.3. Partial Magma Oceans

Giant impacts may not melt the whole planet and could result in a partial magma ocean but the isostatic rebound of the magma could result in a magma ocean the covers the surface [*Tonks and Melosh*, 1993], if it is less dense than the solid. However, the metal silicate emulsion formed by the impact is denser than the already formed mantle of the target proto-Earth and can trigger a Rayleigh-Taylor instability to fall toward the core before a full separation of the emulsion has completed [*Olson and Weeraratne*, 2008]. In this case, a large amount of magma is brought to high pressure at the base of the mantle and, depending on its density, may move back to the surface [*Olson and Weeraratne*, 2008; *King and Olson*, 2011] or stay at the bottom of the mantle hence forming a BMO or joining a preexisting one.

In this scenario, the composition of the magma should be affected by the interaction of metal and silicate up to high pressure and temperature. This could favor a large concentration of Fe in the magma and a high density.

Several scenarios have been sketched in this section and undoubtedly more can be proposed. For example, double diffusive processes are likely to occur in the magma ocean and could result in density stratification [e.g., *Turner and Campbell*, 1986]. Selecting between these scenarios should be possible when a more complete parameterization of the phase diagram of the lower mantle is available. But all of these scenarios share a common feature: the fractional crystallization of a magma ocean will likely create a density stratification that could lead to the formation of a basal magma ocean.

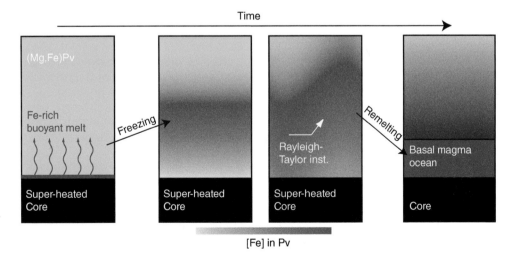

Figure 7.9 Fractional melting of the mantle by a super-heated core. If the melt produced is buoyant and Fe-rich, it can rise and refreeze. This produces an unstable stratification in the solid whose overturn can lead to melting and production of a very Fe-rich basal magma ocean.

7.7. CONCLUSIONS AND OUTSTANDING QUESTIONS

Seismological observations of the lowermost mantle are most easily explained by lateral variation of composition and in particular Fe content, which plays an important role in density variations and therefore dynamics. Additionally, such lateral variations of composition allow the existence of localized pockets of partially molten silicates, the ultra-low velocity zones that are also observed. Even though each observation might be explained, within its error bounds, without requiring composition variations, these are difficult to avoid if one wants to explain the whole picture with a coherently evolutionary model. Similarly, very Fe-rich solid minerals could explain ULVZs, but their solidus is probably lower or very near the CMB temperature. The amount of cooling the core experienced over its history, as evidenced by the long-lasting maintenance of the geomagnetic field, makes the existence of a basal magma ocean in the past inescapable [*Labrosse et al.*, 2007]. This raises several questions regarding the properties of magmas in the conditions of the lower mantle, and in particular their density that needs to be larger than that of the overlying solid.

Most experiments on melting of mantle minerals, as discussed in Section 7.3, considered compositions close to what is expected for the bulk mantle and found contrasting results. However, one should realize that the present situation is the result of 4.5 Gyr of evolution, and the melt present in ULVZ should not be in equilibrium with a solid having the composition of the bulk mantle. The different studies of partitioning during fractional crystallization [*Nomura et al.*, 2011; *Andrault et al.*, 2012; *Tateno et al.*, 2014] all found that Fe is preferentially partitioned in the liquid, even though the precise amounts differ between teams. This means that fractional crystallization will tend to enrich the liquid in Fe and make it denser over time, even if it is less dense than the first solid formed in the magma ocean. Most scenarios of magma ocean crystallization result in a large amount of Fe-rich melt that should end up at the bottom of the mantle. The subsequent fractional crystallization would further increase the Fe content of the magma and its density up to the present day.

In recent years, much progress has been made in the mineral physics relevant to the issues raised by the basal magma ocean scenario. Much effort has been expended, particularly in shock experiments [*Thomas et al.*, 2012; *Thomas and Asimow*, 2013], to extend the compositions to a wider range that is expected to occur during BMO crystallization. A parameterized phase diagram covering the full pressure, temperature, and composition space is still lacking but should be attainable in the near future.

Much work is still necessary on the modeling front to go from the simple parameterized calculation [*Labrosse et al.*, 2007] to the fully dynamical one. A step forward has been performed by *Ulvrová et al.* [2012] who modeled thermal convection in the BMO interacting with a solidification front and showed that classical scalings of Rayleigh-Bénard convection do apply to this case. Future work should also consider convection in the solid part with the moving bottom boundary and include the effect of composition evolution during fractional crystallization and melting.

Even more challenging are the models of magma ocean crystallization needed to go beyond the cartoons presented in Section 7.6 because they require the treatment of two-phase regions (in particular for the slurry regions), and the dynamics in liquid regions is inherently turbulent.

ACKNOWLEDGMENTS

We are thankful to the editors, James Badro and Mike Walter, for convening early Earth sessions at AGU fall meetings and soliciting this chapter. Two anonymous reviewers and Cin-Ty Lee provided thorough and useful reviews.

REFERENCES

Abe, Y. (1993), Thermal evolution and chemical differentiation of the terrestrial magma ocean, in *Evolution of the Earth and Planets*, edited by E. Takahashi, R. Jeanloz, and D. Rubie, Geophysical monograph 74, IUGG 14, pp. 41–54, AGU/IUGG.

Abe, Y. (1997), Thermal and chemical evolution of the terrestrial magma ocean, *Phys. Earth Planet. Inter.*, 100, 27–39.

Agee, C. B. (1998), Crystal-liquid density inversions in terrestrial and lunar magmas, *Phys. Earth Planet. Inter.*, 107 (1–3), 63–74, doi:10.1016/S0031-9201(97)00124-6.

Andrault, D., N. Bolfan-Casanova, G. L. Nigro, M. A. Bouhifd, G. Garbarino, and M. Mezouar (2011), Solidus and liquidus profiles of chondritic mantle: Implication for melting of the Earth across its history, *Earth Planet. Sci. Lett.*, 304, 251–259, doi:10.1016/j.epsl.2011.02.006.

Andrault, D., S. Petitgirard, G. L. Nigro, J.-L. Devidal, G. Veronesi, G. Garbarino, and M. Mezouar (2012), Solid-liquid iron partitioning in Earth's deep mantle, *Nature*, 487 (7407), 354–357.

Andrault, D., G. Pesce, M. A. Bouhifd, N. Bolfan-Casanova, J.-M. H´enot, and M. Mezouar (2014), Melting of subducted basalt at the core-mantle boundary, *Science, 344*, 892–895.

Anzellini, S., A. Dewaele, M. Mezouar, P. Loubeyre, and G. Morard (2013), Melting of iron at Earth's inner core boundary based on fast X-ray diffraction, *Science, 340* (6131), 464–466, doi:10.1126/science.1233514.

Aubert, J., S. Labrosse, and C. Poitou (2009), Modelling the palaeo-evolution of the geodynamo, *Geophys. J. Int.*, 179, 1414–1428, doi:10.1111/j.1365-246X.2009.04361.x.

Becker, T. and L. Boschi (2002), A comparison of tomographic and geodynamic mantle models, *Geochem., Geo-phys., Geosys.*, *3*, 2001GC000,168.

Berryman, J. G. (2000), Seismic velocity decrement ratios for regions of partial melt in the lower mantle, *Geophys. Res. Lett.*, *27*, 421–424.

Boyet, M., and R. W. Carlson (2005), ^{142}Nd Evidence for Early (>4.53 *Ga*) Global Differentiation of the Silicate Earth, *Science*, *309* (5734), 576–581, doi: 10.1126/science.1113634.

Braginsky, S. I. and P. H. Roberts (1995), Equations governing convection in Earth's core and the geodynamo, *Geophys. Astrophys. Fluid Dyn*, *79*, 1–97.

Brandenburg, J. and P. van Keken (2007), Deep storage of oceanic crust in a vigorously convecting mantle, *J. Geophys. Res.*, *112* (B6), B06403.

Burke, K., B. Steinberger, T. Torsvik, and M. Smethurst (2008), Plume Generation Zones at the margins of Large Low Shear Velocity Provinces on the core-mantle boundary, *Earth Planet. Sci. Lett.*, *265*, 49–60.

Burkemper, L. K., C. B. Agee, and K. A. Garcia (2012), Constraints on core formation from molybdenum solubility in silicate melts at high pressure, *Earth Planet. Sci. Lett.*, *335–336* (0), 95–104, doi:10.1016/j.epsl.2012.04.040.

Canup, R. M. (2012), Forming a Moon with an Earth-like Composition via a Giant Impact, *Science*, *338* (6110), 1052–1055, doi:10.1126/science.1226073.

Christensen, U. R. (2010), Dynamo scaling laws and applications to the planets, *Space Sci. Rev.*, *152*, 565–590, doi: 10.1007/s11214-009-9553-2.

Christensen, U. R. and J. Aubert (2006), Scaling properties of convection-driven dynamos in rotating spherical shells and application to planetary magnetic fields, *Geophys. J. Int.*, *166*, 97–114, 10.1111/j.1365-246X.2006.03009.x.

Christensen, U. R. and A. W. Hofmann (1994), Segregation of subducted oceanic crust in the convecting mantle, *J. Geophys. Res.*, *99* (B10), 19,867–19,884.

Cobden, L., S. Goes, M. Ravenna, E. Styles, F. Cammarano, K. Gallagher, and J. Connolly (2009), Thermochemical interpretation of 1-D seismic data for the lower mantle: The significance of nonadiabatic thermal gradients and compositional heterogeneity, *J. Geophys. Res.*, *114*, B11,309, doi: 10.1029/2008JB006262.

Coltice, N. and Y. Ricard (1999), Geochemical observations and one layer mantle convection, *Earth Planet. Sci. Letter*, *74*, 125–137.

Coltice, N., M. Moreira, J. W. Hernlund, and S. Labrosse (2011), Crystallization of a basal magma ocean recorded by helium and neon, *Earth Planet. Sci. Lett.*, *308*, 193–199, doi:10.1016/j.epsl.2011.05.045.

Condie, K. and V. Pease (2008), When did plate tectonics begin on planet Earth?, *Geological Society of America*, pp. 1–294.

Cottaar, S. and B. Romanowicz (2012), An unusually large ULVZ at the base of the mantle near Hawaii, *Earth Planet. Sci. Lett.*, *355–356*, 213–222, doi: 10.1016/j.epsl.2012.09.005.

Ćuk, M. and S. T. Stewart (2012), Making the Moon from a Fast-Spinning Earth: A Giant Impact Followed by Resonant Despinning, *Science*, *338* (6110), 1047–1052, doi: 10.1126/science.1225542.

Davaille, A. (1999a), Two-layer thermal convection in miscible viscous fluids, *J. Fluid Mech.*, *379*, 223–253.

Davaille, A. (1999b), Simultaneous generation of hotspots and superswells by convection in a heterogeneous planetary mantle, *Nature*, *402*, 756–760.

de Koker, N., B. B. Karki, and L. P. Stixrude (2013), Thermodynamics of the MgO–SiO$_2$ liquid system in Earth's lowermost mantle from first principles, *Earth Planet. Sci. Lett.*, *361*, 58–63.

Deschamps, F., L. Cobden, and P. Tackley (2012), The primitive nature of large low shear-wave velocity provinces, *Earth Planet. Sci. Lett.*, *349–350*, 198–208.

Dewaele, A., M. Mezouar, N. Guignot, and P. Loubeyre (2010), High melting points of tantalum in a laser-heated diamond anvil cell, *Phys. Rev. Lett.*, *104*, 255,701, doi: 10.1103/PhysRevLett.104.255701.

Dixon, J. E., L. Leist, C. Langmuir, and J.-G. Schilling (2002), Recycled dehydrated lithosphere observed in plume-influenced mid-ocean-ridge basalt, *Nature*, *420* (6914), 385–389.

Dziewonski, A., B. Hager, and R. O'Connell (1977), Large-scale heterogeneities in the lower mantle, *J. Geophys. Res.*, *82*, 239–255.

Elkins-Tanton, L. (2008), Linked magma ocean solidification and atmospheric growth for Earth and Mars, *Earth Planet. Sci. Lett.*, *271*, 181–191.

Elkins-Tanton, L. T., S. E. Zaranek, E. M. Parmentier, and P. C. Hess (2005), Early magnetic field and magmatic activity on mars from magma ocean cumulate overturn, *Earth Planet. Sci. Lett.*, *236*, 1–12.

Fiquet, G., A. L. Auzende, J. Siebert, A. Corgne, H. Bureau, H. Ozawa, and G. Garbarino (2010), Melting of peridotite to 140 gigapascals, *Science*, *329* (5998), 1516–1518, doi:10.1126/science.1192448.

Flasar, F. M. and F. Birch (1973), Energetics of core formation: a correction, *J. Geophys. Res.*, *78*, 6101–6103.

Ford, S., E. Garnero, and A. McNamara (2006), A strong lateral shear velocity gradient and anisotropy heterogeneity in the lowermost mantle beneath the southern Pacific, *J. Geophys. Res.*, *111*, B03,306, doi:10.1029/2004JB003574.

Frost, D. J., Y. Asahara, D. C. Rubie, N. Miyajima, L. S. Dubrovinsky, C. Holzapfel, E. Ohtani, M. Miyahara, and T. Sakai (2010), Partitioning of oxygen between the Earth's mantle and core, *J. Geophys. Res.*, *115* (B2), B02,202, doi: 10.1029/2009JB006302.

Funamori, N. and T. Sato (2010), Density contrast between silicate melts and crystals in the deep mantle: An integrated view based on static-compression data, *Earth Planet. Sci. Lett.*, *295* (3–4), 435–440, doi: 10.1016/j.epsl.2010.04.021.

Garnero, E. and D. Helmberger (1996), Seismic detection of a thin laterally varying boundary layer at the base of the mantle beneath the central-Pacific, *Geophys. Res. Lett.*, *23*, 977–980.

Garnero, E. and A. McNamara (2008), Structure and dynamics of Earth's lower mantle, *Science*, *320*, 626–628.

Gomi, H., K. Ohta, K. Hirose, S. Labrosse, R. Caracas, M. J. Verstraete, and J. W. Hernlund (2013), The high conductivity of iron and thermal evolution of the Earth's core, *Phys. Earth Planet. Inter.*, *224*, 88–103, doi: 10.1016/j.pepi.2013.07.010.

Gonnermann, H. M., A. M. Jellinek, M. A. Richards, and M. Manga (2004), Modulation of mantle plumes and heat flow at the core mantle boundary by plate-scale flow: results from laboratory experiments, *Earth Planet. Sci. Lett.*, *226* (1–2), 53–67.

Gubbins, D., T. G. Masters, and J. A. Jacobs (1979), Thermal evolution of the Earth's core, *Geophys. J. R. Astr. Soc.*, *59*, 57–99.

Gubbins, D., D. Alfè, G. Masters, G. D. Price, and M. J. Gillan (2004), Gross thermodynamics of 2-component core convection, *Geophys. J. Int.*, *157*, 1407–1414.

Helmberger, D., L. Wen, and X. Ding (1998), Seismic evidence that the source of the Iceland hotspot lies at the core-mantle boundary, *Nature*, *396*, 251–255.

Hernlund, J. and C. Houser (2008), On the distribution of seismic velocities in Earth's deep mantle, *Earth Planet. Sci. Lett.*, *265*, 423–437, doi:10.1016/j.epsl.2007.10.042.

Hernlund, J. and M. Jellinek (2010), Dynamics and structure of a stirred partially molten ultralow-velocity zone, *Earth Planet. Sci. Lett.*, *296*, 1–8, doi:10.1016/j.epsl.2010.04.027.

Hernlund, J., and A. McNamara (2015), The core-mantle boundary region, in *Treatise on Geophyiscs, vol. 7*, edited by G. Schubert and D. Bercovici, pp. 461–519, Elsevier.

Hernlund, J. and P. Tackley (2007), Some dynamical consequences of partial melting at the base of Earth's mantle, *Phys. Earth Planet. Inter.*, *162*, 149–163.

Hernlund, J. W., C. Thomas, and P. J. Tackley (2005), Phase boundary double crossing and the structure of Earth's deep mantle, *Nature*, *434*, 882–886, doi:10.1038/nature03472.

Hess, P. C. and E. M. Parmentier (1995), A model for the thermal and chemical evolution of the Moon's interior: implications for the onset of mare volcanism, *Earth Planet. Sci. Lett.*, *134*, 501–514.

Hier-Majumder, S. (2008), Influence of contiguity on seismic velocities of partially molten aggregates, *J. Geophys. Res.*, *113* (B12), doi:10.1029/2008JB005662.

Hirose, K. and Y. Fei (2002), Subsolidus and melting phase relations of basaltic composition in the uppermost lower mantle, *Geochim. Cosmochim. Acta.*, *66* (12), 2099–2108, doi:10.1016/S0016-7037(02)00847-5.

Hirose, K., S. Labrosse, and J. W. Hernlund (2013), Composition and state of the core, *Ann. Rev. Earth Planet. Sci.*, *41*, 657–691, doi:10.1146/annurev-earth-050212-124007.

Hopkins, M., T. Harrison, and C. Manning (2008), Low heat flow inferred from >4 gyr zircons suggests hadean plate boundary interactions, *Nature*, *456*, 493–496.

Houser, C., G. Masters, P. Shearer, and G. Laske (2008), Shear and compressional velocity models of the mantle from cluster analysis of long-period waveforms, *Geophys. J. Int.*, *174* (1), 195–212, doi:10.1111/j.1365-246X.2008.03763.x.

Hutko, A., T. Lay, J. Revenaugh, and E. Garnero (2008), Anticorrelated seismic velocity anomalies from post-perovskite in the lowermost mantle, *Science*, *320*, 1070–1074, doi:10.1126/science.1155822.

Ishii, M. and J. Tromp (2004), Constraining large-scale mantle heterogeneity using mantle and inner-core sensitive modes, *Phys. Earth Planet. Int.*, *146*, 113–124.

Ito, E., A. Kubo, T. Katsura, and M. Walter (2004), Melting experiments of mantle materials under lower mantle conditions with implications for magma ocean differentiation, *Phys. Earth Planet. Inter.*, *143–144*, 397–406, doi:10.1016/j.pepi.2003.09.016.

Kiefer, B., L. Stixrude, and R. Wentzcovitch (2002), Elasticity of (Mg,Fe)SiO$_3$ perovskite at high pressures, *Geophys. Res. Lett.*, *29*, 1539–1539.

King, C. and P. L. Olson (2011), Heat partitioning in metal-silicate plumes during Earth differentiation, *Earth Planet. Sci. Lett.*, *304* (3–4), 577–586, doi: 10.1016/j.epsl.2011.02.037.

Labrosse, S. (2003), Thermal and magnetic evolution of the Earth's core, *Phys. Earth Planet. Inter.*, *140* (1–3), 127–143.

Labrosse, S. (2015), Thermal evolution of the core with a high thermal conductivity, *Phys. Earth Planet. Inter.*, doi:10.1016/j.pepi.2015.02.002.

Labrosse, S., J.-P. Poirier, and J.-L. Le Mouël (1997), On cooling of the Earth's core, *Phys. Earth Planet. Inter.*, *99*, 1–17.

Labrosse, S., J. W. Hernlund, and N. Coltice (2007), A crystallizing dense magma ocean at the base of Earth's mantle, *Nature*, *450* (7171), 866–869.

Lay, T. and E. Garnero (2011), Deep mantle seismic modeling and imaging, *Ann. Rev. Earth Planet. Sci.*, *39*, 91–123.

Lay, T., E. Garnero, and Q. Williams (2004), Partial melting in a thermo-chemical boundary layer at the base of the mantle, *Phys. Earth Planet. Int.*, *146*, 441–467.

Lay, T., J. Hernlund, E. J. Garnero, and M. S. Thorne (2006), A Post-Perovskite Lens and D" Heat Flux Beneath the Central Pacific, *Science*, *314* (5803), 1272–1276.

Le Bars, M. and A. Davaille (2002), Stability of thermal convection in two superimposed miscible viscous fluids, *J. Fluid Mech.*, *471*, 339–363.

Lekic, V., S. Cottaar, A. Dziewonski, and B. Romanowicz (2012), Cluster analysis of global lower mantle tomography: A new class of structure and implications for chemical heterogeneity, *Earth Planet. Sci. Lett.*, *357–358*, 68–77, doi:10.1016/j.epsl.2012.09.014.

Li, J. and C. B. Agee (1996), Geochemistry of mantle-core differentiation at high pressure, *Nature*, *381* (6584), 686–689.

Li, M. and A. McNamara (2013), The difficulty for subducted oceanic crust to accumulate at the Earth's core-mantle boundary, *J. Geophys. Res.*, pp. 1–10.

Li, M., A. McNamara, and E. Garnero (2014), Chemical complexity of hotspots caused by cycling oceanic crust through mantle reservoirs, *Nature Geosci.*, doi:10.1038/ngeo2120.

Liebske, C. and D. J. Frost (2012), Melting phase relations in the MgO-MgSiO$_3$ system between 16 and 26 GPa: Implications for melting in Earth's deep interior, *Earth Planet. Sci. Lett.*, *345*, 159–170, doi:10.1016/j.epsl.2012.06.038.

Lister, J. R. and B. A. Buffett (1995), The strength and efficiency of the thermal and compositional convection in the geodynamo, *Phys. Earth Planet. Inter.*, *91*, 17–30.

Luo, S., S. Ni, and D. Helmberger (2001), Evidence for a sharp lateral variation of velocity at the core-mantle boundary from multipathed PKPab, *Earth Planet. Sci. Lett.*, *189*, 155–164.

Manners, U. (2008), Investigating the structure of the core-mantle boundary using S and P diffracted waves, Ph.D. thesis, University of California, San Diego.

Mao, W., G. Shen, V. Prakapenka, Y. Meng, A. Cambell, D. Heinz, J. Shu, R. Hemley, and H. Mao (2004), Ferromagnesian postperovskite silicates in the D" layer of the Earth, *Proc. Natl. Acad. Sci. USA*, *101*, 15,867–15,869.

Mao, W. L., H.-K. Mao, W. Sturhahn, J. Zhao, V. B. Prakapenka, Y. Meng, J. Shu, Y. Fei, and R. J. Hemley (2006), Iron-rich post-perovskite and the origin of Ultralow-Velocity Zones, *Science*, *312* (5773), 564–565.

Marty, B. (2012), The origins and concentrations of water, carbon, nitrogen and noble gases on Earth, *Earth Planet. Sci. Lett.*, *313–314* (C), 56–66, doi:10.1016/j.epsl.2011.10.040.

McDonough, W. F. (2003), Compositional model for the Earth's core, *Treatise on Geochemistry*, *2*, 547–568, doi: 10.1016/B0-08-043751-6/02015-6.

McNamara, A. and S. Zhong (2004), Thermochemical structures within a spherical mantle: Superplumes or piles?, *J. Geophys. Res. B*, *109*, B07, 402.

McNamara, A. K., E. J. Garnero, and S. Rost (2010), Tracking deep mantle reservoirs with ultra-low velocity zones, *Earth Planet. Sci. Lett.*, *299* (1–2), 1–9, doi: 10.1016/j.epsl.2010.07.042.

Monnereau, M. and D. Yuen (2007), Topology of the postperovskite phase transition and mantle dynamics, *Proc. Natl. Acad. Sci.*, *104*, 9156–9161.

Mukhopadhyay, S. (2012), Early differentiation and volatile accretion recorded in deep-mantle neon and xenon, *Nature*, *486* (7401), 101–4.

Nakagawa, T. and P. Tackley (2005), The interaction between the post-perovskite phase change and a thermo-chemical boundary layer near the core-mantle boundary, *Earth Planet. Sci. Lett.*, *238*, 204–216.

Nakagawa, T. and P. J. Tackley (2004), Effects of thermo-chemical mantle convection on the thermal evolution of the Earth's core, *Earth Planet. Sci. Lett.*, *220*, 107–119.

Nakajima, M. and D. J. Stevenson (2014), Investigation of the initial state of the moon-forming disk: Bridging SPH simulations and hydrostatic models, *Icarus*, *233*, 259–267, doi:10.1016/j.icarus.2014.01.008.

Nimmo, F., G. D. Price, J. Brodholt, and D. Gubbins (2004), The influence of potassium on core and geodynamo evolution, *Geophys. J. Int.*, *156*, 263–376.

Nomura, R., H. Ozawa, S. Tateno, K. Hirose, J. W. Hernlund, S. Muto, H. Ishii, and N. Hiraoka (2011), Spin crossover and iron-rich silicate melt in the Earth's deep mantle, *Nature*, *473* (7346), 199–202.

Nomura, R., K. Hirose, K. Uesugi, Y. Ohishi, A. Tsuchiyama, A. Miyake, and Y. Ueno (2014), Low core-mantle boundary temperature inferred from the solidus of pyrolite, *Science*, *343* (6170), 522–525, doi:10.1126/science.1248186.

Olson, P. L. and D. Weeraratne (2008), Experiments on metal-silicate plumes and core formation, *Phil. Trans. R. Soc. A*, *366* (1883), 4253–4271, doi:10.1038/nature01459.

Ozawa, H., K. Hirose, M. Mitome, Y. Bando, N. Sata, and Y. Ohishi (2009), Experimental study of reaction between perovskite and molten iron to 146 GPa and implications for chemically distinct buoyant layer at the top of the core, *Phys. Chem. Minerals*, *36*, 355–363, doi:10.1007/s00269-008-0283-x.

Pozzo, M., C. J. Davies, D. Gubbins, and D. Alfè (2012), Thermal and electrical conductivity of iron at Earth's core conditions, *Nature*, *485*, 355–358.

Rama Murthy, V., W. van Westrenen, and Y. Fei (2003), Radioactive heat sources in planetary cores: Experimental evidence for potassium, *Nature*, *423*, 163–165.

Reasoner, C. and J. Revenaugh (2000), ScP constraints on ultralow-velocity zone density and gradient thickness beneath the Pacific, *J. Geophys. Res. B*, *105*, 173–182.

Ricard, Y., M. Richards, C. Lithgow-Bertelloni, and Y. Le Stunff (1993), A geodynamic model of mantle density heterogeneity, *J. Geophys. Res.*, *98* (B12), 21,895–21,909.

Ricard, Y., O. Šrámek, and F. Dubuffet (2009), A multi-phase model of runaway core-mantle segregation in planetary embryos, *Earth Planet. Sci. Lett.*, *284* (1–2), 144–150, doi: 10.1016/j.epsl.2009.04.021.

Rost, S. and J. Revenaugh (2003), Small scale ultralow velocity zone structure imaged by ScP, *J. Geophys. Res. B*, *108*, 2056.

Rost, S., E. J. Garnero, Q. Williams, and M. Manga (2005), Seismological constraints on a possible plume root at the core-mantle boundary, *Nature*, *435*, 666–669, doi: 10.1038/nature03620.

Rost, S., E. Garnero, and Q. Williams (2006), Fine-scale ultralow-velocity zone structure from high-frequency seismic array data, *J. Geophys. Res.*, *111*, B09, 310, doi: 10.1029/2005JB004088.

Samuel, H. and D. Bercovici (2006), Oscillating and stagnating plumes in the Earth's lower mantle, *Earth Planet. Sci. Lett.*, *248* (1–2), 90–105, doi:10.1016/j.epsl.2006.04.037.

Schuberth, B. S., C. Zaroli, and G. Nolet (2012), Synthetic seismograms for a synthetic Earth: long-period P- and S-wave traveltime variations can be explained by temperature alone, *Geophys. J. Int.*, *188* (3), 1393–1412, doi: 10.1111/j.1365-246X.2011.05333.x.

Sha, X. and R. E. Cohen (2011), First-principles studies of electrical resistivity of iron under pressure, *J. Phys.: Condens Matter*, *23* (7), 075,401.

Shen, G. and P. Lazor (1995), Measurement of melting temperatures of some minerals under lower mantle pressures, *J. Geophys. Res.*, *100* (B9), 17,699–17,713, doi: 10.1029/95JB01864.

Siebert, J., J. Badro, D. Antonangeli, and F. J. Ryerson (2012), Metal–silicate partitioning of Ni and Co in a deep magma ocean, *Earth Planet. Sci. Lett.*, *321–322*, 189–197, doi:10.1016/j.epsl.2012.01.013.

Simmons, N., A. Forte, and S. Grand (2006), Constraining mantle flow with seismic and geodynamic data: A joint approach, *Earth Planet. Sci. Lett.*, *246*, 109–124, doi: 10.1029/2006GL028009.

Sleep, N. H. (2000), Evolution of the mode of convection within terrestrial planets, *J. Geophys. Res.*, *105* (E7), 17,563–17,578.

Solomatov, V. S. (2000), Fluid Dynamics of a Terrestrial Magma Ocean, in *Origin of the Earth and Moon*, edited by R. Canup and K. Righter, pp. 323–338, University of Arizona Press, Tucson.

Solomatov, V. S. (2007), Magma oceans and primordial mantle differentiation, *Treatise on Geophysics*, *9*, 91–120.

Šrámek, O., L. Milelli, Y. Ricard, and S. Labrosse (2012), Thermal evolution and differentiation of planetesimals and planetary embryos, *Icarus*, *217*, 339–354, doi: 10.1016/j.icarus.2011.11.021.

Stacey, F. D. (1991), Effects on the core of structure within D", *Geophys. Astrophys. Fluid Dyn*, *60*, 157–163.

Stevenson, D. (1981), Models of the Earth core, *Science*, *214* (4521), 611–619.

Stevenson, D. J. (1990), Fluid dynamics of core formation, in *Origin of the Earth*, edited by H. E. Newsom and J. H. Jones, pp. 231–249, Oxford University Press, New York.

Stewart, A. J., M. W. Schmidt, W. van Westrenen, and C. Liebske (2007), Mars: A new core-crystallization regime, *Science*, *316* (5829), 1323–1325.

Stixrude, L. and C. Lithgow-Bertelloni (2005), Thermodynamics of mantle minerals: 1. Physical properties, *Geophys. J. Int.*, *162*, 610–632, doi:10.1111/j.1365- 246X.2005.02642.x.

Stixrude, L., N. de Koker, N. Sun, M. Mookherjee, and B. B. Karki (2009), Thermodynamics of silicate liquids in the deep Earth, *Earth Planet. Sci. Lett.*, *278* (3–4), 226–232, doi:10.1016/j.epsl.2008.12.006.

Stolper, E., D. Walker, B. H. Hager, and J. F. Hays (1981), Melt segregation from partially molten source regions: The importance of melt density and source region size, *J. Geophys. Res.*, *86* (B7), 6261–6271, doi: 10.1029/JB086iB07p06261.

Sun, D., E. Tan, D. Helmberger, and M. Gurnis (2007), Seismological support for the metastable superplume model, sharp features, and phase changes within the lower mantle, *Proc. Nat. Acad. Sci.*, *104*, 9151–9155.

Tackley, P. (1998), Three-dimensional simulations of mantle convection with a thermo-chemical basal boundary layer: D"?, in *The Core-Mantle Boundary Region*, edited by M. Gurnis, M. Wysession, E. Knittle, and B. Buffett, pp. 231–253, American Geophysical Union Monograph.

Tackley, P. (2002), Strong heterogeneity caused by deep mantle layering, *Geochem. Geophys. Geosys.*, *3*, 1–22, doi: 10.1029/2001GC000167.

Tan, E., W. Leng, S. Zhong, and M. Gurnis (2011), On the location and mobility of thermo-chemical structures with high bulk modulus in the 3-D compressible mantle, *Geochem. Geophys. Geosys.*, *12*, Q07,005.

Tateno, S., K. Hirose, and Y. Ohishi (2014), Melting experiments on peridotite to lowermost mantle conditions, *J. Geophys. Res.*, *119*, 4684–4694, doi:10.1002/2013JB010616.

Thomas, C. W., and P. D. Asimow (2013), Direct shock compression experiments on premolten forsterite and progress toward a consistent high-pressure equation of state for CaO-MgO-Al2O3-SiO2-FeO liquids, *J. Geophys. Res.*, *118* (11), 5738–5752, doi:10.1002/jgrb.50374.

Thomas, C. W., Q. Liu, C. B. Agee, P. D. Asimow, and R. A. Lange (2012), Multi-technique equation of state for Fe2SiO4 melt and the density of Fe-bearing silicate melts from 0 to 161 GPa, *J. Geophys. Res.*, *117* (B10), B10,206.

Thorne, M., E. Garnero, and S. Grand (2004), Geographic correlation between hot spots and deep mantle lateral shear-wave velocity gradients, *Phys. Earth Planet. Int.*, *146*, 47–63.

Thorne, M., Y. Zhang, and J. Ritsema (2013), Evaluation of 1-D and 3-D seismic models of the Pacific lower mantle with S, SKS, and SKKS traveltimes and amplitudes, *J. Geophys. Res.*, *118*, 50,054, doi:10.1002/jgrb.50054.

To, A., B. Romanowicz, Y. Capdeville, and N. Takeuchi (2005), 3D effects of sharp boundaries at the borders of the African and Pacific superplumes: Observation and modeling, *Earth Planet. Sci. Lett.*, *233* (1–2), 137–153.

Tolstikhin, I. and A. Hofmann (2005), Early crust on top of the Earth's core, *Phys. Earth Planet. Inter.*, *148*, 109–130.

Tonks, W. B. and H. J. Melosh (1993), Magma ocean formation due to giant impacts, *J. Geophys. Res.*, *98* (E3), 5319–5333.

Trønnes, R. G. and D. J. Frost (2002), Peridotite melting and mineral–melt partitioning of major and minor elements at 22–24.5 GPa, *Earth Planet. Sci. Lett.*, *197* (1–2), 117–131, doi:10.1016/S0012-821X(02)00466-1.

Turner, J. and I. Campbell (1986), Convection and mixing in magma chambers, *Earth-Science Reviews*, *23* (4), 255–352, doi:10.1016/0012-8252(86)90015-2.

Turner, S., T. Rushmer, M. Reagan, and J. -F. Moyen (2014), Heading down early on? *Start of subduction on Earth, Geology*, *42*, 139–142, doi:10.1130/G34886.1.

Ulvrová, M., S. Labrosse, N. Coltice, P. Råback, and P. J. Tackley (2012), Numerical modeling of convection interacting with a melting and solidification front: application to the thermal evolution of the basal magma ocean, *Phys. Earth Planet. Inter.*, *206–207*, 51–66.

Usui, Y., J. A. Tarduno, M. Watkeys, A. Hofmann, and R. D. Cottrell (2009), Evidence for a 3.45-billion-year-old magnetic remanence: Hints of an ancient geodynamo from conglomerates of south africa, *Geochem. Geophys. Geosyst.*, *10*, Q09Z07, doi:10.1029/2009GC002496.

Wang, Y. and L. Wen (2004), Mapping the geometry and geographic distribution of a very low velocity province at the base of the Earth's mantle, *J. Geophys. Res.*, *109*, B10,305.

Wen, L. (2001), Seismic evidence for a rapidly-varying compositional anomaly at the base of the Earth's mantle beneath the Indian Ocean, *Earth Planet. Sci. Lett.*, *194*, 83–95.

Wen, L. (2006), A compositional anomaly at the Earth's core–mantle boundary as an anchor to the relatively slowly moving surface hotspots and as source to the DUPAL anomaly, *Earth Planet. Sci. Lett.*, *246* (1–2), 138–148, doi:10.1016/j.epsl.2006.04.024.

Wicks, J. K., J. M. Jackson, and W. Sturhahn (2010), Very low sound velocities in iron-rich (Mg,Fe)O: Implications for the core-mantle boundary region, *Geophys. Res. Lett.*, *37* (15), L15,304, doi:10.1029/2010GL043689.

Williams, Q. and E. J. Garnero (1996), Seismic evidence for partial melt at the base of the Earth's mantle, *Science*, *273*, 1528.

Williams, Q., J. Revenaugh, and E. Garnero (1998), A correlation between ultra-low basal velocities in the mantle and hot spots, *Science*, *281*, 546–549.

Wookey, J., S. Stackhouse, J. Kendall, J. Brodholt, and G. Price (2005), Efficacy of the post-perovskite phase as an explanation of lowermost mantle seismic properties, *Nature*, *438*, 1004–1008.

Xie, S. and P. Tackley (2004), Evolution of helium and argon isotopes in a convecting mantle, *Phys. Earth Planet. Inter.*, *146*, 417–439.

Zerr, A. and R. Boehler (1993), Melting of (Mg, Fe)SiO3-perovskite to 625 kilobars: Indication of a high melting temperature in the lower mantle, *Science*, *262* (5133), 553–555, doi:10.1126/science.262.5133.553.

Zhang, J. and C. Herzberg (1994), Melting experiments on anhydrous peridotite KLB-1 from 5.0 to 22.5 GPa, *J. Geophys. Res.*, *99*, 17,729–17,742.

8

Early Differentiation and Its Long-Term Consequences for Earth Evolution

Richard W. Carlson[1], Maud Boyet[2], Jonathan O'Neil[3], Hanika Rizo[1,4,5], and Richard J. Walker[4]

ABSTRACT

Several first order features of Earth owe their origin to processes occurring before, during, and within a few hundred million years of Earth formation. Arguably the most significant expression of these early events is the bulk composition of Earth. Earth's depletion in some volatile elements likely was inherited from the materials from which it formed. This is most easily attributed to Earth's accumulation from planetesimals formed in the inner Solar System where the temperatures were hot enough, for long enough, to keep many volatile elements in the gas phase until after the solids had accumulated into at least planetesimal-sized objects. Improved understanding of the processes of planetary accretion makes it increasingly clear that the main fraction of Earth's mass was accumulated through violent collisions with large planetesimals, not by gentle accumulation of primitive bodies. The accreted planetesimals likely had already experienced global differentiation to separate core from mantle and crust, and suffered additional volatile loss by gravitational escape of any atmosphere formed through this early differentiation on the small planetesimal. The short-lived ^{182}Hf-^{182}W system indicates that the metal-silicate separation associated with core formation began on planetesimals within a million years or less and on Earth within tens of millions of years of the start of Solar System formation. Metal-silicate separation left Earth's mantle deficient in siderophile elements relative to their abundances in bulk chondrites. Mantle abundances of moderately siderophile elements suggest high-pressure and temperature equilibrium between metal and silicate, consistent with metal-silicate segregation occurring during largely or entirely molten stages of early Earth history. By contrast, the mantle abundances of highly siderophile elements are most easily reconciled with addition of approximately half a percent of Earth's mass of material with chondritic composition after chemical exchange between mantle and core had stopped. Evidence for early differentiation of the silicate Earth, as would be expected for a terrestrial magma ocean, is remarkably subdued, but is now being extracted from information provided by short-lived radioactive systems such as ^{129}I-^{129}Xe, ^{146}Sm-^{142}Nd, and ^{182}Hf-^{182}W. For example, ^{129}Xe and ^{142}Nd heterogeneities in the mantle point to a major terrestrial differentiation event occurring between circa 4.4 and 4.45 Ga, which is most easily attributed to the time of the Moon-forming giant impact. What little evidence remains for the nature of Earth's crust that formed immediately after the resulting magma ocean suggests the presence of a primitive mafic crust that did not become reworked into substantial felsic continental crust until 3.8 to 4.0 Ga.

[1]*Department of Terrestrial Magnetism, Carnegie Institution for Science, Washington, DC, USA*

[2]*Laboratoire Magmas et Volcans, Universite Blaise Pascal, CNRS UMR 6524, Clermont-Ferrand, France*

[3]*Department of Earth and Environmental Sciences, University of Ottawa, Ottawa, Ontario, Canada*

[4]*Department of Geology, University of Maryland, College Park, MD, USA*

[5]*Now at Department des sciences de la Terre et de l'atmosphere, Universite du Quebec á Montreal, Montreal, Canada*

The Early Earth: Accretion and Differentiation, Geophysical Monograph 212, First Edition.
Edited by James Badro and Michael Walter.
© 2015 American Geophysical Union. Published 2015 by John Wiley & Sons, Inc.

8.1. INTRODUCTION

Abundant energy sources accompanied Earth formation including the heat produced by decay of short-lived radioactive isotopes such as ^{26}Al, the gravitational potential energy released by core formation, and the kinetic energy imparted by the large collisions involved in the terminal stages of planet growth. This energetic birth drove extensive differentiation of Earth during and shortly following its formation, resulting in the development of its core, mantle, crust, and atmosphere. There is ample evidence that similar global-scale differentiation processes acted on the other terrestrial bodies, including the Moon and Mars, as well as on differentiated asteroids. Despite our general knowledge of large-scale planetary differentiation, numerous questions remain regarding how Earth acquired its bulk composition and, once formed, the nature of the differentiation processes that led to the Earth we know now. One such question, the focus of this chapter, concerns the relative role of differentiation processes accompanying Earth formation compared to longer-acting processes, such as crust-mantle differentiation driven by plate tectonics throughout much of Earth history. Evidence for early-Earth differentiation has begun to emerge from a variety of geological and geochemical studies, yet the interpretation of this evidence has proven difficult because the continuing geologic activity of Earth has complicated the signal of earlier chemical processing of the planet.

Despite nature's best attempts to obscure the consequences of early-Earth differentiation, significant strides have been made in the past decade in deciphering the relative role of early differentiation processes involved in the establishment of first order compositional features of Earth. The following factors are contributing to these advances: (a) detection of mass independent isotope variability between Earth and its potential building blocks, as represented by meteorites, that reflects nucleosynthetic heterogeneity in the solar nebula inherited by Earth [*Dauphas et al.*, 2002; *Dauphas et al.*, 2004; *Andreasen and Sharma*, 2007; *Carlson et al.*, 2007; *Trinquier et al.*, 2007; *Simon et al.*, 2009; *Trinquier et al.*, 2009; *Chen et al.*, 2010; *Qin et al.*, 2010; *Burkhardt et al.*, 2011]; (b) mass-dependent isotope variability caused by chemical processing of elements before and during Earth formation [*Thiemens and Heidenreich*, 1983; *Clayton*, 2003; *Poitrasson et al.*, 2004; *Williams et al.*, 2006; *Georg et al.*, 2007; *Fitoussi et al.*, 2009; *Shahar et al.*, 2009; *Simon and DePaolo*, 2010; *Paniello et al.*, 2012; *Valdes et al.*, 2014]; (c) evidence conveyed by isotopic heterogeneity in short-lived radioactive systems that was not completely erased by longer-term geologic processes, such as crustal recycling and mantle convection [*Wetherill*, 1975; *Harper and Jacobsen*, 1992; *Lee and Halliday*, 1995, 1996; *Ozima and Podosek*, 1999;

Kleine et al., 2002; *Yin et al.*, 2002; *Boyet and Carlson*, 2006; *Caro et al.*, 2006; *Willbold et al.*, 2011; *Touboul et al.*, 2012; *Rizo et al.*, 2013]; and (d) improvements in geochronologic and petrogenetic interrogation that allow peering through the complex history of ancient rocks to better understand their conditions of formation during the first billion years of Earth history [e.g., *Froude et al.*, 1983; *Bowring et al.*, 1990; *Harrison*, 2009; *O'Neil et al.*, 2012; *Cawood et al.*, 2013; *Nutman et al.*, 2013].

This chapter examines the evidence for the nature and timing of differentiation processes that occurred during Earth formation through to the point in Earth history when it can be demonstrated that crust formation and recycling processes, similar to those occurring today, took over as the main driver of chemical differentiation of Earth's interior. Distinguishing which of Earth's present day features owe their origin to differentiation events occurring during early Earth history, as opposed to >4.5 Ga of geologic processing will improve our understanding not only of the consequences of the still-running crust-mantle differentiation on Earth but also provide an important look at what characteristics of Earth may have been established at birth.

8.2. VOLATILE DEPLETION

Models of Earth's bulk composition [*Jagoutz et al.*, 1979; *McDonough and Sun*, 1995; *Lyubetskaya and Korenaga*, 2007; *Palme and O'Neill*, 2014] usually start with the composition of the Sun, as the Sun constitutes over 99% of the mass of the Solar System. Representing solar composition for this purpose are CI chondrites [*Anders and Grevesse*, 1989]. CI chondrites have compositions similar to the Sun for all elements but the most volatile, and those (e.g., Li) that are consumed by nuclear processes in the Sun. Compared to CI chondrites, however, Earth is depleted in volatile elements, even in some elements classified as moderately volatile (Fig. 8.1a). Among the volatile elements depleted in Earth are many members of both short- and long-lived radioactive decay schemes, including U-Th-Pb, Rb-Sr, Mn-Cr, Pd-Ag, and I-Pu-Xe. These radioisotope systems provide a means to estimate the time when Earth acquired its volatile depleted characteristic and suggest that at least for the moderately volatile elements, Earth most likely formed from volatile-depleted materials rather than became volatile-depleted as the result of volatile-loss events after its formation.

In the U-Th-Pb system, U and Th are both refractory (50% condensation temperatures [Tc] of 1610 and 1659 K, respectively [*Lodders*, 2003]) lithophile elements, and thus, should not have been fractionated from Earth due to any volatile loss mechanism, nor should they have been removed from the silicate Earth during core formation. In contrast, Pb is both moderately volatile (Tc = 727 K) and

(a)

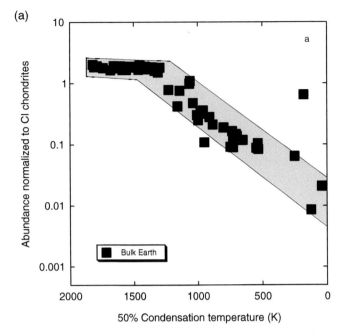

(b)

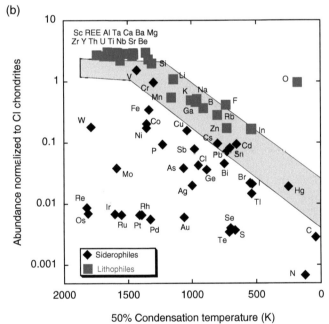

Figure 8.1 Element abundance normalized to CI chondrite in (a) the bulk Earth and (b) separately for lithophile and siderophile element abundances in the mantle [*Palme and O'Neill*, 2014] relative to the temperature at which 50% of each element would condense from a low-pressure gas of Solar composition [*Lodders*, 2003].

chalcophile, which could have led to its depletion in the bulk-silicate-Earth (BSE) by volatile loss and/or by partitioning into the core. Bulk-silicate-Earth is defined as the average composition of the mantle plus crust. Lead is depleted in the BSE by a factor of ~14 compared to CI

abundances [*Palme and O'Neill*, 2014]. As a result, the $^{238}U/^{204}Pb$ ratio of the BSE is of order 8 compared to a CI ratio of 0.2. This large change in U/Pb ratio left an indelible imprint in the Pb isotopic evolution of the silicate Earth. On a Pb isotope isochron diagram, connecting an estimate of the initial Solar System Pb isotopic composition, determined from measurements of sulfides in iron meteorites, with the Pb isotopic composition of modern oceanic sediments provides a correlation with a slope corresponding to an age of 4.55 ± 0.07 Ga. This age, as interpreted by Clair Patterson, was the first modern determination of the age of the Earth [*Patterson*, 1956]. In order for this Pb isotope tie line to approximate the age of the Earth requires that the changes in U/Pb ratio experienced over the entire history of the materials that make up modern oceanic sediments were small relative to the large step between Solar and mantle U/Pb ratio. Patterson's "age of the Earth" [*Patterson*, 1956] adjusts to 4.48 Ga using modern values for U decay constants [*Begemann et al.*, 2001] and a $^{238}U/^{235}U = 137.84$ [*Weyer et al.*, 2008; *Condon et al.*, 2010; *Richter et al.*, 2010]. This result shows that the high U/Pb ratio of Earth was present very early in Earth history, but understanding exactly what event the average Pb isotopic composition of the Earth dates is complicated due to uncertainty in how much of this U/Pb ratio increase was due to Pb volatility [*Azbel et al.*, 1993; *Allègre et al.*, 2008] as opposed to partitioning of Pb into the core [*Wood and Halliday*, 2005, 2010]. Additional complicating factors in the interpretation of the U-Pb age of Earth include the potential sensitivity of this model age to addition of Pb by late-arriving, volatile-rich planetesimals [*Albarède*, 2010], and that the relatively high concentration of Pb in the continental crust makes the U-Pb result very sensitive to the history of continent formation [*Yin and Jacobsen*, 2006].

The Xe isotopic composition of Earth received contributions from two short-lived radioisotopes, the moderately volatile ^{129}I (Tc = 535 K [*Lodders*, 2003]; half-life = 15.7 Ma), and refractory ^{244}Pu (fission half-life = 82 Ma). The presence of about a 7% excess in ^{129}Xe in Earth's atmosphere, compared to solar Xe, was attributed to the decay of ^{129}I in the first use of short-lived radionuclides to document Earth differentiation events occurring during the lifetime of these isotopes [*Reynolds*, 1960]. Simple models of atmosphere outgassing translate the I-Xe systematics of Earth's atmosphere into atmosphere formation ages of ~100 Ma after the 4.568 Ga start of Solar System formation [*Wetherill*, 1975; *Ozima and Podosek*, 1983; *Pepin and Porcelli*, 2006]. This age of atmosphere formation, thus, is similar to that of the U-Pb "age of the Earth," both of which imply a relatively long formation interval for Earth. As with U-Pb, however, understanding exactly what event the apparent I-Pu-Xe age dates is not straightforward. Of primary concern is the relatively poorly known original abundances of I and Xe in Earth, and how much Xe has been lost from Earth's atmosphere [*Tolstikhin and O'Nions*, 1994], or added by mantle outgassing or late arriving, volatile-rich, accreting planetesimals [*Dauphas*, 2003]. Earth's atmosphere is deficient in Xe compared to Kr by more than a factor of ten compared to their relative abundance in carbonaceous chondrites (C-chondrite) [*Ozima and Podosek*, 1983]. The cause of the relatively low Xe abundance in Earth's atmosphere is not clear. By contrast, Earth's Xe/Kr ratio is similar to that of Venus and Mars, which are both similar to the solar ratio [*Pepin and Porcelli*, 2002]. Collectively, this may suggest that C-chondrites are anomalous rather than Earth. If it is Earth that has the anomalous relative abundance of Xe, alternative explanations include selectively poor outgassing of Xe from Earth's interior [*Ozima and Podosek*, 1983], absorption of Xe onto various minerals or rocks [*Bernatowicz et al.*, 1984; *Matsuda and Matsubara*, 1989], entrainment of Xe by H_2 during its hydrodynamic escape from the atmosphere [*Tolstikhin and O'Nions*, 1994], or loss of ionized xenon from the atmosphere [*Bernatowicz and Hagee*, 1987; *Marty*, 2012]. In addition to the sub-C-chondrite Xe/Kr ratio, terrestrial Xe shows substantial isotopic mass fractionation, and is enriched in heavy isotopes of Xe by about 3.5% per AMU compared to solar or chondritic Xe [*Pepin and Porcelli*, 2002]. As with the low apparent Xe abundance, the cause of this fractionation is not well understood. Recent evidence suggests that the isotopic mass fractionation of Xe was lower in the past, with Archean atmospheric Xe approaching the stable isotopic composition of chondritic Xe [*Pujol et al.*, 2011]. This is important for estimating the time of volatile loss from Earth because reconciling the atmosphere closure age determined from ^{129}I-^{129}Xe with that indicated by ^{244}Pu-^{136}Xe is complicated due to the large enrichment in ^{136}Xe caused by the mass fractionation. If the mass fractionation of atmospheric Xe is a feature that developed over most of Earth history, then I-Xe "ages" for Earth's atmosphere provide only a lower limit to the initial volatile loss of Earth. Models that include substantial post-Earth formation loss of Xe from the atmosphere push the I-Pu-Xe age closer to the beginning of Solar System formation, for example within 50–70 Ma [*Ozima and Podosek*, 1999].

Unlike the U-Pb and I-Pu-Xe systems, the bulk-Earth abundances of moderately volatile Rb (Tc = 800 K) and refractory Sr (Tc = 1464 K) are well constrained because neither are likely to have been incorporated in the core or lost from the atmosphere. The bulk-Earth's Rb/Sr ratio (^{87}Rb/^{86}Sr ~0.09) is about a factor of ten lower than that in CI chondrites, reflecting the non-chondritic abundance of volatile Rb in Earth. A planetary body with a CI chondrite ^{87}Rb/^{86}Sr = 0.925 [*Palme and O'Neill*, 2014]

experiences an increase in ^{87}Sr/^{86}Sr of approximately 0.0013 every 100 Ma. Refractory calcium-aluminum-rich inclusions (CAI) in primitive carbonaceous chondrites that would normally be used to pin the initial ^{87}Sr/^{86}Sr of the Solar System because of their ancient age and low Rb/Sr ratios, show a range of about 1.3 parts in 10,000 in their calculated initial ^{87}Sr/^{86}Sr [*Gray et al.*, 1973; *Podosek et al.*, 1991; *Hans et al.*, 2013], likely reflecting a combination of nucleosynthetic heterogeneity and stable Sr isotope mass fractionation [*Moynier et al.*, 2012] along with disturbance in the Rb-Sr system of the CAIs [*Podosek et al.*, 1991]. With a solar Rb/Sr ratio, this range in ^{87}Sr/^{86}Sr corresponds to 6.5 Ma of evolution if the range in ^{87}Sr/^{86}Sr were due purely to the ingrowth of ^{87}Sr via ^{87}Rb decay, establishing the precision limit with which Rb-Sr systematics can be used to track volatile depletion in planetary materials. Using the Solar System initial ^{87}Sr/^{86}Sr = 0.698975 ± 0.000008 suggested by *Hans et al.* [2013], a body with solar Rb/Sr ratio would have evolved to a ^{87}Sr/^{86}Sr >0.702, a typical low value seen in modern mid-ocean ridge basalts (MORB), by 4.35 Ga. This age is only a lower limit on the age of volatile depletion because the Sr isotopic composition of modern MORB reflects >4.5 billion years of evolution of the mantle's Sr isotopic composition with a non-zero ^{87}Rb/^{86}Sr ratio. An earlier sample of mantle Sr, for example the Sr contained in an Archean (3450 Ma) hydrothermal barite from Australia, provides the lowest initial ^{87}Sr/^{86}Sr determined for a terrestrial rock of 0.700502 [*McCulloch*, 1994]. This value is similar to that calculated for the initial Sr isotopic composition of mineral separates from 3.45 Ga komatiites from the Onverwacht Group in South Africa [*Jahn and Shih*, 1974]. Compared to the CAI initial Sr isotopic composition, the barite Sr provides a lower limit of 4.46 Ga to the time when Earth's Rb/Sr ratio became markedly lower than in CI chondrites (Fig. 8.2). Once again, this age overlaps with the simple U-Pb and I-Xe model ages for Earth's volatile depletion, but, as with I-Xe, most likely represents only a lower limit to the true time of volatile depletion. A potential measure of ^{87}Sr/^{86}Sr earlier in Earth history might be obtained from the Moon if the Moon formed primarily from Earth via a giant impact. If the lowest ^{87}Sr/^{86}Sr for a lunar rock (0.698987, [*Nyquist et al.*, 1974; *Carlson and Lugmair*, 1988]) provides an estimate for Earth's Sr isotopic composition earlier in its history than does the Archean barite, then Earth, or more likely its precursor materials, had to have been depleted in Rb compared to CI chondrites within less than a million years of Solar System formation (Fig. 8.2).

Potentially more precise estimates for the time of Earth's volatile depletion can be extracted from the short-lived ^{53}Mn-^{53}Cr decay scheme. ^{53}Mn has a half-life of 3.7 Ma. Different groups of meteorites display a correlated variability in ^{53}Cr/^{52}Cr vs. Mn/Cr ratio (Fig. 8.3) [*Lugmair*

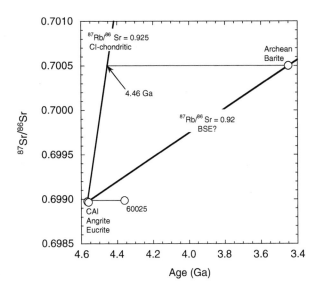

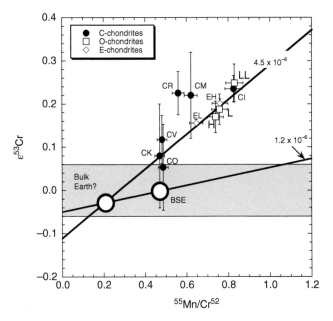

Figure 8.2 Sr isotope evolution diagram. Volatile-rich CI chondrites with $^{87}Rb/^{86}Sr = 0.925$ [*Palme and O'Neill*, 2014] evolve ^{87}Sr rapidly rising above $^{87}Sr/^{86}Sr = 0.701$ by 4.42 Ga starting from the Solar System initial Sr measured in calcium-aluminum-rich inclusions (CAIs) in primitive meteorites [*Hans et al.*, 2013]. Terrestrial $^{87}Sr/^{86}Sr$ at circa 3.45 Ga, as determined in barites [*McCulloch*, 1994] and calculated for komatiite minerals [*Jahn and Shih*, 1974], provide an estimate of BSE $^{87}Rb/^{86}Sr = 0.092$ that provides a lower limit to the time of Earth's volatile depletion of 4.46 Ga. If lunar rocks, such as the ancient crustal anorthosite 60025 [*Carlson and Lugmair*, 1988], provide an earlier measure of terrestrial Sr isotopic composition, then Earth must have obtained its low Rb/Sr ratio within 1 Ma of Solar System formation. Magmatic rocks from volatile depleted planetesimals such as the angrite and eucrite parent bodies also indicate that their volatile depletion was established within a million years or less of Solar System formation [*Hans et al.*, 2013].

Figure 8.3 Mn-Cr isochron diagram showing data for different groups of meteorites [*Trinquier et al.*, 2008; *Qin et al.*, 2010]. Excluding the CM, CR, and CK meteorite groups where there is concern about complete dissolution of refractory presolar grains [*Trinquier et al.*, 2008], the best fit line through the remaining data define a slope corresponding to a $^{53}Mn/^{55}Mn$ ratio of $(4.5 \pm 1.5) \times 10^{-6}$, which in turn corresponds to an age of 4565 ± 2 Ma. The large circles show the estimated bulk mantle $^{55}Mn/^{53}Cr$ ratio [*Palme and O'Neill*, 2014] at $\varepsilon^{53}Cr = 0$, and where the bulk-Earth (including core and mantle) would have to plot for its $^{55}Mn/^{52}Cr$ ratio [*Palme and O'Neill*, 2014] In order to fall on the meteorite isochron. The line connecting the large circles defines a slope that corresponds to an age approximately 7 Ma younger than the meteorite isochron.

and Shukolyukov, 1998; *Moynier et al.*, 2007; *Trinquier et al.*, 2008; *Qin et al.*, 2010]. The range in Mn/Cr ratio likely reflects the more volatile nature of Mn (Tc = 1158 K) compared to Cr (Tc = 1296 K [*Lodders*, 2003]). Mn-Cr data for chondrites define a slope of $(4.5 \pm 1.5) \times 10^{-6}$ on the Mn-Cr isochron diagram of Figure 8.3. Taking the D'Orbigny angrite, which has an initial $^{53}Mn/^{55}Mn = (3.24 \pm 0.04) \times 10^{-6}$ [*Glavin et al.*, 2004] and U-Pb age of 4563.37 ± 0.25 Ma [*Brennecka and Wadhwa*, 2012] as a reference point for Solar System $^{53}Mn/^{55}Mn$ evolution, the slope of the chondrite correlation shown in Figure 8.3 corresponds to an age of 4565 ± 2 Ma [*Qin et al.*, 2010], compared to a Solar System formation age of 4567.6 ± 0.4 Ma [*Jacobsen et al.*, 2008] to 4568.2 ± 0.4 Ma [*Bouvier and Wadhwa*, 2010; *Brennecka and Wadhwa*, 2012]. Earth's mantle has a $^{53}Cr/^{52}Cr$ ratio below that measured for chondrites, suggesting that Earth's volatile depletion (e.g., low Mn/Cr ratio in this case) was already present

while ^{53}Mn was extant. To use the Mn-Cr system to place a robust age on this volatile depletion, the possibility of Mn-Cr fractionation during core formation must be considered [*Palme and O'Neill*, 2014]. The near chondritic Na/Mn ratio of Earth's mantle suggests that very little of Earth's Mn resides in the core [*McDonough*, 2003]. The Cr/Mg ratio in the mantle (~0.01), however, is about 40% lower than that typical of chondrites (~0.024) implying that about 60% of Earth's Cr is in the core [*Palme and O'Neill*, 2014]. The BSE Mn/Cr ratio is estimated to be 0.417 ($^{55}Mn/^{52}Cr = 0.471$), which is higher than the $^{55}Mn/^{52}Cr$ ratio of 0.26 needed for the terrestrial mantle $^{53}Cr/^{52}Cr$ to fall on the chondrite isochron. The bulk-Earth's (including core) Mn/Cr ratio is 0.17 [*McDonough*, 2003]. Whether the core's $^{53}Cr/^{52}Cr$ ratio is the same as the mantle's depends on whether core formation occurred while ^{53}Mn was still extant. For the bulk-Earth to lie on the meteorite isochron with its Mn/Cr ratio of 0.17 ($^{55}Mn/^{52}Cr = 0.192$), it would have to have an $\varepsilon^{53}Cr$ of -0.03, whereas the present mantle, by definition, has $\varepsilon^{53}Cr = 0$. Given that the measurement precision on these numbers is of order ± 0.05 ε-unit, the mantle and bulk Earth do not have resolvable differences in Cr isotopic compositions. The Mn-Cr isochron slope connecting this point to the mantle point is 1.2×10^{-6}, nominally 7 Ma younger than the meteorite isochron in Figure 8.3. If the slope of the bulk-Earth – BSE tie line in Figure 8.3 has any age significance, it most likely reflects a model age for core-mantle differentiation, but Earth's volatile depletion would then predate this event. As with Rb-Sr, the Mn-Cr systematics of Earth, in comparison to chondrites, suggests that Earth's volatile depletion was established within, at most, a few million years of Solar System formation. Given that Earth may have taken several tens of millions of years to grow to most of its current mass [Chambers, 2004], our interpretation of these apparent ages is that volatile depletion was a characteristic of the building blocks of Earth. This suggests that the accreted planetesimals that make up most of Earth's mass grew in a hot part of the inner nebular where condensation of moderately volatile elements was very limited. Whether there are volatile loss events from Earth after it formed, for example during giant impacts that may be responsible for affecting relative noble gas abundances in the mantle [Tucker and Mukhopadhyay, 2014], must be evaluated in the context that Earth formed volatile depleted, and likely never had CI or Solar abundances of even moderately volatile elements.

8.3. CORE FORMATION

In addition to the evidence for volatile element depletion early in Earth history, the consequences of core formation on mantle abundances of core-soluble (siderophile) elements also is clear. Compared to chondritic abundances, Earth's mantle is depleted in siderophile elements by factors of a few to as much as a few hundred (Fig. 8.1b). Most siderophile elements in the mantle, however, are present in abundances that are higher than would be predicted for relatively low temperature and pressure metal-silicate equilibration. Consequently, the issue of whether or not the core and mantle are in chemical equilibrium has been the subject of considerable debate [Righter, 2011; Rubie et al., 2011]. Much of the debate hinges on predictions originally made by Ringwood [1977] and later by Murthy [1991], that metal-silicate partition coefficients for siderophile elements are likely to be functions of temperature and pressure. Numerous experimental determinations of siderophile element metal-silicate partition coefficients at increasingly higher temperatures and pressures have shown that mantle abundances for some siderophile elements can be accounted for by metal-silicate equilibration at high temperatures (> 2700 K) and pressures (> 30 GPa) [Bouhifd and Jephcoat, 2011; Righter

et al., 2014], but this conclusion is mainly true for the so-called moderately siderophile elements that include Co, Ni, W, Ag, and Mo. These are elements that for 1 atmosphere experiments are characterized by metal silicate distribution coefficients of 1 to 1000. The effects of potentially changing oxygen fugacities [Wade and Wood, 2005; Siebert et al., 2013], and the presence of different light elements (e.g., C, S, Si) in the core [Willis and Goldstein, 1982; Jones and Walker, 1991; Chabot et al., 2010] also have been experimentally explored for siderophile elements. The now rich literature on this subject is reviewed in more detail in Chapter 6.

The highly siderophile elements (HSE) include the platinum-group elements Ir, Os, Ru, Pt, Rh, and Pd, along with Au and Re. These elements have extremely high 1 atmosphere metal-silicate partition coefficients of $> 10^4$ [Borisov and Palme, 1997]. Under some high P-T experimental conditions, the distribution coefficients for Au, Pd, and Pt have been shown to decrease sufficiently to account for their mantle abundances by metal-silicate partitioning [Cottrell and Walker, 2006; Righter et al., 2008; Wheeler et al., 2010]. Other HSE, such as Os, Ir, and Ru, however, do not show sufficient decreases in their metal-silicate partition coefficients to account for their mantle abundances in a similar manner [Brenan and McDonough, 2009; Mann et al., 2012]. Perhaps more problematical to accounting for their unexpectedly high abundances in the mantle by high P-T metal-silicate partitioning is the observation that the HSE are present in the mantle at relative abundances very similar to their relative abundances in chondrites. In detail, there are small differences in Ru/Ir and Pd/Ir ratios between chondrites and estimates of mantle HSE abundances (Fig. 8.4a) that are also seen in impact melts from the Moon (Fig. 8.4b) indicating that the late impactor flux accreted to Earth and Moon did not have precisely chondritic HSE patterns [Becker et al., 2006].

The most stringent test of a "chondritic" mantle in terms of HSE abundances is provided by the ^{187}Re-^{187}Os isotopic system. The decay of ^{187}Re to ^{187}Os provides a time-integrated measure of the Re/Os ratio in the mantle that is unaffected by the elemental fractionation that accompanies recent partial melting events [Shirey and Walker, 1998; Carlson, 2005]. The estimated present day $^{187}Os/^{188}Os$ ratio of the primitive mantle is 0.1296 ± 0.0010 [Meisel et al., 2001], which is within the range measured for ordinary and enstatite chondrites [Meisel et al., 2001; Walker, 2009]. This observation requires that the Re/Os ratio of Earth's mantle has been within about 5% of an average chondritic ratio for most of Earth history. To explain the generally chondritic relative abundances of the HSE in the mantle by metal-silicate partitioning, therefore, would require that metal-silicate distribution coefficients converge to an essentially single value for all

(a)

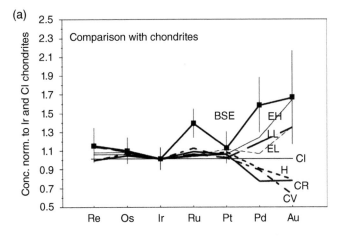

(b)

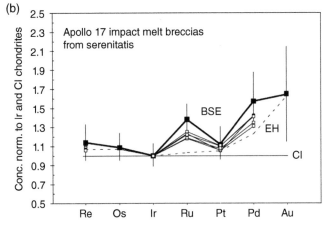

Figure 8.4 CI-chondrite and Ir normalized HSE abundances in (a) different types of meteorites and the BSE, and (b) in Apollo 17 impact melt breccias. Both figures after *Becker et al.* [2006] with the data in part b from *Norman et al.* [2002].

HSE at a certain temperature-pressure condition. To date, no experiments indicate that this may be the case, and some experiments indicate that it is not the case [*Mann et al.*, 2012]. Collectively these results suggest that the mantle abundances of the moderately and highly siderophile elements were established by different processes.

A clue to the origin of the HSE in the mantle comes from comparisons of the Earth with other planetary bodies. The absolute and relative abundances of the HSE for Mars and the Moon are, unfortunately, much more poorly constrained than for Earth. No *bona fide* mantle samples for either Mars or the Moon exist in our collections, so far as is known. Consequently, inferences about bulk mantle HSE abundances on these planetary bodies must be derived from the analysis of basalts. For Earth, fractionation of the HSE during partial melting of mantle lithologies can be severe, depending on factors such as the degree of melting and oxidation state, so extrapolation from basalt abundances to mantle source characteristics

is problematic. Nevertheless, chondrite normalized HSE pattern shapes for martian and lunar basalts are generally similar to those of terrestrial basalts with comparable MgO [*Day et al.*, 2007; *Brandon et al.*, 2012]. This suggests that the relative partitioning of the HSE between mantle and melt are similar in these two bodies with the implication that the HSE contents of their mantles are in broadly chondritic relative abundances. More importantly, the Os isotopic compositions of the mantles of both the Moon [*Birck and Allègre*, 1994; *Walker et al.*, 2004; *Day et al.*, 2007] and Mars [*Birck and Allègre*, 1994; *Brandon et al.*, 2000] appear to be mostly within the range of chondritic meteorites, like the terrestrial primitive mantle. This requires that the long-term Re/Os ratios of portions of both mantles have been precisely chondritic. Further, HSE abundances in martian basalts suggest that the martian mantle has very similar absolute abundances to the terrestrial mantle [*Brandon et al.*, 2012]. By contrast, the lunar mantle appears to be depleted in HSE by a factor of ≥20, compared to the terrestrial mantle [*Day et al.*, 2007].

One other planetary body that may shed light on the siderophile issue is the asteroid Vesta, from which the howardite-eucrite-diogenite meteorite suite likely derives. Of this meteorite suite, the diogenites sample either crustal cumulates from mafic melts or mantle materials. Diogenites are characterized by highly variable HSE absolute abundances [*Dale et al.*, 2012; *Day et al.*, 2012]. Samples with high HSE abundances are characterized by chondritic relative abundances of HSE (and chondritic $^{187}Os/^{188}Os$), whereas the meteorites with lowest HSE abundances are characterized by fractionated, non-chondritic HSE patterns (and suprachondritic $^{187}Os/^{188}Os$). Although some of the HSE-rich rocks could have been modified by impact-derived contamination, *Day et al.* [2012] interpreted both high and low abundance HSE signatures to mainly reflect compositions resulting from primary differentiation. The apparently chondritic relative abundances of HSE in the martian, lunar, and possibly vestan mantles make appealing to high P-T metal-silicate partitioning problematic. The P-T conditions needed to reconcile moderately siderophile element partition coefficients with mantle abundances in Earth are much higher than would be present during core formation on Mars, the Moon, or particularly Vesta. Thus, another mechanism is necessary to establish the HSE in the mantles of these planets, moons, and planetesimals.

Understanding the origin of the siderophile element abundances in the mantle is critical to the application of the ^{182}Hf-^{182}W system to determine the timing of Earth's core formation. The Hf-W system is particularly amenable to constraining the timing of metal-silicate separation because W is siderophile, but Hf is not, so metal-silicate separation will create a metal phase with an Hf/W ratio near zero, and hence eventually unradiogenic W, whereas the complementary silicate will have a high Hf/W ratio and will evolve radiogenic W. Two important conclusions for the discussion here that come from Hf-W study results are that (1) metal-silicate separation began on planetesimals within 1 to 1.5 Ma of the start of Solar System formation as evidenced by Hf-W ages for weakly irradiated magmatic iron meteorites [*Kruijer et al.*, 2012; *Kruijer et al.*, 2014], and (2) the 200 ppm higher $^{182}W/^{184}W$ ratio in Earth's mantle compared to chondrites indicates that core formation on Earth occurred while ^{182}Hf (8.9 Ma half-life) was still extant. The issue of equilibration between metal and silicate also significantly affects the interpretation of Hf-W ages for core formation. For example, if during an accretionary event to Earth, there was no equilibration between metal and silicate as the impacting planetesimal's iron merged with Earth's core, the Hf-W age would reflect the mean age of metal-silicate separation on the accreted planetesimals, not on Earth. On the other hand, if reequilibration between Earth's core and mantle occurred with every major impact into Earth, then the Hf-W age deduced from the W isotopic composition of Earth's mantle would reflect the time of the last major impact. Reality likely lies between these extremes, so the question of how much W isotope exchange there was between the metal in accreting planetesimals and Earth's mantle as the newly delivered metal found its way into Earth's core has been a subject of active investigation, both for its consequences for Hf-W chronology, but also the process of core formation on Earth [*Halliday and Kleine*, 2005; *Jacobsen*, 2005; *Nimmo and Agnor*, 2006; *Kleine et al.*, 2009; *Rudge et al.*, 2010]. A more recent update on these issues can be found in Chapter 5.

8.4. LATE ACCRETION

The near chondritic relative abundances of the HSE in the terrestrial mantle spawned alternative explanations to equilibrium partitioning between mantle and core that include inefficient removal of iron metal from the mantle during core formation [*Jones and Drake*, 1986] and accretion of materials with chondritic bulk HSE compositions after chemical exchange between core and mantle had stopped [*Kimura et al.*, 1974; *Chou et al.*, 1983; *Wanke et al.*, 1984]. The latter explanation has been termed "late accretion" or "late veneer," and is often incorrectly confused with "late heavy bombardment" that refers to lunar basin forming impacts occurring circa 3.9 to 4.0 Ga. The idea of a late spike in accretion rate derived originally from the dominance of ~3.9 Ga ages in lunar impact rocks [*Tera et al.*, 1974]. More recent considerations of lunar impact history suggest that the impact record is biased to young ages due to younger impacts covering the deposits created by older impacts [*Chapman et al.*, 2007;

Norman et al., 2010] and that the crustal and thermal structure of the Moon may have caused similar size impacts to create larger basins on the Earth-facing side of the Moon than on the far side [*Miljkovic et al.*, 2013]. Both results work in the direction of reducing the estimate for the amount of impacting material arriving during the period of the late heavy bombardment. As a result, the late heavy bombardment, if there even was one, almost certainly contributed too little mass to have any consequence for the bulk composition of Earth or the Moon.

If the majority of HSE in Earth's mantle were added by accreting planetesimals with chondritic bulk HSE compositions that arrived after chemical communication between core and mantle had become insignificant, then mantle HSE abundances can be used to constrain the mass of late accretion. For example, in the extreme case where the mantle was totally stripped of Ir by core formation, addition of 2.35×10^{25} g of average chondrite having 600 ng/g Ir [*Horan et al.*, 2003] to the 4×10^{27}g of mantle would leave the mantle with its observed 3.5 ng/g Ir concentration. This "late veneer," thus, need only represent the accretion of about half a percent of the mass of Earth after core formation was complete in order to explain both the abundances and the abundance ratios of the HSE in Earth's mantle. The variable HSE abundances in different meteorites [*Horan et al.*, 2003] and the unknown efficiency of transfer of HSE from impactors to Earth's core during the late accretion led to uncertainty in the exact mass of late accreting material, but the HSE constraint shows that a small mass of late accreting material can have a big effect on mantle HSE abundances.

With the exception of the improved precision in the estimates of mantle HSE abundances introduced by Re-Os isotope analyses, the arguments presented above supporting a late addition of material with bulk chondritic compositions to Earth after core formation was complete, have changed little over the last 40 years. Recently, however, two measurements have allowed new insight into the question of the origin of the HSE in the terrestrial mantle.

8.4.1. Isotopic Composition of the Late Veneer

Improved precision in isotopic analysis over about the last decade has revealed a number of elements that display isotopic variability of nucleosynthetic origin that distinguish different groups of meteorites from the isotopic composition of the terrestrial planets [*Trinquier et al.*, 2007; *Trinquier et al.*, 2009; *Qin et al.*, 2010; *Steele et al.*, 2011; *Warren*, 2011]. Two of the elements that fall into this category are the siderophile elements Mo [*Dauphas et al.*, 2002; *Burkhardt et al.*, 2011] and Ru [*Dauphas et al.*, 2004; *Chen et al.*, 2010]. Of these, Ru is one of the HSE. As such, if late accretion accounts for most of the HSE in the mantle, the Ru isotopic composition of the mantle will be dominated by the material added late in Earth's accretion history. Compared to Earth, some C-chondrites, and most iron meteorites, show deficits in ^{100}Ru (Fig. 8.5) that reflect depletion in the isotopes produced by s-process nucleosynthesis [*Dauphas et al.*, 2004; *Chen et al.*, 2010]. Other meteorite groups are less distinct in Mo and Ru isotopic composition compared to Earth, and some (the CM chondrite Murchison) show slight excesses in ^{100}Ru [*Chen et al.*, 2010]. With the limited data available at this time, the only meteorite group with Ru (and Mo) isotopic composition that overlaps that of Earth's mantle are the IAB iron meteorites [*Chen et al.*, 2010; *Fischer-Gödde et al.*, 2013]. This observation, as will be discussed later, extends beyond the siderophile elements and may rule out C-chondrites as the primary constituent of material added to Earth late during its accretionary history, but a conclusive statement on this point awaits more Ru isotope data for C-chondrites. A similar conclusion was suggested previously by the observation that the ^{187}Os/^{188}Os isotopic composition of the modern mantle overlaps that of enstatite and ordinary chondrites, but is slightly higher than seen in C-chondrites [*Walker et al.*, 2002].

Assessing whether or not C-chondrites were substantively involved in late accretion is important. Carbonaceous chondrites have played particularly important roles in

some models of Earth accretion because, not only can they potentially explain the mantle abundances of HSE, but also because some C-chondrites contain substantial quantities of water and other volatile elements. Thus, late accretion is a possible means for adding water and organic-rich materials to the planet. The mass of water in Earth's oceans is about 1.4 x 10^{24}g. For this mass of water to be added by the amount of late accreting material suggested by the HSE constraints would require the late accreting material to have on average ~6 wt% H$_2$O. This much water is within the range seen in C-chondrites, but it is much higher than typical of other groups of chondrites. Adding Earth's water through modest late accretion is an attractive option because of the evidence discussed previously that Earth likely formed volatile element depleted. In addition, the hydrogen and nitrogen isotopic compositions of Earth overlap those of C-chondrites [*Alexander et al.*, 2012], as do the ratios of S/Se and Se/Te [*Wang and Becker*, 2013], both of which support the argument that much of Earth's volatile content was added by post-core formation accretion of a small amount of volatile-rich material such as C-chondrites.

Other water-rich objects, such as comets, also are an often-invoked source of Earth's water. At their very high water concentrations (tens of weight percent), the mass of comet that would have to be added to explain Earth's ocean mass would be small, and hence would have little effect on other element abundances, including the HSE. The prime evidence against comets being the source of Earth's water is their distinct hydrogen isotopic composition [*Villanueva et al.*, 2009; *Drake*, 2005]. Recently, comet 103P/Hartley 2 was found to have a deuterium to hydrogen ratio similar to Earth [*Hartogh et al.*, 2011], so a cometary source of volatiles remains an option, although this comet has a nitrogen isotopic composition unlike Earth's atmosphere, which appears to limit the cometary contribution to Earth's volatile abundance to no more than a few percent [*Marty*, 2012]. The other possibility is that Earth's oceans reflect efficient outgassing of water from Earth's interior [*Elkins-Tanton*, 2008]. Adding 1.4 × 10^{24} g of water into the 4 × 10^{27} g of Earth's mantle raises its water concentration by 350 ppm, and with a water concentration of 140 ppm in the mantle source of mid-ocean ridges basalts [*Saal et al.*, 2002], would place the BSE water concentration at ~500 ppm, which is within the range predicted by various other approaches [*Dixon and Clague*, 2001; *Palme and O'Neill*, 2014]. Whether significant water was added to Earth by late accretion remains an open question, but the nucleosynthetic anomalies in Ru, Mo, and other elements are providing increasingly strong evidence that at least C-chondrites were not a major, or perhaps even minor, component of Earth's building blocks.

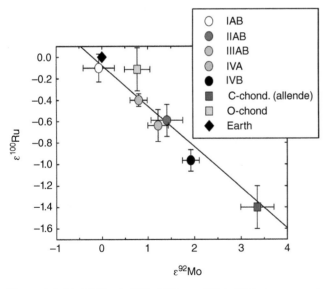

Figure 8.5 Variability in ^{100}Ru/^{101}Ru and ^{92}Mo/^{96}Mo (expressed in parts in 10,000 difference (ε) from the terrestrial standard) in different groups of meteorites. The line shows the expected mixing line between s-process and r-process rich end members. Figure originally from *Dauphas et al.* [2004] updated with data from *Burkhardt et al.* [2011] and *Chen et al.* [2010].

8.4.2. W Isotope Variability

Recent work that pushes the precision of the $^{182}W/^{184}W$ ratio measurements into the few ppm range have revealed that a number of Archean crustal rocks display excesses of about 5 to 20 ppm in $^{182}W/^{184}W$ ratio, compared to the modern mantle (Fig. 8.6) [*Willbold et al.*, 2011; *Touboul et al.*, 2012; *Touboul et al.*, 2014]. The two plausible explanations for these excesses are (a) the Archean/Hadean rocks formed from a more radiogenic (in ^{182}W) portion of the mantle that remained after core formation, but before late accreting, chondritic, material was homogenized into the mantle, leading to a drawing down of mantle $^{182}W/^{184}W$, and (b) the W isotope signal was not solely controlled by core formation, but also by Hf-W fractionation occurring during the lifetime of ^{182}Hf, caused by crystal-liquid fractionation in a terrestrial magma ocean. If the former, a correlation between ^{182}W excess and depletion in HSE abundances might be expected in the source of the Archean crustal rocks. An analysis of HSE

abundances in komatiites erupted throughout the Archean suggested that mantle HSE abundances increased between the Eoarchean and Neoarchean [*Maier et al.*, 2009]. In contrast, the komatiite suite for which elevated $^{182}W/^{184}W$ has been measured does not have HSE abundances that would suggest their derivation from an HSE-depleted source. For example, the 2.82 Ga Kostomuksha komatiites have an average $\mu^{182}W$ of +15 ppm, compared to 3.47 Ga Komati komatiites that show no resolved excess in ^{182}W ($\mu^{182}W = 2.7 \pm 4.1$; Fig. 8.6), yet *Touboul et al.* [2012] concluded that the mantle sources for both the komatiites contained >80% of modern HSE abundances.

8.5. DIFFERENTIATION OF THE SILICATE EARTH

8.5.1. Establishing the Isotopic Baseline to Detect Differentiation Events

The nucleosynthetic anomalies in Ru and Mo that were mentioned previously are just two elements out of many that are now known to show isotopic distinctions between meteorites and Earth. Among other elements displaying isotopic differences attributable to distinct nucleosynthetic components are Ca [*Simon et al.*, 2009; *Huang et al.*, 2012], Ti [*Trinquier et al.*, 2009], Cr [*Trinquier et al.*, 2007; *Qin et al.*, 2010], Sr [*Qin et al.*, 2011; *Moynier et al.*, 2012; *Hans et al.*, 2013], Ba, Nd, Sm [*Andreasen and Sharma*, 2007; *Carlson et al.*, 2007; *Gannoun et al.*, 2011a; *Qin et al.*, 2011; *Boyet and Gannoun*, 2013], and Hf [*Qin et al.*, 2011]. The most likely explanation for the isotopic variability in these elements is that the different meteorite parent bodies, and the terrestrial planets, sample different mixtures of the various stellar contributions to the Solar nebula, reflecting their formation in different regions in the nebula. An open question, however, is whether, and how, these small isotopic variations in mostly rare elements translate into compositional characteristics of the terrestrial planets. Early attempts to convert the observed isotopic variability into proportions of different meteorite building blocks in the formation of Mars and Earth highlight the relatively minor role C-chondrites may have played in terrestrial planet formation [*Warren*, 2011].

One isotope system that has been extensively used for investigating the compositional evolution of the terrestrial planets is Sm-Nd. The combination of the long-lived ^{147}Sm-^{143}Nd ($t_{1/2}$ = 106 Ga) and short-lived ^{146}Sm-^{142}Nd ($t_{1/2}$ = 103 Ma or 68 Ma as proposed by *Kinoshita et al.* [2012]) decay schemes coupled with the predictable geochemical behavior of these two neighboring rare earth elements (REE) has provided significant insight into the various processes that have affected mantle composition [*DePaolo and Wasserburg*, 1976; *Jacobsen*

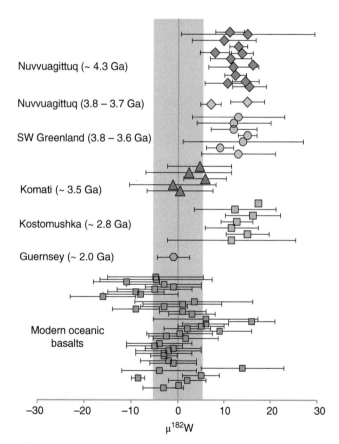

Figure 8.6 Differences in $^{182}W/^{184}W$ compared to the terrestrial standard (gray field), reported in parts per million μ. Data from *Willbold et al.* [2011]; *Touboul et al.* [2012, 2014].

and Wasserburg, 1979; Allègre et al., 1983b]. Boyet and Carlson [2005] first showed that the ^{142}Nd/^{144}Nd ratios of all modern Earth rocks are displaced to higher values than seen in the majority of chondritic meteorites. Excess ^{142}Nd can be imparted either by decay of ^{146}Sm coupled with a superchondritic Sm/Nd ratio, or by an excess in s-process nucleosynthetic products. Leaching/selective dissolution experiments on C-chondrites [*Qin et al.*, 2011] and enstatite chondrites (E-chondrites) [*Boyet and Gannoun*, 2013] have greatly expanded the range of known nucleosynthetic anomalies in Nd (Fig. 8.7). Various Solar System materials plot along the mixing line that extends from nearly pure s-process Nd in pre-solar SiC grains [*Hoppe and Ott*, 1997] to rare CAIs that display significant excesses in r-process Nd isotopes [*Lugmair et al.*, 1978; *McCulloch and Wasserburg*, 1978]. As with Ru and Mo isotope variation, terrestrial Nd lies at one end of the mixing line of r-, s-process produced Nd, where C-chondrites occupy the other end (Fig. 8.8). Most groups of meteorites, including those derived from differentiated asteroids, such as eucrites and angrites, lie somewhere between C-chondrites and Earth and are similar to the isotopic composition of ordinary chondrites (O-chondrites) (Fig 8.8). Of the meteorite groups for which data exist, only about a third of E-chondrites overlap in ^{142}Nd/^{144}Nd with Earth, with the remainder having lower ^{142}Nd/^{144}Nd similar to the ratios measured in O-chondrites [*Gannoun et al.*, 2011a].

Whether the offset of Earth from the majority of the chondrite data also reflects a different nucleosynthetic mixture potentially can be solved by a comparison of ^{142}Nd vs. ^{148}Nd variability, as ^{148}Nd will be affected by s-, r-process mixing, but not ^{146}Sm decay (Fig. 8.8). The s-, r-process mixing lines for leaches of E- and C-chondrites

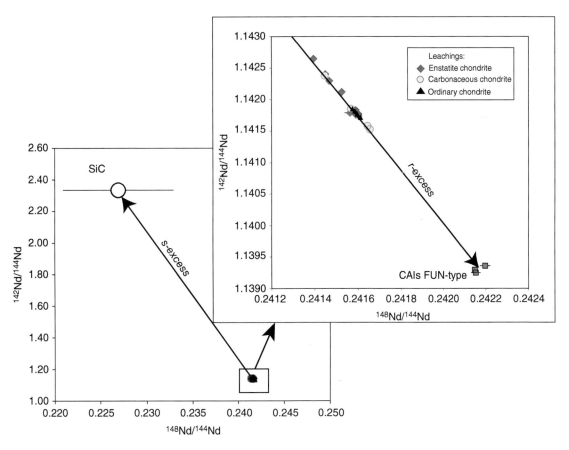

Figure 8.7 Nucleosynthetic variability in Nd isotopic composition reflecting mixing between an almost pure s-process end member, presolar SiC [*Hoppe and Ott*, 1997], with r-process enriched "FUN type" CAIs (grey squares [*Lugmair et al.*, 1978; *McCulloch and Wasserburg*, 1978]). Along the mixing line between these two s- and r-process enriched end members are leaches and residues from C- and O-chondrites [*Qin et al.*, 2011] and E-chondrites [*Boyet and Gannoun*, 2013]. ^{142}Nd/^{144}Nd ratios are corrected for the radiogenic decay of ^{146}Sm over the age of the Solar System based on a common evolution and a chondritic ^{147}Sm/^{144}Nd ratio of 0.196 [*Bouvier et al.*, 2008].

(a)

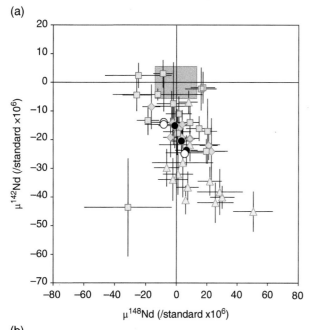

(b)

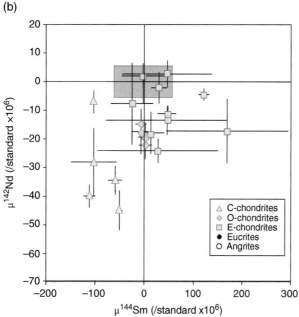

Figure 8.8 Whole rock measurements of Nd and Sm isotopic composition in a variety of chondrites and the eucrite and angrite magmatic meteorites. Ratios include $^{142}Nd/^{144}Nd$, $^{148}Nd/^{144}Nd$ and $^{144}Sm/^{152}Sm$, all expressed in parts per million deviation from the terrestrial standard. Gray box reflects the range of values measured in samples derived from the modern terrestrial mantle. [Data from *Boyet and Carlson*, 2005; *Andreasen and Sharma*, 2006, 2007; *Carlson et al.*, 2007; *Gannoun et al.*, 2011a; *Sanborn et al.*, 2014].

approach is that at terrestrial $^{148}Nd/^{144}Nd$, the chondrite data show a wide range in $^{142}Nd/^{144}Nd$ that may reflect the additional contribution to ^{142}Nd from direct p-process synthesis of ^{142}Nd and ^{146}Sm (Fig. 8.8b). Carbonaceous chondrites show a deficit in pure p-process ^{144}Sm [*Andreasen and Sharma*, 2007; *Carlson et al.*, 2007], and hence, likely both pure p-process ^{146}Sm and the p-process contribution to ^{142}Nd that translates to additional lowering of the $^{142}Nd/^{144}Nd$ in C-chondrites. By contrast, the majority of E-chondrites have elevated $^{144}Sm/^{152}Sm$ compared to terrestrial Sm [*Gannoun et al.*, 2011a]. Data for O-chondrites, all of which have terrestrial $^{148}Nd/^{144}Nd$ and $^{144}Sm/^{152}Sm$, have $^{142}Nd/^{144}Nd$ between 15 and 25 ppm lower than terrestrial. Given the lack of evidence for a nucleosynthetic distinction between O-chondrites (and eucrites/angrites) in stable Nd and Sm isotopic composition, the offset in $^{142}Nd/^{144}Nd$ between O-chondrites and Earth likely is best explained by ^{146}Sm decay in a portion of the mantle that had superchondritic Sm/Nd ratio during the lifetime of ^{146}Sm. The precision of both ^{142}Nd, but particularly ^{148}Nd and ^{144}Sm data for meteorites, however, is not sufficient at this time to convincingly rule out a nucleosynthetic origin for the ^{142}Nd variability. As with the I-Xe, Hf-W, and Mn-Cr evidence discussed previously, the ^{146}Sm-^{142}Nd system appears to point to significant Earth differentiation driven by events occurring during the first few hundred million years of Earth history.

8.5.2. A Layered or Nonchondritic Earth?

On the assumption that the circa 20 ppm elevated $^{142}Nd/^{144}Nd$ ratio of the modern terrestrial mantle, in comparison to O-chondrites, reflects the decay of ^{146}Sm at a superchondritc Sm/Nd ratio, *Boyet and Carlson* [2005, 2006] developed a model for an early-formed compositional reservoir–the "early depleted reservoir" (EDR)–that is characterized by a Sm/Nd ratio about 6% higher than chondritic. If formed at 4.568 Ga, the EDR, to evolve the 20 ppm excess in ^{142}Nd, would need a Sm/Nd ratio that would lead it to have a present day $\varepsilon^{143}Nd$ of +7 (model 1 in Figure 8.9), which is similar to the most common value seen for intraplate oceanic basalts [*Zindler and Hart*, 1986], and just slightly below the average of all mid-ocean ridge basalts (MORB) [*Gale et al.*, 2013]. Delaying the time when the EDR obtained this superchondritic Sm/Nd ratio by as much as 30 Ma would require the EDR to have a higher Sm/Nd ratio in order to produce the chondrite-Earth offset in $^{142}Nd/^{144}Nd$. This would result in a present day $\varepsilon^{143}Nd$ in excess of values typical of MORB, the major reservoir in the modern mantle that has the highest $\varepsilon^{143}Nd$ (model 4 in Figure 8.9). As pointed out by *Bourdon et al.* [2008], considering the additional increase in the mantle's Sm/Nd ratio caused by extraction of continental crust over Earth history would

cross the terrestrial value for $^{148}Nd/^{144}Nd$ at a $^{142}Nd/^{144}Nd$ ratio between 25 ppm (C-chondrites [*Qin et al.*, 2011] and 6 ppm (E-chondrites [*Boyet and Gannoun*, 2013]) lower than the modern terrestrial value. Complicating this

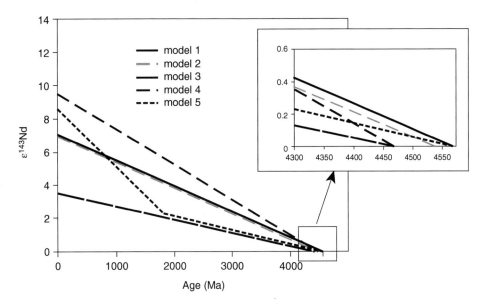

Figure 8.9 Different evolutionary models for the Sm/Nd ratio. Model 1: One stage model: $^{147}Sm/^{144}Nd=0.208$ since 4.568 Ga reaches present day $^{142}Nd/^{144}Nd$ 20 ppm higher than O-chondrite. Model 2: Two stage model: $^{147}Sm/^{144}Nd = 0.196$ (chondritic) during the first 30 Ma, then 0.208 until today. In this model, the present day $^{142}Nd/^{144}Nd$ of this reservoir is only 15 ppm higher than O-chondrite. Models 3 to 5 lead to 10-ppm excesses in $^{142}Nd/^{144}Nd$ compared to O-chondrites (assumes 10 ppm of the O-chondrite—Earth difference comes from nucleosynthetic anomalies). Model 3: Two stage model: chondritic evolution during the first 100 Ma then $^{147}Sm/^{144}Nd=0.2019$ until today. Model 4: Two stage model: chondritic evolution during the first 100 Ma then $^{147}Sm/^{144}Nd=0.2125$ until today. Model 5: Two stage model: $^{147}Sm/^{144}Nd=0.2025$ until 1.8 Ga then $^{147}Sm/^{144}Nd=0.223$ until today.

push this age limit back in time by an amount that depends on the details of continent formation, particularly its mean age and the volume of mantle affected by continent extraction.

Compositional estimates for the portion of Earth's mantle that either can be sampled directly or that provides melts that form the crust, referred to here as "accessible Earth," traditionally start from the assumption that refractory lithophile elements, such as Sm and Nd, are present at chondritic relative abundances [*Jagoutz et al.*, 1979; *McDonough and Sun*, 1995; *Palme and O'Neill*, 2014]. Changing this baseline bulk-Earth compositional approximation to account for the ^{142}Nd evidence for a superchondritic Sm/Nd ratio propagates into the broader compositional estimates for the mantle. This is because these estimates often use ratios of elements of similar compatibility, or deduced from radiogenic isotope constraints, to fill in the abundance estimates for elements that cannot be predicted from chondritic starting models [e.g., *Salters and Stracke*, 2004]. The compositional consequences of a superchondritic Sm/Nd ratio for the accessible Earth have been explored by numerous papers and are summarized in Table 8.1 [*Boyet and Carlson*, 2005, 2006; *Carlson and Boyet*, 2008; *Caro and Bourdon*, 2010; *Jackson et al.*, 2010; *Caro*, 2011; *Jackson and Carlson*, 2011; *Jackson and Jellinek*, 2013]. Some of the major implications of a non-chondritic composition for the

accessible Earth include (a) the abundances of the main heat-producing radioactive elements U, Th, and K in the EDR are about 60% of those predicted from chondritic primitive mantle models (Table 8.1), implying substantially reduced internal radiogenic heating of the mantle with consequences for the thermal evolution of Earth's interior; (b) most of Earth's radiogenic ^{40}Ar is in the atmosphere, rather than ~50% being retained in the mantle, as currently assumed [*Allègre et al.*, 1983a]; (c) the source of the oceanic intraplate basalts characterized by high $^3He/^4He$ ratios may be remaining portions of the EDR, less affected by continent extraction than the source of MORB; (d) the volume of the mantle source depleted by continent extraction occupies not a small fraction, but most of the mantle; and (e) the mantle source for MORB is less depleted than estimates (Table 8.1) that assume that all of its incompatible element depletion is due to extraction of a continental crust of circa 2 Ga mean age [*Workman and Hart*, 2005].

The degree to which the chondrite-Earth difference in $^{142}Nd/^{144}Nd$ affects estimates of mantle composition depends on how much of this difference is the result of different mixtures of nucleosynthetic products, as opposed to ^{146}Sm decay in an Earth with superchondritic Sm/Nd ratio. For example, if Earth began with only a circa 10 ppm excess in $^{142}Nd/^{144}Nd$ compared to O-chondrites, as it would if it were isotopically similar to E-chondrites

Table 8.1 Composition of Major Terrestrial Reservoirs

Reservoir Reference	BSE M&S	DMM W&H	DMM S&S	DMM B&C	EDR B&C	SCHEM C&B	NCM J&J	EER (26%) CB	EER (4%) CB
SiO_2	45.0	44.7	44.9		45.0		44.9	44.9	44.5
TiO_2	0.2	0.13	0.19		0.14		0.17	0.35	0.13
$A_{l2}O_3$	4.4	4.0	4.3		4.3		4.29	4.6	6.1
FeO	8.1	8.2	8.1		8.1		8.09	8.0	7.3
MgO	37.8	38.7	38.2		38.0		38.14	37.3	33.4
CaO	3.5	3.2	3.5		3.5		3.44	3.6	4.1
Na_2O	0.4	0.13	0.29		0.31		0.28	0.50	1.4
(ppm)									
Rb	0.60	0.05	0.088	0.10	0.38		0.33	1.23	5.53
Ba	6.60	0.563	1.20	1.37	3.25		4.4	16.1	65.8
Th	0.080	0.0079	0.014	0.016	0.047		0.055	0.17	0.80
U	0.020	0.0032	0.0047	0.0054	0.014		0.014	0.039	0.19
K	240	50	60	68	150		166	501	2177
Nb	0.66	0.15	0.21	0.24	0.30		0.65	1.68	9.23
La	0.65	0.19	0.234	0.26	0.38		0.47	1.40	6.83
Ce	1.68	0.55	0.772	0.87	1.12		1.26	3.27	14.5
Pb	0.15	0.018	0.0232	0.035	0.095		0.10	0.31	1.32
Sr	20	7.66	9.80	11	12.9		15.5	40	182
Nd	1.25	0.581	0.713	0.806	0.921		1.01	2.19	8.89
Hf	0.28	0.157	0.199	0.217	0.24		0.24	0.41	1.34
Sm	0.41	0.239	0.270	0.293	0.318		0.35	0.657	2.51
Eu	0.15	0.096	0.107	0.115	0.12		0.14	0.24	0.91
Gd	0.54	0.358	0.395	0.422	0.44		0.49	0.83	2.91
Er	0.44	0.348	0.371	0.386	0.40		0.41	0.55	1.39
Yb	0.44	0.365	0.401	0.401	0.41		0.42	0.52	1.16
Lu	0.07	0.058	0.063	0.063	0.065		0.064	0.075	0.14
$^{87}Rb/^{86}Sr$	0.088	0.0072	0.0260	0.0263	0.0857	0.061	0.0609	0.0905	0.0891
$^{147}Sm/^{144}Nd$	0.1963	0.2487	0.2289	0.2200	0.2085	0.2082	0.2081	0.1817	0.1709
$^{176}Lu/^{177}Hf$	0.0339	0.0524	0.0449	0.0413	0.0387	0.0375	0.0377	0.0259	0.0147
Th/U	3.92	2.47		2.90	3.46			4.37	4.21
$^{238}U/^{204}Pb$	8.68	11.51	13.12	9.93	9.22		9.07	8.19	9.14
$^{87}Sr/^{86}Sr$	0.70483	0.70263	0.7027	0.70295	0.70466	0.7030	0.7030	0.70498	0.70489
$^{143}Nd/^{144}Nd$	0.51264	0.51313	0.5131	0.51316	0.51300	0.51299	0.51300	0.51219	0.51186
$^{176}Hf/^{177}Hf$	0.28284	0.28326	0.28328	0.28337	0.28327	0.28313	0.28314	0.28213	0.28113
$^{206}Pb/^{204}Pb$	18.25	18.275		19.07	18.80			17.75	18.74
$^{207}Pb/^{204}Pb$	15.84	15.486		16.26	16.18			15.53	16.14
$^{208}Pb/^{204}Pb$	38.36		37.9	37.50	37.82			38.84	39.56

Abbreviations: BSE = Bulk-Silicate-Earth, DMM = Depleted MORB mantle, EDR = Early-Depleted Reservoir, EER = Early-Enriched Reservoir, SCHEM = Super Chondritic Earth Model, NCM = Non-Chondritic Mantle, M&S = *McDonough and Sun* [1995], W&H = *Workman and Hart* [2005]. S&S = *Salters and Stracke* [2004], B&C = *Boyet and Carlson* [2006], C&B = *Caro and Bourdon* [2010], J&J = *Jackson and Jellinek* [2013], CB = *Carlson and Boyet* [2008]. Numbers under EER give the percentage of mantle occupied by these reservoirs.

[*Gannoun et al.*, 2011a], or to the values suggested by lunar studies [*Sprung et al.*, 2013; *Carlson et al.*, 2014], then the putative EDR would have to have a $^{147}Sm/^{144}Nd$ ratio of at least 0.2019, which would then lead to a minimum present day $\varepsilon^{143}Nd$ of +3.5 (model 3 in Figure 8.9). In this case, the composition of this non-chondritic reservoir would lie about halfway between those calculated assuming a bulk-Earth with chondritic relative abundances of refractory lithophile elements [*McDonough and Sun*, 1995; *Palme and O'Neill*, 2014], and those assuming a 20 ppm offset between terrestrial and O-chondrite initial $^{142}Nd/^{144}Nd$ (Table 8.1) [*Boyet and Carlson*, 2005, 2006; *Carlson and Boyet*, 2008; *Caro and Bourdon*, 2010; *Jackson et al.*, 2010; *Jackson and Jellinek*, 2013]. Starting from this initial $^{142}Nd/^{144}Nd$, if Earth evolved with a chondritic Sm/Nd ratio for 100 Ma after Solar System formation, and then formed an EDR with $^{147}Sm/^{144}Nd = 0.2125$, the EDR today would have a 20 ppm excess in $^{142}Nd/^{144}Nd$ compared to O-chondrites, and an $\epsilon^{143}Nd = +9.5$ (model 5 in Figure 8.9), on the high side, but within the range seen in modern MORB [Gale et al., 2013]. Alternatively, if Earth formed with a slightly superchondritic Sm/Nd ratio (e.g., $^{147}Sm/^{144}Nd = 0.2025$), the modern day average $\epsilon^{143}Nd = +8.6$ of MORB [Gale et al., 2013] would be reached if this starting $^{147}Sm/^{144}Nd$ were increased to 0.223 at 1.8 Ga (model 5 in Figure 8.9), the mean age typically assumed for continent extraction [Jacobsen and Wasserburg, 1979; Goldstein et al., 1984]. Estimates for the Sm/Nd ratio in the mantle source of MORB range widely because of the need to correct measured Sm/Nd ratios in MORB for the fractionation that occurs during partial melting. Recent estimates of the $^{147}Sm/^{144}Nd$ in the MORB mantle include <0.21 [Gale et al., 2013], 0.220 [Boyet and Carlson, 2006], 0.229 [Salters, 2004], 0.249 [Workman and Hart, 2005] and between 0.236 and 0.269 [Huang et al., 2013]. Given this large range in the estimated composition of the MORB source mantle, the Sm-Nd characteristics of the MORB source can be equally well satisfied whether it was created by continent extraction from either an EDR or chondritic starting mantle composition.

Because Sm and Nd are both refractory lithophile elements, the most likely means to form a reservoir with non-chondritic Sm/Nd ratio is by partial melting and fractional crystallization. Andreasen et al. [2008] suggested that the core could be a reservoir of fractionated Sm-Nd, but there currently is no evidence that the REE become siderophile at high P-T [Bouhifd et al., 2014]. The REE can be concentrated into sulfides (e.g., oldhamite) in very reduced bodies such as enstatite chondrites [Crozaz and Lundberg, 1995], but Gannoun et al. [2011b] showed that oldamites have superchondritic Sm/Nd ratios, and thus, if segregated into Earth's core, could not be the chemical complement of the EDR. Labrosse et al. [2007], using the possibility of a density cross over between silicate liquid and solid in the deep mantle [Mosenfelder et al., 2007; Stixrude and Lithgow-Bertelloni, 2012], suggested that a terrestrial magma ocean that extended all the way to the core could crystallize from the middle to both shallower and greater depths. The density contrast between solid and melt in the deep mantle remains a widely debated topic [Nomura et al., 2011; Andrault et al., 2012]. If a terrestrial magma ocean extended into the lower mantle and thereby allowed the fractional crystallization of perovskite, both Mg- and Ca-perovskite have REE distribution coefficients that would lead them to crystallize with superchondritic Sm/Nd ratios if formed from a magma with chondritic Sm/Nd ratio [Corgne et al., 2005]. In their modeling of the

incompatible lithophile element concentrations in the EDR, Boyet and Carlson [2005, 2006] noted that the EDR trace element pattern was most consistent with the distribution coefficient pattern of clinopyroxene, which would be true only if the main fractionation that produced the EDR occurred at low pressures. Their model followed a lunar analog where a terrestrial magma ocean crystallized from the bottom up, but unlike the Moon, ended up with an Fe-rich basaltic crust strongly enriched in incompatible elements. That crust, the EER or Early Enriched Reservoir, was buoyantly unstable and foundered, eventually ending up in the lowermost mantle. In this type of model, the BSE has chondritic relative abundances of at least the refractory lithophile elements but is separated into two chemically complementary reservoirs, one depleted in incompatible elements with superchondritic Sm/Nd ratio (EDR), and the other enriched in incompatible elements with low Sm/Nd (EER). The former has been the source of crustal rocks on Earth throughout most of Earth history. The latter has not yet been sampled on Earth.

An alternative to the idea of chemically complementary reservoirs within Earth is that the bulk Earth simply does not have chondritic relative abundances of refractory lithophile elements [Caro and Bourdon, 2010]. The only mechanism proposed by which this could come about goes by the term "collisional erosion" [O'Neill and Palme, 2008]. In this model, planetary differentiation starts on small planetesimals, likely through global-scale melting, driven by the heat released by ^{26}Al decay. As these planetesimals collide, accretion is not a perfectly efficient process [Asphaug et al., 2006]. During collisions, the crusts of the colliding planetesimals are the most likely component to be ejected at escape velocities from the gravity well of the colliding objects. Compositionally, the collisional erosion model has similar consequences to the magma ocean model of Boyet and Carlson [2005] but with the distinction that the "crustal" component is lost to space rather than to Earth's deep mantle. The collisional erosion model also is subject to the same type of mass-balance constraints. The estimated Sm and Nd concentrations in the BSE in the chondritic model are 0.42 and 1.3 ppm, respectively [McDonough and Sun, 1995] compared to 0.32 and 0.92 ppm for the EDR (Table 8.1). If the whole mantle has the composition estimated for the EDR, Earth has lost 1.53×10^{21} g of Nd and 0.413×10^{21} g of Sm ($^{147}Sm/^{144}Nd = 0.163$ for materials removed) compared to what it would have in the chondritic model. One example of a melt on planetesimals is eucrites, which have about ten times chondritic REE abundances and Sm/Nd ratios similar to chondritic [Consolmagno and Drake, 1977]. For a typical eucrite Nd concentration, formation of Earth's EDR would require the loss of 2.5×10^{26} g, or about 4% of Earth's mass, but the relatively unfractionated Sm/Nd ratio of eucrites would not provide the degree of Sm-Nd fractionation calculated for the EDR. A more effective option would be an incompatible element rich, and more fractionated component such as the KREEP component of the Moon that is estimated to have about 180 ppm Nd [*Warren and Wasson*, 1979] and a ^{147}Sm/^{144}Nd = 0.17 [*Lugmair and Carlson*, 1978]. Losing the equivalent of 0.15% of Earth's mass of a material such as KREEP could create a BSE with EDR like Sm and Nd characteristics. Analysis of the Lu-Hf and Sm-Nd systematics of lunar basalts, however, suggests that the bulk-Moon has chondritic Lu/Hf and Sm/Nd ratios [*Sprung et al.*, 2013], which, in turn argues for a bulk Earth with chondritic ratios of these and other refractory lithophile elements.

8.5.3. The Aftermath of Giant Impacts

The last stages of planet formation involved violent collisions between large planetary embryos. The transformation of kinetic energy to heat after these large collisions likely melted large portions of the impacted planet. Thus, giant impacts played an important role in the differentiation of the rocky planets. The putative Moon-forming giant impact is commonly considered to be the last very energetic event experienced by Earth. Two questions are still unresolved: (1) When did this event occur? and (2) How did the dynamics of the impact result in a Moon that shares some strong isotopic similarities with Earth? A growing number of geochemical observations show a very close match in the isotopic composition of Moon and Earth, with both being different from meteorites and other terrestrial planets [*Wiechert et al.*, 2001; *Zhang et al.*, 2012]. Only extremely small differences in oxygen isotopic composition [*Herwartz et al.*, 2014] and in the isotopic composition of volatile zinc [*Paniello et al.*, 2012] distinguish lunar and terrestrial materials. The isotopic similarity of Earth and Moon suggests that the Moon formed primarily from Earth materials ejected when a large planetary embryo collided with the proto-Earth. Until recently, dynamic simulations of a giant impact could only place enough material into Earth orbit to form the Moon if the impact was a glancing blow that resulted in the majority of Moon-forming fragments deriving from the impactor, not the Earth [*Canup and Asphaug*, 2001; *Canup*, 2004]. These models were constrained to match the orbital angular momentum of the current Earth-Moon system. Two recent models [*Canup*, 2012; *Ćuk and Stewart*, 2012] relax that requirement by appealing to a transfer of angular momentum to the Sun. These models then allow for Moon formation either through the impact of like-sized planetary embryos [*Canup*, 2012] or with a relatively small impactor that hits a rapidly rotating proto-Earth [*Ćuk and Stewart*, 2012].

Another approach to increasing the fraction of Earth in a moon formed by impact is through a glancing blow on the proto-Earth by a high relative velocity impactor where a fraction of the angular momentum of the resulting system is carried off by material escaping Earth orbit [*Reufer et al.*, 2012]. The resulting moon in all of these models is either made primarily of Earth materials or of such an intimate mixture of proto-Earth and impactor that the Moon and Earth end up compositionally indistinguishable. The new impact models have quite different consequences for the proto Earth. Both would cause extensive melting of Earth, but the impact of a smaller embryo [*Ćuk and Stewart*, 2012] could allow a portion of the proto-Earth to remain unmelted after the collision. In this case, portions of Earth could retain memory of differentiation events occurring prior to the last giant impact. Even if Earth were entirely melted during the giant impact, a sufficient density contrast between layers, for example between core and mantle, could allow preservation of differentiated reservoirs created before the giant impact. Consequently, while Hf-W model ages clearly indicate that Earth's core formation occurred during the narrow, early, time window when ^{182}Hf was still extant, an ancient core formation age does not necessarily deny the possibility that the giant impact occurred much later.

Dating the giant impact that formed the Moon would be helpful for better understanding the final stages of the accretion of the Earth. If the last giant impact in Earth history occurred early in the sequence of Earth accretion, keeping the amount of post-giant impact accreting material to as small a mass as suggested by the mantle HSE constraints would be difficult. For example, numerical simulations of planet formation that try to match the HSE constraints arrive at a Moon formation age of 95 ± 32 Ma after the start of Solar System formation [*Jacobson et al.*, 2014]. Determining the age of the Moon by direct analysis has not proven easy. More than 40 years after the lunar sample return by Apollo missions, the age of the Moon remains a subject of debate. One approach to estimating the Moon's age is to find the oldest lunar crustal rocks, particularly the group of highland rocks known as ferroan-anorthosites (FANs) [*Dowty et al.*, 1974]. In the magma ocean model for initial lunar differentiation [*Wood et al.*, 1970], the FANs are interpreted to be flotation cumulates from the magma ocean. Until the thick anorthositic crust forms, a lunar magma ocean will crystallize rapidly, of order a million years or less [*Solomatov*, 2000; *Elkins-Tanton*, 2011]. Thus, if FANs indeed formed as flotation cumulates, their age should be within a million years of lunar magma ocean formation, and they should have formed over a very short time window. FANs have proven notoriously difficult to date as they do not contain zircons, they have very low abundances of the incompatible elements that make up the radioactive dating systems, most FANs have experienced shock metamorphism, and many FANs have extensive cosmic ray exposure histories that further complicate isotope systematics. Ages have been reported for only six Apollo FAN samples, all from the Apollo 16 site, and one lunar meteorite [*Nyquist et al.*, 2006]. The ages range from 4.56 ± 0.24 Ga for 67016c [*Alibert et al.*, 1994] to 4.29 ± 0.06 Ga for 62236 [*Borg et al.*, 1999]. Only two FANs have provided concordant ages from more than one radioisotope system: 60025 has Pb-Pb and Sm-Nd ages averaging 4.360 ± 0.003 Ga [*Borg et al.*, 2011], and 67075 has Rb-Sr and Sm-Nd ages averaging 4.48 ± 0.05 Ga [*Nyquist et al.*, 2010]. The available FAN age data thus do not obviously support the idea that FANs formed over a short time interval, but the question of how much of the age variability reflects shock resetting of the isotope systematics remains.

The oldest lunar zircon is dated at 4.417 ± 0.006 Ga [*Nemchin et al.*, 2009], but the main peak in lunar zircon ages is at 4.32 Ga [*Grange et al.*, 2013], which is similar to the age peak [*Carlson et al.*, 2014] obtained for the other type of lunar crustal rock, the Mg-suite [*James*, 1980]. Other approaches to determine the time of earliest lunar differentiation include the following: (a) Sm-Nd and Lu-Hf model ages of 4.36 to 4.44 Ga for the incompatible-element-rich KREEP component of the Moon [*Lugmair and Carlson*, 1978; *Gaffney and Borg*, 2014; *Sprung et al.*, 2013]; (b) ^{146}Sm-^{142}Nd model ages from 4.32 to 4.42 Ga for the source of mare basalts [*Nyquist et al.*, 1995; *Boyet and Carlson*, 2007; *Brandon et al.*, 2009]; (c) the 4.42 Ga U-Pb model age for the lunar highlands [*Tera and Wasserburg*, 1974]; and (d) an upper limit of <4.5 Ga from ^{182}Hf-^{182}W systematics of lunar samples [*Touboul et al.*, 2007]. Most of these ages are closer to 4.4 than 4.57 Ga and are thus consistent with the idea that Moon formation occurred relatively late in the history of planetary accretion.

The Moon-forming impact must have had a major impact on Earth evolution as well. A number of ages between 4.35 and 4.45 Ga, overlapping the ages discussed above for the Moon, can be found in estimates of early Earth differentiation. These include the following: (a) 4.45 Ga U-Pb "age of the Earth" [*Patterson*, 1956; *Allègre et al.*, 2008]; (b) 4.45 Ga I-Xe age of Earth's atmosphere [*Staudacher and Allègre*, 1982; *Pepin and Porcelli*, 2006; *Mukhopadhyay*, 2012]; (c) variable $^{142}Nd/^{144}Nd$ in Archean/Hadean rocks from SW Greenland [*Harper and Jacobsen*, 1992; *Boyet et al.*, 2003; *Caro et al.*, 2003; *Boyet and Carlson*, 2006; *Caro et al.*, 2006; *Bennett et al.*, 2007; *Rizo et al.*, 2011; *Rizo et al.*, 2013], the Yilgarn Craton in western Australia [*Bennett et al.*, 2007b], the Nuvvuatittuq greenstone belt in northern Quebec, Canada [*O'Neil et al.*, 2008; *O'Neil et al.*, 2012; *O'Neil et al.*, 2013; *Roth et al.*, 2013], the Acasta gneisses of northern Canada

[*Roth et al.*, 2014], and the Abitibi greenstone belt, Canada [*Debaille et al.*, 2013] that testify to Earth differentiation events in the range of 4.35 to 4.45 Ga; and (d) 4.36 to 4.40 Ga ages for detrital zircons from western Australia [*Holden et al.*, 2009; *Valley et al.*, 2014]. Although the meaning of each of these ages is subject to debate, the fact that the maximum ages obtained from both Earth and lunar materials are in the range of 4.35 to 4.45 Ga instead of 4.57 Ga supports a late formation for the Moon, some 100–200 Ma after the start of Solar System formation.

In the last few years, elevated $^{182}W/^{184}W$ ratios compared to the modern mantle (Fig. 8.6) have been detected in the Nuvvuagittuq [*Touboul et al.*, 2014] and SW Greenland [*Willbold et al.*, 2011; *Touboul et al.*, 2012; *Rizo et al.*, 2015] rocks and in 2.8 Ga komatiites from Russia [*Touboul et al.*, 2012]. The ^{182}W variability must be caused by the decay of ^{182}Hf and hence, must reflect differentiation events occurring within the first few tens of millions of years of Solar System history. As discussed in Section 8.4, the fractionation in Hf/W is most often interpreted as reflecting metal-silicate separation [*Halliday and Kleine*, 2005], but W is substantially more incompatible than Hf during silicate differentiation [*Righter and Shearer*, 2003], so the Hf/W ratio also can be strongly fractionated during crystal-liquid fractionation in a terrestrial magma ocean. If the ^{182}W excesses in Archean rocks track differentiation of a terrestrial magma ocean, rather than core formation, then one might expect a correlation between $^{182}W/^{184}W$, and that of another short-lived radiogenic isotope system that is sensitive to mantle differentiation, such as ^{146}Sm-^{142}Nd. At the present, there simply are too few localities for which both high precision $^{142}Nd/^{144}Nd$ and $^{182}W/^{184}W$ ratios are available to make a firm conclusion on this point. At least in the rocks of the Nuvvuagittuq greenstone belt (Quebec), the elevated $^{182}W/^{184}W$ [*Touboul et al.*, 2014] are accompanied by deficits in $^{142}Nd/^{144}Nd$ [*O'Neil et al.*, 2012], the opposite of what would be expected for magma ocean differentiation because both W and Nd are the more incompatible elements in these systems. In addition, if caused by magma ocean differentiation, the very early differentiation event suggested by the W data appears to conflict with the ^{142}Nd evidence discussed above for late Moon formation and consequent Earth differentiation. One option is that there was more than one giant impact/magma ocean event on Earth, which is likely according to planetary accumulation models [*Jacobson et al.*, 2014]. For example, the magnitude of $^{3}He/^{22}Ne$ fractionation observed in the present-day mantle is suggested to reflect at least two magma ocean episodes [*Tucker and Mukhopadhyay*, 2014]. Preservation of a ^{182}W (and ^{142}Nd) signal from an earlier magma ocean requires that the later magma ocean did not completely homogenize the differentiated products of the earlier magma ocean. The fact that W does not record a ~4.4 Ga event, as suggested by ^{142}Nd is not surprising because ^{182}Hf was extinct by that time. Combined ^{142}Nd, ^{147}Sm-^{143}Nd and ^{176}Lu-^{176}Hf observations suggest that Nd and Hf isotopic evolution were not correlated in the Hadean/Eoarchean as they have been ever since [*Rizo et al.*, 2011]. The decoupling of these two isotope systems can be explained by the involvement of high-pressure mineral phases, such as Mg- and Ca-perovskite because, unlike most upper mantle phases, the perovskites strongly fractionate the high-field-strength elements, such as Hf, from the REE [*Corgne et al.*, 2005]. This supports the idea that the early Sm/Nd fractionation recorded by the ^{142}Nd anomalies is most likely associated with the crystallization of a terrestrial magma ocean that extended to lower mantle pressure conditions.

8.5.4. Onset of Early Crust Formation

The timing of early crust formation and the nature of the primordial crust that formed after the Moon-forming giant impact are still poorly constrained due to the scarcity of preserved early crustal samples. The Hadean Eon, spanning from 4.568 Ga to 4.0 Ga is almost completely devoid of rock records, except for a few small remnants. Hadean samples include the Acasta gneisses (Northwest Canada) that are the oldest rocks dated by U-Pb on zircons at 4.03 Ga [*Bowring and Williams*, 1999], the Nuvvuagittuq greenstone belt (Northeast Canada) possibly as old as 4.4 Ga [*O'Neil et al.*, 2012], and the Jack Hills (Western Australia) detrital zircons also as old as 4.36 to 4.40 Ga [*Wilde et al.*, 2001; *Holden et al.*, 2009; *Valley et al.*, 2014]. Other occurrences of early crust only emerge at ≤ 3.9 Ga, including the Eoarchean gneiss complex of Southwest Greenland [*Nutman et al.*, 1996; *Polat et al.*, 2011; *Nutman et al.*, 2013], the Saglek Gneiss Complex (Labrador, Canada) [*Collerson and Bridgewater*, 1979; *Collerson et al.*, 1989], the Napier Complex (Antarctica) [*Harley and Black*, 1997; *Kelly and Harley*, 2005], the Anshan area (Northeastern China) [*Song et al.*, 1996; *Liu et al.*, 2008], the Narryer Gneiss Terrane (Western Australia) [*Myers*, 1988; *Pidgeon and Wilde*, 1998], and the Ancient Gneiss Complex (Swaziland) [*Hunter et al.*, 1984; *Kroner et al.*, 2013]. Despite our limited knowledge of the exact processes that led to the formation of the primordial crust, ages as old as 4.4 Ga suggest that crust could form and be stabilized, at least to some extent, very quickly after the Moon-forming impact.

Despite the fact that the Jack Hills detrital zircons provide invaluable information about the primitive crust, their host rocks have been destroyed, leaving no direct samples of Earth's primordial crust and limiting our ability to put compositional constraints on the Hadean crust. The initial Hf isotopic compositions of the Jack Hills zircons suggest that their host rock was of broadly felsic composition from the tonalite-trondhjemite-granodiorite series (TTG) [*Blichert-Toft and Albarède*, 2008]. Felsic rocks however, cannot be directly produced by melting of mantle peridotite but must instead be derived from extensive fractional crystallization of, or the melting of, an older mafic precursor. A mafic precursor for the Jack Hills zircon-bearing TTGs is supported by the εHf-zircon vs. time trend defined by the least disturbed Jack Hills detrital zircons (Fig. 8.10) [*Kemp et al.*, 2010]. The Acasta Gneiss Complex includes orthogneisses as old as 4.03 Ga [*Bowring and Williams*, 1999]. These felsic TTGs must also have an older basaltic precursor. One Acasta zircon with a xenocrystic core as old as 4.2 Ga [*Iizuka et al.*, 2006] clearly shows that the Acasta gneisses include a component of reworked Hadean crust, as does the recent detection of variable ^{142}Nd/^{144}Nd in Acasta gneisses [*Roth et al.*, 2014]. Figure 8.10 shows that the Acasta Hadean to Eoarchean zircons follow a similar εHf secular evolution to the least disturbed Jack Hills detrital zircons consistent with remelting of a comparable Hadean basaltic protolith to produce the Acasta TTG.

Felsic rocks are the most likely host rocks for the zircons that provide robust geochronological constraints for old polymetamorphosed rocks, which have focused much work on Archean crust on the felsic component. The mafic "greenstone" components of granite-greenstone terranes also have seen abundant geochemical study, particularly those in SW Greenland [e.g., *Polat et al.*, 2011]. The mafic component of the 3.7 to 3.8 Ga Greenland crust bears chemical signatures similar to those of modern magmas found in convergent margins. The same is true of the ancient mafic rocks in the Nuvvuagittuq greenstone belt of the Northeastern Superior Province of Canada, possibly as old as 4.3 Ga, that is primarily composed of highly metamorphosed mafic rocks whose protoliths range in composition from basalt to basaltic andesite [*O'Neil et al.*, 2011]. The geochemical composition and stratigraphic compositional relationships of the Nuvvuagittuq mafic rocks, called the Ujaraaluk unit, are consistent with a protolith consisting of a hydrothermally altered basaltic oceanic-type crust perhaps formed in an intraoceanic convergent margin setting [*O'Neil et al.*, 2011; *Turner et al.*, 2014]. Both greenstone belts show variability in ^{142}Nd. The SW Greenland Eoarchean samples show excesses in ^{142}Nd compared to the modern terrestrial value, but no correlation between Sm/Nd and ^{142}Nd/^{144}Nd ratios, which is consistent with their formation at circa 3.8 Ga, after ^{146}Sm was virtually extinct [*Caro et al.*, 2003; *Bennett et al.*, 2007; *Rizo et al.*, 2013]. In contrast, the mafic rocks from the Nuvvuagittuq greenstone belt show considerable variability in ^{142}Nd/^{144}Nd

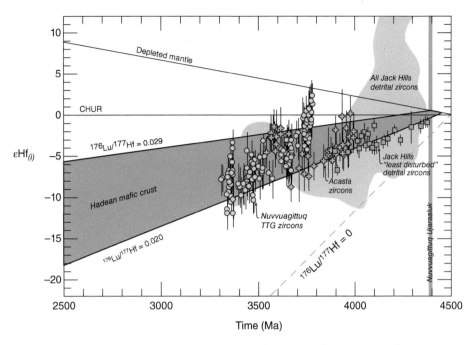

Figure 8.10 Hf isotope evolution for zircons from three Eoarchean to Hadean terranes, the Nuvvuagittuq (yellow circles [*O'Neil et al.*, 2013]), Acasta (green crosses [*Iizuka et al.*, 2009]) and Jack Hills. The blue squares show the "least disturbed" zircons based on the criteria used by *Kemp et al.* [2010] whereas the green field shows all data for Jack Hills zircons [*Amelin et al.*, 1999; *Harrison et al.*, 2005; *Blichert-Toft and Albarède*, 2008; *Harrison et al.*, 2008]. The gray band shows the Hf isotope evolution expected for Hadean mafic crust with Lu/Hf ratios within the range observed for the Ujaraaluk mafic rocks of the Nuvvuagittuq Greenstone Belt.

($\mu^{142}Nd$ = +8 to −18), with most samples having a deficit in ^{142}Nd compared to the modern terrestrial value [*O'Neil et al.*, 2008; *O'Neil et al.*, 2012; *Roth et al.*, 2013]. Together with ultramafic co-genetic cumulates, these rocks exhibit a correlation between $^{142}Nd/^{144}Nd$ and Sm/Nd ratios that has been interpreted by *O'Neil et al.* [2008, 2012] to be an isochron giving a formation age of $4.388\,^{+15}_{-17}$ Ga using the ^{146}Sm half-life of 68 Ma from *Kinoshita et al.* [2012] that would make them the oldest preserved crustal remnant on Earth. The age is 4289 Ma using the previously proposed 103 Ma half-life for ^{146}Sm. If this age indeed dates their eruption, the Nuvvuagittuq mafic rocks are the oldest, Hadean, crustal remnant preserved on Earth. In this case, the Nuvvuagittuq greenstone belt may be an example of the beginning of crust formation in the Hadean, to be followed some half-billion years later by the formation of the preserved Eoarchean terranes that mark the onset of the Archean crustal record.

The Hadean formation age of the Nuvvuagittuq rocks, however, has been challenged because the oldest U-Pb ages on zircons found in rare intrusive trondhjemitic rocks from the Nuvvuagittuq greenstone belt are ~3.8 Ga [*Cates and Mojzsis*, 2007; *David et al.*, 2008; *O'Neil et al.*, 2013]. An alternative model suggests formation of the Nuvvuagittuq mafic rocks by mixing between mantle-derived melts in the Eoarchean with an

incompatible element enriched reservoir formed in the Hadean that contributed the ^{142}Nd deficits of the Nuvvuagittuq mafic rocks [*Guitreau et al.*, 2013; *Roth et al.*, 2013]. Despite the fact that the long-lived ^{147}Sm-^{143}Nd isotopic system in the Ujaraaluk rocks shows evidence of significant disturbance in the Neoarchean, large massive gabbroic sills intrusive into the Ujaraaluk provide a Sm/Nd vs. $^{143}Nd/^{144}Nd$ correlation that yields a slope corresponding to an age of 4.12 ± 0.10 Ga, consistent with an older Hadean age for the intruded Ujaraaluk unit [*O'Neil et al.*, 2012]. The lack of igneous zircons in the mafic Ujaraaluk hinders the confirmation of their Hadean crystallization age using U-Pb geochronologic techniques. Nevertheless, the ^{142}Nd isotopic composition of the Ujaraaluk unit is consistent with it being either a Hadean mafic protocrust or an Eoarchean volcanic province that involved extensive interaction with a preexisting Hadean incompatible element enriched crustal component that has identical compositional and isotopic characteristics to the Ujaraaluk.

In spite of the evidence for a mafic primitive crust on Earth (and Mars and Mercury), Archean cratons are primarily composed of much more evolved TTG, suggesting that early geologic processes on Earth rapidly reworked the primitive basaltic crust to produce second-stage felsic rocks. Despite the large number of zircons dated in

Archean terrains, only the Jack Hills conglomerates and the Acasta gneisses contain zircons (or cores) older than 4.0 Ga (Fig. 8.11). The scarcity of Hadean zircons is consistent with either a long-lived (zircon-free) mafic primitive crust or nearly complete destruction of an early felsic crust, likely through subduction and recycling into the mantle. The zircon age distribution in itself suggests a long quiescent period of over half a billion years (until ~3.8 Ga) before the ~4.4 Ga mafic primitive crust was extensively reworked to produce a significant amount of felsic rocks. This is also consistent with what is observed in the Nuvvuagittuq greenstone belt. The Nuvvuagittuq mafic rocks are surrounded and intruded by multiple generations of Eoarchean TTG produced at 3.76 Ga, 3.66 Ga, 3.51 Ga, and 3.35 Ga [O'Neil et al., 2013]. The secular Hf isotopic evolution of the zircons from most of

these TTG is consistent with derivation from a ~4.4 Ga mafic crust and is similar to what is seen for the Jack Hills and Acasta zircons (Fig. 8.10). Therefore, the same type of mafic primordial crust also could have produced the Acasta gneisses, as well as the TTGs that served as sources for the Jack Hills Hadean/Eoarchean zircons. Most Nuvvuagittuq TTGs also contain ^{142}Nd deficits [O'Neil et al., 2013]. Their Eoarchean zircon ages clearly show that these rocks were produced after the extinction of ^{146}Sm, so their low $^{142}Nd/^{144}Nd$ ratios can only be inherited from a >4.1 Ga precursor. Both the compositional characteristics [Adam et al., 2012] and the ^{142}Nd deficits in the Nuvvuagittuq TTG are thus also consistent with them being derived from the melting of a Hadean Ujaraaluk-type mafic crust. Recycling of this basaltic protocrust for an extended period of time, producing

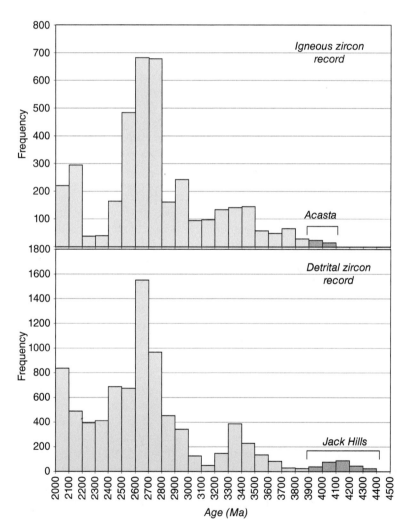

Figure 8.11 Histogram of zircon ages comparing the igneous rock record to the detrital sedimentary zircon record. In both cases, zircons become most abundant over a billion years after the oldest zircon. One component of this age gap almost certainly reflects imperfect preservation of the oldest rock record, but another component could be that there was a substantial hiatus between the formation of the primordial crust and its extensive reworking into zircon-rich felsic crust. Data from *Condie et al.* [2009] and references therein.

TTG over several hundred million years, may have played a significant role in the stabilization of the Archean cratons and appears to have been an important process of crustal formation only after ~3.8 Ga.

8.6. CONCLUSIONS

Peering through 4.5 Ga of geologic processing of Earth has made it difficult to disentangle the chemical consequences of the operation of plate tectonics and its role in creating crust and recycling it back into the mantle. First order features of Earth, notably its volatile depletion and the separation of core from mantle, clearly date to the very earliest epoch of Earth evolution (Fig. 8.12). Our understanding of the timescale of core formation on both Earth and the planetesimals from which Earth was made was dramatically improved through application of the Hf-W radiometric system. Although evidence that Earth formed volatile depleted, rather than became volatile depleted, is becoming more clear, much remains to be understood about the process of volatile depletion, in particular the relative roles of a simple lack of condensation

of volatile elements in the materials that make up Earth, compared to loss of highly volatile elements from atmospheres formed on planetesimals that were not large enough to gravitationally retain their atmospheres. Whether Earth grew with something approaching its current volatile inventory or whether key volatiles, such as water, were added slowly by later accreting material remains an active area of investigation with major consequences for our understanding of the compositional development of Earth's atmosphere. Surprisingly, perhaps the most difficult aspect of understanding Earth's early differentiation involves its most accessible part–its crust. The extremely limited amount of crust preserved from prior to 3.8 Ga leaves many open questions about the nature of Earth's first crust, the process(es) that produced the early crust, and particularly the rate at which it was created and recycled into the mantle. As isotopic techniques continue to improve, the ability to interrogate the history of Earth's oldest rocks improves with the prospect of providing a clearer picture of how the earliest differentiation events experienced by Earth may have influenced or determined the characteristics of the modern Earth.

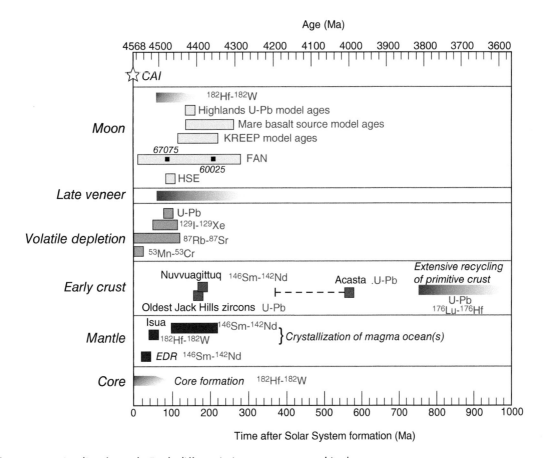

Figure 8.12 Timeline for early Earth differentiation events covered in the text.

REFERENCES

Adam, J., T. Rushmer, J. O'Neil, and D. Francis (2012), Hadean greenstones from the Nuvvuagittuq fold belt and the origin of the Earth's early continental crust, *Geology*, *40*, 363–366.

Albarède, F. (2010), Volatile accretion history of the terrestrial planets and dynamic implications, *Nature*, *461*, 1227–1233.

Alexander, C. M. O. D., R. Bowden, M. L. Fogel, K. T. Howard, C. D. K. Herd, and L. R. Nittler (2012), The provenances of asteroids, and their contributions to the volatile inventories of the terrestrial planets, *Science*, *337*, 721–723.

Alibert, C., M. D. Norman, and M. T. McCulloch (1994), An ancient Sm-Nd age for a ferroan noritic anorthosite clast from lunar breccia 67016, *Geochimica et Cosmochimica Acta.*, *58*, 2921–2926.

Allègre, C., T. Staudacher, P. Sarda, and M. Kurz (1983a), Constraints on evolution of Earth's mantle from rare gas systematics, *Nature*, *303*, 762–766.

Allègre, C. J., S. R. Hart, and J.-F. Minster (1983b), Chemical structure and evolution of the mantle and continents determined by inversion of Nd and Sr isotopic data, I. Theoretical methods, *Earth and Planetary Science Letters*, *66*, 177–190.

Allègre, C. J., G. Mahnes, and C. Göpel (2008), The major differentiation of the Earth at ~4.45 Ga, *Earth and Planetary Science Letters*, *267*, 353–364.

Amelin, Y., D.-C. Lee, A. N. Halliday, and R. T. Pidgeon (1999), Nature of the Earth's earliest crust from hafnium isotopes in single detrital zircons, *Nature*, *399*, 252–255.

Anders, E. and N. Grevesse (1989), Abundances of the elements: Meteoritic and solar, *Geochimica et Cosmochimica Acta.*, *53*, 197–214.

Andrault, D., S. Petitgirard, S. L. Nigro, J.-L. Devidal, G. Garbarino, and M. Mezouar (2012), Solid-liquid iron partitioning in Earth's deep mantle, *Nature*, *487*, 354–357.

Andreasen, R. and M. Sharma (2006), Solar nebula heterogeneity in p-process samarium and neodymium isotopes, *Science*, *314*(5800), 806–809, doi:10.1126/science.1131708.

Andreasen, R. and M. Sharma (2007), Mixing and homogenization in the early solar system: clues from Sr, Ba, Nd, and Sm isotopes in meteorites, *The Astrophysical Journal*, *665*, 874–883.

Andreasen, R., M. Sharma, K. V. Subbarao, and S. G. Viladkar (2008), Where on Earth is the enriched Hadean reservoir?, *Earth and Planetary Science Letters*, *266*, 14–28.

Asphaug, E., C. B. Agnor, and Q. Williams (2006), Hit-and-run planetary collisions, *Nature*, *439*, 155–160, doi:10.1038/nature04311.

Azbel, I. Y., I. N. Tolstikhin, J. D. Kramers, G. V. Pechernikova, and A. V. Vityazev (1993), Core growth and siderophile element depletion of the mantle during homogeneous Earth accretion, *Geochimica et Cosmochimica Acta.*, *57*, 2889–2898.

Becker, H., M. F. Horan, R. J. Walker, S. Gao, J.-P. Lorand, and R. L. Rudnick (2006), Highly siderophile element composition of the Earth's primitive upper mantle: constraints from new data on peridotite massifs and xenoliths, *Geochimica et Cosmochimica Acta.*, *70*, 4528–4550.

Begemann, F., K. R. Ludwig, G. W. Lugmair, K. Min, L. E. Nyquist, P. J. Patchett, P. R. Renne, C.-Y. Shih, I. M. Villa, and R. J. Walker (2001), Call for an improved set of decay constants for geochronological use, *Geochimica et Cosmochimca Acta.*, *65*(1), 111–121.

Bennett, V. C., A. D. Brandon, and A. P. Nutman (2007), Coupled ^{142}Nd-^{143}Nd isotopic evidence for Hadean mantle dynamics, *Science*, *318*(5858), 1907–1910, doi:10.1126/science.1145928.

Bernatowicz, T. J. and B. E. Hagee (1987), Isotopic fractionation of Kr and Xe implanted in solids at very low energies, *Geochimica et Cosmochimica Acta.*, *51*, 1599–1611.

Bernatowicz, T. J., F. A. Podosek, M. Honda, and F. E. Kramer (1984), The atmospheric inventory of xenon and noble gases in shales: the plastic bag experiment, *Journal of Geophysical Research*, *89*, 4597–4611.

Birck, J. L. and C. J. Allègre (1994), Contrasting Re/Os magmatic fractionation in planetary basalts, *Earth and Planetary Science Letters*, *124*, 139–148.

Blichert-Toft, J. and F. Albarède (2008), Hafnium isotopes in Jack Hills zircons and the formation of the Hadean crust, *Earth and Planetary Science Letters*, *265*, 686–702.

Borg, L. E., J. N. Connelly, M. Boyet, and R. W. Carlson (2011), Chronological evidence that the Moon is either young or did not have a global magma ocean, *Nature*, *477*, 70–73.

Borg, L. E., M. Norman, L. E. Nyquist, D. D. Bogard, G. A. Snyder, L. A. Taylor, and M. M. Lindstrom (1999), Isotopic studies of ferroan anorthosite 62236: a young lunar crustal rock from a light rare-earth-element-depleted source, *Geochimica et Cosmochimca Acta.*, *63*, 2679–2691.

Borisov, A. and H. Palme (1997), Experimental determination of the solubility of platinum in silicate melts, *Geochimica et Cosmochimca Acta.*, *61*, 4349–4357.

Bouhifd, M. A. and A. P. Jephcoat (2011), Convergence of Ni and Co metal-silicate partition coefficients in the deep magma-ocean and coupled silicon-oxygen solubility in iron melts at high pressures, *Earth and Planetary Science Letters*, *307*, 341–348.

Bouhifd, M. A., M. Boyet, C. Cartier, T. Hammouda, N. Bolfan-Casanova, J.-L. Devidal, and D. Andrault (2015), Superchondritic Sm/Nd ratio of the Earth: impact of Earth's core formation, *Earth and Planetary Science Letters*, *413*, 158–166.

Bourdon, B., M. Touboul, G. Caro, and T. Kleine (2008), Early differentiation of the Earth and Moon, *Philosophical Transactions Royal Society of London A*, *366*, 4105–4128.

Bouvier, A., J. D. Vervoort, and P. J. Patchett (2008), The Lu-Hf and Sm-Nd isotopic composition of CHUR: constraints from unequilibrated chondrites and implications for the bulk composition of terrestrial planets, *Earth and Planetary Science Letters*, *273*, 48–57.

Bouvier, A. and M. Wadhwa (2010), The age of the solar system redefined by the oldest Pb-Pb age of a meteoritic inclusion, *Nature Geoscience*, *3*, 637–641.

Bowring, S. A., T. B. Housh, and C. E. Isachsen (1990), The Acasta gneisses: remnant of Earth's early crust, in *The Origin of the Earth*, edited by H. E. Newsom and J. H. Jones, pp. 319–343, Oxford University Press, New York.

Bowring, S. A. and I. S. Williams (1999), Priscoan (4.00–4.03 Ga) orthogneisses from northwestern Canada, *Contributions to Mineralogy and Petrology*, *134*, 3–16.

Boyet, M., J. Blichert-Toft, M. Rosing, M. Storey, P. Telouk, and F. Albarede (2003), ^{142}Nd evidence for early Earth differentiation, *Earth and Planetary Science Letters*, *214*, 427–442.

Boyet, M. and R. W. Carlson (2005), ^{142}Nd evidence for early (>4.53 Ga) global differentiation of the silicate Earth, *Science*, *309*, 576–581.

Boyet, M. and R. W. Carlson (2006), A new geochemical model for the Earth's mantle inferred from ^{146}Sm-^{142}Nd systematics, *Earth and Planetary Science Letters*, *250*, 254–268.

Boyet, M. and R. W. Carlson (2007), A highly depleted moon or a non-magma ocean origin for the lunar crust?, *Earth and Planetary Science Letters*, *262*, 505–516.

Boyet, M. and A. Gannoun (2013), Nucleosynthetic Nd isotope anomalies in primitive enstatite chondrites, *Geochimica et Cosmochimca Acta.*, *121*, 652–666.

Brandon, A. D., T. J. Lapen, V. Debaille, B. L. Beard, K. Rankenburg, and C. Neal (2009), Re-evaluating ^{142}Nd/^{144}Nd in lunar mare basalts with implications for the early evolution and bulk Sm/Nd of the Moon, *Geochimica et Cosmochimica Acta.*, *73*, 6421–6445.

Brandon, A. D., I. S. Puchtel, R. J. Walker, J. M. D. Day, A. J. Irving, and L. A. Taylor (2012), Evolution of the martian mantle inferred from the ^{187}Re-^{187}Os isotope and highly siderophile element abundance systematics of shergottite meteorites, *Geochimica et Cosmochimca Acta.*, *76*, 206–235.

Brandon, A. D., R. J. Walker, J. W. Morgan, and G. G. Goles (2000), Re–Os isotopic evidence for early differentiation of the Martian mantle, *Geochimica et Cosmochimica Acta.*, *64*, 4083–4095.

Brenan, J. M. and W. F. McDonough (2009), Core formation and metal-silicate fractionation of osmium and iridium from gold, *Nature Geoscience*, *2*, 798–801.

Brennecka, G. A. and M. Wadhwa (2012), Uranium isotope compositions of the basaltic angrite meteorites and the chronological implications of the early Solar System, *Proceedings of the National Academy of Sciences*, *109*, 9299–9303.

Burkhardt, C., T. Kleine, F. Oberli, A. Pack, B. Bourdon, and R. Wieler (2011), Molybdenum isotope anomalies in meteorites: Constraints on solar nebula evolution and origin of the Earth, *Earth and Planetary Science Letters*, *312*, 390–400.

Canup, R. M. (2004), Simulations of a late lunar-forming impact, *Icarus*, *168*, 433–456.

Canup, R. M. (2012), Forming a moon with an Earth-like composition via a giant impact, *Science*, *338*, 1052–1055.

Canup, R. M. and E. Asphaug (2001), Origin of the Moon in a giant impact near the end of the Earth's formation, *Nature*, *412*, 708–712.

Carlson, R. W. (2005), Application of the Pt-Re-Os isotopic systems to mantle geochemistry and geochronology, *Lithos*, *82*, 249–272.

Carlson, R. W. and M. Boyet (2008), Composition of Earth's interior: the importance of early events, *Philosophical Transactions Royal Society of London A*, *366*, 4077–4103.

Carlson, R. W. and G. W. Lugmair (1988), The age of ferroan anorthosite 60025: oldest crust on a young Moon?, *Earth and Planetary Science Letters*, *90*, 119–130.

Carlson, R. W., L. E. Borg, A. M. Gaffney, and M. Boyet (2014), Rb-Sr, Sm-Nd and Lu-Hf isotope systematics of the lunar Mg-suite: the age of the lunar crust and its relation to the time of Moon formation, *Philosophical Transactions Royal Society of London A*, *372*, 20130246.

Carlson, R. W., M. Boyet, and M. Horan (2007), Chondrite barium, neodymium, and samarium isotopic heterogeneity and early earth differentiation, *Science*, *316*, 1175–1178.

Caro, G. (2011), Early Silicate Earth Differentiation, *Annual Reviews of Earth and Planetary Science*, *39*(1), 31–58, doi:10.1146/annurev-earth-040610-133400.

Caro, G. and B. Bourdon (2010), Non-chondritic Sm/Nd ratio in the terrestrial planets: Consequences for the geochemical evolution of the mantle–crust system, *Geochimica et Cosmochimica Acta.*, *74*(11), 3333–3349, doi:10.1016/j.gca.2010.02.025.

Caro, G., B. Bourdon, J.-L. Birck, and S. Moorbath (2006), High-precision ^{142}Nd/^{144}Nd measurements in terrestrial rocks: Constraints on the early differentiation of the Earth's mantle, *Geochimica et Cosmochimica Acta.*, *70*(1), 164–191, doi:10.1016/j.gca.2005.08.015.

Caro, G., B. Bourdon, J. L. Birck, and S. Moorbath (2003), ^{146}Sm-^{142}Nd evidence from Isua metamorphosed sediments for early differentiation of the Earth's mantle, *Nature*, *423*, 428–432.

Cates, N. L. and S. J. Mojzsis (2007), Pre-3750 Ma supracrustal rocks from the Nuvvagittuq supracrustal belt, northern Quebec, *Earth and Planetary Science Letters*, *255*, 9–21.

Cawood, P. A., C. J. Hawkesworth, and B. Dhuime (2013), The continental record and the generation of continental crust, *Geological Society of America, Bulletin*, *125*, 14–32.

Chabot, N. L., T. M. Safko, and W. F. McDonough (2010), Effect of silicon on trace element partitioning in iron-bearing metalic melts, *Meteoritics and Planetary Science*, *45*, 1–15.

Chambers, J. E. (2004), Planetary accretion in the inner Solar System, *Earth and Planetary Science Letters*, *223*(3–4), 241–252, doi:10.1016/j.epsl.2004.04.031.

Chapman, C. R., B. A. Cohen, and D. H. Grinspoon (2007), What are the real constraints on the existence and magnitude of the late heavy bombardment?, *Icarus*, *189*, 233–245.

Chen, J. H., D. A. Papanastassiou, and G. J. Wasserburg (2010), Ruthenium endemic isotope effects in chondrites and differentiated meteorites, *Geochimica et Cosmochimca Acta.*, *74*, 3851–3862.

Chou, C.-L., D. M. Shaw, and J. H. Crocket (1983), Siderophile trace elements in the Earth's oceanic crust and upper mantle, *Journal of Geophysical Research*, *88*, A507–A518.

Clayton, R. N. (2003), Oxygen isotopes in meteorites, in *Treatise on Geochemistry*, edited by A. M. Davis, pp. 129–142, Elsevier, Amsterdam.

Collerson, K. D. and D. Bridgewater (1979), Metamorphic development of early Archaean tonalitic and trondhjemitic gneisses, Saglek area, Labrador, in *Trondhjemites, Dacites and Related Rocks*, edited by F. Barker, p. 659, Elsevier, Amsterdam.

Collerson, K. D., M. T. McCulloch, and A. P. Nutman (1989), Sr and Nd isotope systematics of polymetamorphic Archean gneisses from southern West Greenland and northern Labrador, *Canadian Journal of Earth Science*, *27*, 446–466.

Condie, K. C., E. Belousova, W. L. Griffin, and K. N. Sircombe (2009), Granitoid events in space and time: constraints from igneous and detrital zircon age spectra, *Gondwana Research*, *15*, 228–242.

Condon, D. J., N. McLean, S. R. Noble, and S. A. Bowring (2010), Isotopic composition ($^{238}U/^{235}U$) of some commonly used uranium reference materials, *Geochimica et Cosmochimica Acta.*, *74*(24), 7127–7143, doi:10.1016/j.gca.2010.09.019.

Consolmagno, G. J. and M. J. Drake (1977), Composition and evolution of the eucrite parent body: evidence from rare earth elements, *Geochimica et Cosmochimca Acta.*, *41*, 1271–1282.

Corgne, A., C. Liebske, B. J. Wood, D. C. Rubie, and D. J. Frost (2005), Silicate perovskite-melt partitioning of trace elements and geochemical signature of a deep perovskitic reservoir, *Geochimica et Cosmochimica Acta.*, *69*(2), 485–496, doi:10.1016/j.gca.2004.06.041.

Cottrell, E. and D. Walker (2006), Constraints on core formation from Pt partitioning in mafic silicate liquids at high temperatures, *Geochimica et Cosmochimca Acta.*, *70*, 1656–1580.

Crozaz, G. and L. L. Lundberg (1995), The origin of oldhamite in unequilibrated enstatite chondrites, *Geochimica et Cosmochimca Acta*, *59*, 3817–3831.

Ćuk, M. and S. T. Stewart (2012), Making the Moon from a fast-spinning Earth: A giant impact followed by resonant despinning, *Science*, *338*, 1047–1052.

Dale, C. W., K. W. Burton, R. C. Greenwood, A. Gannoun, J. Wade, B. J. Wood, and D. G. Pearson (2012), Late accretion on the earliest planetesimals revealed by the highly siderophile elements, *Science*, *336*, 72–75.

Dauphas, N. (2003), The dual origin of the terrestrial atmosphere, *Icarus*, *165*, 326–339.

Dauphas, N., A. M. Davis, B. Marty, and L. Reisberg (2004), The cosmic molybdenum-ruthenium isotope correlation, *Earth and Planetary Science Letters*, *226*, 465–475.

Dauphas, N., B. Marty, and L. Reisberg (2002), Molybdenum evidence for inherited planetary scale isotope heterogeneity of the protosolar nebula, *The Astrophysical Journal*, *565*, 640–644.

David, J., L. Godin, R. Stevenson, J. O'Neil, and D. Francis (2008), U-Pb ages (3.8–2.7 Ga) and Nd isotope data from the newly identified Eoarchean Nuvvuagittuq supracrustal belt, Superior Craton, Canada, *Geological Society of America Bulletin, doi:10.1130/B26369.1.*

Day, J. M. D., D. G. Pearson, and L. A. Taylor (2007), Highly siderophile element constraints on accretion and differentiation of the Earth-Moon system, *Science*, *315*, 217–219.

Day, J. M. D., R. J. Walker, L. P. Qin, and D. Rumble (2012), Late accretion as a natural consequence of planetary growth, *Nature Geoscience*, *5*, 614–617.

Debaille, V., C. O'Neill, A. D. Brandon, P. Haenecour, Q. -Z. Yin, N. Mattielli, and A. H. Treiman (2013), Stagnant-lid tectonics in early Earth revealed by 142Nd variations in late Archean rocks, *Earth and Planetary Science Letters*, *373*, 83–92.

DePaolo, D. J. and G. J. Wasserburg (1976), Nd isotopic variations and petrogenetic models, *Geophysical Research Letters*, *3*, 249–252.

Dixon, J. E. and D. A. Clague (2001), Volatiles in basaltic glasses from Loihi Seamount, Hawaii: evidence for a relatively dry plume component, *Journal of Petrology*, *42*, 627–654.

Dowty, E., M. Prinz, and K. Keil (1974), Ferroan anorthosite: a widespread and distinctive lunar rock type, *Earth and Planetary Science Letters*, *24*, 15–25.

Drake, M. J. (2005), Origin of water in the terrestrial planets, *Meteoritics and Planetary Science*, *40*, 519–527.

Elkins-Tanton, L. T. (2008), Linked magma ocean solidification and atmospheric growth for Earth and Mars, *Earth and Planetary Science Letters*, *271*, 181–191.

Elkins-Tanton, L. T. (2011), Magma oceans in the inner solar system, *Annual Reviews of Earth and Planetary Science*, *40*, 113–139.

Fischer-Gödde, M., C. Burkhardt, and T. Kleine (2013), Origin of the late veneer inferred from Ru isotope systematics, *44th Lunar and Planetary Science Conference*, Abstract, *2876*.

Fitoussi, C., B. Bourdon, T. Kleine, F. Oberli, and B. C. Reynolds (2009), Si isotope systematics of meteorites and terrestrial peridotites: implications for Mg/Si fractionation in the solar nebula and for Si in the Earth's core, *Earth and Planetary Science Letters*, *287*(1–2), 77–85, doi:10.1016/j.epsl.2009.07.038.

Froude, D. O., T. R. Ireland, P. D. Kinny, I. S. Williams, W. Compston, I. R. Williams, and J. S. Myers (1983), Ion microprobe identification of 4100–4200 Myr old terrestrial zircons, *Nature*, *304*, 616–618.

Gaffney, A. M. and L. E. Borg (2014), A young solidification age for the lunar magma ocean, *Geochimica et Cosmochimica Acta*, *140*, 227–240.

Gale, A., C. A. Dalton, C. H. Langmuir, Y. Sun, and J.-G. Schilling (2013), The mean composition of ocean ridge basalts, *Geochemistry Geophysics Geosystems*, *14*(3), 489–518, doi:10.1029/2012GC004334.

Gannoun, A., M. Boyet, H. Rizo, and A. E. Goresy (2011a), ^{146}Sm-^{142}Nd systematics measured in enstatite chondrites reveals a heterogeneous distribution of ^{142}Nd in the solar nebula, *Proceedings of the National Academy of Sciences*, *108*, 7693–7697.

Gannoun, A., M. Boyet, A. E. Goresy, and B. Devouard (2011b), REE and actinide microdistribution in Sahara 97072 and ALHA77295 EH3 chondrites: a combined cosmochemical and petrologic investigation, *Geochimica et Cosmochimica Acta.*, *75*, 3269–3289.

Georg, R. B., A. N. Halliday, E. A. Schauble, and B. C. Reynolds (2007), Silicon in the Earth's core, *Nature*, *447*(7148), 1102–1106, doi:10.1038/nature05927.

Glavin, D. P., A. Kubny, E. Jagoutz, and G. W. Lugmair (2004), Mn-Cr isotope systematics of the D'Orbigny angrite, *Meteoritics and Planetary Science*, *39*, 693–700.

Goldstein, S. L., R. K. O'Nions, and P. J. Hamilton (1984), A Sm-Nd isotopic study of atmospheric dusts and particulates from major river systems, *Earth and Planetary Science Letters*, *70*, 221–236.

Grange, M. L., A. A. Nemchin, R. T. Pidgeon, R. E. Merle, and N. E. Timms (2013), What lunar zircon ages can tell?, *Lunar and Planetary Science*, *44*, 1884.

Gray, C. M., D. A. Papanastassiou, and G. J. Wasserburg (1973), The identification of early condensates from the solar nebula, *Icarus*, 20, 213–239.

Guitreau, M., J. Blichert-Toft, S. J. Mojzsis, A.S.G. Roth, and B. Bourdon (2013), A legacy of Hadean silicate differentiation inferred from Hf isotopes in Eoarchean rocks of the Nuvvuagittuq supracrustal belt (Quebec, Canada), *Earth and Planetary Science Letters*, 362, 171–181.

Halliday, A. N. and T. Kleine (2005), Meteorites and the timing, mechanisms, and conditions of terrestrial planet accretion and early differentiation, in *Meteorites and the Early Solar System II*, edited by D. S. Lauretta and H. Y. McSween, pp. 775–801, The University of Arizona Press, Tucson.

Hans, U., T. Kleine, and B. Bourdon (2013), Rb-Sr chronology of volatile depletion in differentiated protoplanets: BABI, ADOR and ALL revisited, *Earth and Planetary Science Letters*, 374, 204–214.

Harley, S. L. and L. P. Black (1997), A revised Archaean chronology for the Napier Complex, Enderby Land, from SHRIMP ion-microprobe studies, *Antarctic Science*, 9, 74–91.

Harper, C. L. and S. B. Jacobsen (1992), Evidence from coupled ^{147}Sm-^{143}Nd and ^{146}Sm-^{142}Nd systematics for very early (4.5-Gyr) differentiation of the Earth's mantle, *Nature*, 360, 728–732.

Harrison, T. M. (2009), The Hadean crust: Evidence from >4 Ga zircons, *Annual Reviews of Earth and Planetary Science*, 37(1), 479–505, doi:10.1146/annurev.earth.031208.100151.

Harrison, T. M., J. Blichert-Toft, W. Muller, F. Albèrede, P. Holden, and S. J. Mojzsis (2005), Heterogeneous Hadean hafnium: evidence for continental crust at 4.4 to 4.5 Ga, *Science*, 310, 1947–1950.

Harrison, T. M., A. K. Schmitt, M. T. McCulloch, and O. M. Lovera (2008), Early (4.5 Ga) formation of terrestrial crust: Lu-Hf, $\delta^{18}O$, and Ti thermometry results for Hadean zircons, *Earth and Planetary Science Letters*, 268, 476–486.

Hartogh, P., et al. (2011), Ocean-like water in the Jupiter-family comet 103P/Hartley 2, *Nature*, 478, 218–220.

Herwartz, D., A. Pack, B. Friedrichs, and A. Bischoff (2014), Identification of the giant impactor Theia in lunar rocks, *Science*, 344, 1146–1150.

Holden, P., P. Lanc, T. R. Ireland, T. M. Harrison, J. J. Foster, and Z. Bruce (2009), Mass-spectrometric mining of Hadean zircons by automated SHRIMP multi-collector and single-collector U/Pb zircon age dating: The first 100,000 grains, *International Journal of Mass Spectrometry*, 286, 53–63.

Hoppe, P. and U. Ott (1997), Mainstream silicon carbide grains from meteorites, in *Astrophysical Implications of the Laboratory Study of Presolar Materials*, edited by T. J. Bernatowicz and E. K. Zinner, pp. 27–58, American Institute of Physics, Woodbury.

Horan, M. F., R. J. Walker, J. W. Morgan, J. N. Grossman, and A. E. Rubin (2003), Highly siderophile elements in chondrites, *Chemical Geology*, 196(1–4), 27–42, doi:10.1016/s0009-2541(02)00405-9.

Huang, S., J. Farkas, G. Yu, M. I. Petaev, and S. B. Jacobsen (2012), Calcium isotopic ratios and rare earth element abundances in refractory inclusions from the Allende CV3 chondrite, *Geochimica et Cosmochimca Acta.*, 77, 252–265.

Huang, S., S. B. Jacobsen, and S. Mukhopadhyay (2013), ^{147}Sm-^{143}Nd systematics of Earth are inconsistent with a superchondritic Sm/Nd ratio, *Proceedings of the National Academy of Sciences*, 110, 4929–4934.

Hunter, D. R., F. Barker, and H. T. Millard (1984), Geochemical investigation of Archaean bimodal and Dwalile metamorphic suites, Ancient Gneiss Complex, Swaziland, *Precambrian Research*, 24, 131–155.

Iizuka, T., K. Horie, T. Komiya, S. Maruyama, T. Hirata, H. Hidaka, and B. F. Windley (2006), 4.2 Ga zircon xenocryst in an Acasta gneiss from northwestern Canada: evidence for early continental crust, *Geology*, 34, 245–248.

Iizuka, T., T. Komiya, S. P. Johnson, Y. Kon, S. Maruyama, and T. Hirata (2009), Reworking of Hadean crust in the Acasta gneisses, northwestern Canada: evidence from in situ Lu-Hf isotope analysis of zircon, *Chemical Geology*, 259, 230–239.

Jackson, M. G. and R. W. Carlson (2011), An ancient recipe for flood-basalt genesis, *Nature*, 476, 316–318.

Jackson, M. G., R. W. Carlson, M. D. Kurz, P. D. Kempton, D. Francis, and J. Blusztajn (2010), Evidence for the survival of the oldest terrestrial mantle reservoir, *Nature*, 466, 853–856.

Jackson, M. G. and A. M. Jellinek (2013), Major and trace element composition of the high ^{3}He/^{4}He mantle: Implications for the composition of a nonchondritic Earth, *Geochemistry Geophysics Geosystems*, 14(8), 2954–2976, doi:10.1002/ggge.20188.

Jacobsen, B., Q.-Z. Yin, F. Moynier, Y. Amelin, A. N. Krot, K. Nagashima, I. D. Hutcheon, and H. Palme (2008), ^{26}Al-^{26}Mg and ^{207}Pb-^{206}Pb systematics of Allende CAIs: Canonical solar initial ^{26}Al/^{27}Al ratio reinstated, *Earth and Planetary Science Letters*, 272(1–2), 353–364, doi:10.1016/j.epsl.2008.05.003.

Jacobsen, S. B. (2005), The Hf-W isotopic system and the origin of the Earth and Moon, *Annual Reviews of Earth and Planetary Science*, 33, 531–570.

Jacobsen, S. B. and G. J. Wasserburg (1979), The mean age of mantle and crustal reservoirs, *Journal of Geophysical Research*, 84, 7411–7427.

Jacobson, S. A., A. Morbidelli, S. N. Raymond, D. P. O'Brien, K. J. Walsh, and D. C. Rubie (2014), Highly siderophile elements in Earth's mantle as a clock for the Moon-forming impact, *Nature*, 508, 84–87.

Jagoutz, E., H. Palme, H. Baddenhausen, K. Blum, M. Cendales, G. Dreibus, B. Spettel, V. Lorenz, and H. Wanke (1979), The abundances of major, minor and trace elements in the Earth's mantle as derived from primitive ultramafic nodules, *Proc. Lunar Sci. Conf. 10th*, 2031–2050.

Jahn, B.-M. and C. Shih (1974), On the age of the Overwacht Group, Swaziland Sequence, South Africa, *Geochimica et Cosmochimca Acta.*, 38, 873–885.

James, O. B. (1980), Rocks of the early lunar crust, *Proceedings of the 11th Lunar and Planetary Science Conference*, 365–393.

Jones, J. H. and M. J. Drake (1986), Geochemical constraints on core formation in the Earth, *Nature (London)*, 322(6076), 221–228.

Jones, J. H. and D. Walker (1991), Partitioning of siderophile elements in the Fe-Ni-S system; 1 bar to 80 kbar, *Earth and Planetary Science Letters*, 105(1–3), 127–133.

Kelly, N. M. and S. L. Harley (2005), An integrated microtextural and chemical approach to zircon geochronology: refining the Archean history of the Napier Complex, east Antarctica, *Contributions to Mineralogy and Petrology*, 149, 57–84.

Kemp, A. I. S., S. A. Wilde, C. J. Hawkesworth, C. D. Coath, A. Nemchin, R. T. Pidgeon, J. D. Vervoort, and S. A. DuFrane (2010), Hadean crustal evolution revisited: New constraints from Pb–Hf isotope systematics of the Jack Hills zircons, *Earth and Planetary Science Letters*, 296(1–2), 45–56, doi:10.1016/j.epsl.2010.04.043.

Kimura, K., R. S. Lewis, and E. Anders (1974), Distribution of gold and rhenium between nickel-iron and silicate melts: implications for the abundances of siderophile elements on the Earth and Moon, *Geochimica et Cosmochimca Acta.*, 38, 683–701.

Kinoshita, N., et al. (2012), A shorter ^{142}Nd half-life measured and implications for ^{146}Sm-^{142}Nd chronology in the solar system, *Science*, 335, 1614–1617.

Kleine, T., C. Munker, K. Mezger, and H. Palme (2002), Rapid accretion and early core formation on asteroids and the terrestrial planets from Hf-W chronometry, *Nature*, 418, 952–955.

Kleine, T., M. Touboul, B. Bourdon, F. Nimmo, K. Mezger, H. Palme, S. B. Jacobsen, Q. Z. Yin, and A. N. Halliday (2009), Hf-W chronology of the accretion and early evolution of asteroids and terrestrial planets, *Geochimica et Cosmochimca Acta.*, 73, 5150–5188.

Kroner, A., J. E. Hoffmann, H. Xie, F. Wu, C. Munker, E. Hegner, J. Wong, Y. Wan, and D. Liu (2013), Generation of early Archaean felsic greenstone volcanic rocks through crustal melting in the Kaapvaal craton, southern Africa, *Earth and Planetary Science Letters*, 381, 188–197.

Kruijer, T. S., P. Sprung, T. Kleine, I. Leya, C. Burkhardt, and R. Wieler (2012), Hf-W chronometry of core formation in planetesimals inferred from weakly irradiated iron meteorites, *Geochimica et Cosmochimca Acta.*, 99, 287–304.

Kruijer, T. S., M. Touboul, M. Fischer-Godde, K. R. Bermingham, R. J. Walker, and T. Kleine (2014), Protracted core formation and rapid accretion of protoplanets, *Science*, 344, 1150–1154.

Labrosse, S., J. W. Herlund, and N. A. Coltice (2007), A crystallizing dense magma ocean at the base of the Earth's mantle, *Nature*, 450, 866–869.

Lee, D.-C. and A. N. Halliday (1995), Hafnium-tungsten chronometry and the timing of terrestrial core formation, *Nature*, 378, 771–774.

Lee, D.-C. and A. N. Halliday (1996), Hf-W isotopic evidence for rapid accretion and differentiation in the early solar system, *Science*, 274, 1876–1879.

Liu, D., S. A. Wilde, Y. Wan, J. Wu, H. Zhou, C. Dong, and X. Yin (2008), New U-Pb and Hf isotopic data confirm Anshan as the oldest preserved segment of the North China Craton, *Amer. J. Sci.*, 308, 200–231.

Lodders, K. (2003), Solar system abundances and condensation temperatures of the elements, *Astrophysical Journal*, 591, 1220–1247.

Lugmair, G. W. and R. W. Carlson (1978), The Sm-Nd history of KREEP, *Proceedings of the Lunar and Planetary Science Conference*, 9, 689–704.

Lugmair, G. W. and A. Shukolyukov (1998), Early solar system timescales according to ^{53}Mn-^{53}Cr systematics, *Geochimica et Cosmochimica Acta.*, 62, 2863–2886.

Lugmair, G. W., K. Marti, and N. B. Scheinin (1978), Incomplete mixing of products from R-, P-, and S-process nucleosynthesis: Sm-Nd systematics in Allende inclusion EK 1-04-1, paper presented at Lunar and Planetary Science, Lunar and Planetary Institute, Houston.

Lyubetskaya, T. and J. Korenaga (2007), Chemical composition of Earth's primitive mantle and its variance: 1. Methods and results, *Journal of Geophysical Research*, 112(B03211), doi:10.1029/2005JB004223.

Maier, W. D., S. J. Barnes, I. H. Campbell, M. L. Fiorentini, P. Peltonen, S. J. Barnes, and R. H. Smithies (2009), Progressive mixing of meteoritic veneer into the early Earth's deep mantle, *Nature*, 460, 620–623.

Mann, U., D. J. Frost, D. C. Rubie, H. Becker, and A. Audetat (2012), Partitioning of Ru, Rh, Pd, Re, Ir and Pt between liquid metal and silicate at high pressures and high temperatures—implications for the origin of the highly siderophile element concentrations in Earth's mantle, *Geochimica et Cosmochimca Acta.*, 84, 593–613.

Marty, B. (2012), The origins and concentrations of water, carbon, nitrogen and noble gases on Earth, *Earth and Planetary Science Letters*, 313–314, 56–66.

Matsuda, J. and K. Matsubara (1989), Noble gases in silica and their implications for the terrestrial "missing" Xe, *Geophysical Research Letters*, 16, 81–84.

McCulloch, M. T. (1994), Primitive ^{87}Sr/^{86}Sr from an Archean barite and conjecture on the Earth's age and origin, *Earth and Planetary Science Letters*, 126, 1–13.

McCulloch, M. T. and G. J. Wasserburg (1978), Barium and neodymium isotopic anomalies in the Allende meteorite, *The Astrophysical Journal*, 220, L15–L19.

McDonough, W. F. (2003), Compositional model for the Earth's core, in *Treatise on Geochemistry*, edited by R. W. Carlson, pp. 547–568, Elsevier, Amsterdam.

McDonough, W. F. and S.-s. Sun (1995), The composition of the Earth, *Chemical Geology*, 120, 223–253.

Meisel, T., R. J. Walker, A. J. Irving, and J. P. Lorand (2001), Osmium isotopic compositions of mantle xenoliths: A global perspective, *Geochimica et Cosmochimica Acta.*, 65, 1311–1323.

Miljkovic, K., M. A. Wieczorek, G. S. Collins, M. Laneuville, G. A. Neumann, H. J. Melosh, S. C. Solomon, R. J. Phillips, D. E. Smith, and M. T. Zuber (2013), Asymmetric distribution of lunar impact basins caused by variations in target properties, *Science*, 342, 724–726.

Mosenfelder, J. L., P. D. Asimow, and T. J. Ahrens (2007), Thermodynamic properties of Mg_2SiO_4 liquid at ultra-high pressures from shock measurements to 200 GPa on forsterite and wadsleyite, *Journal of Geophysical Research*, 112(B6), doi:10.1029/2006jb004364.

Moynier, F., J. M. D. Day, W. Okui, T. Yokoyama, A. Bouvier, R. J. Walker, and F. A. Podosek (2012), Planetary-scale strontium isotopic heterogeneity and the age of volatile depletion of early Solar System materials, *Astrophysical Journal*, 758, 45–52, doi:10.1088/0004-637X/758/1/45.

Moynier, F., Q.-Z. Yin, and B. Jacobsen (2007), Dating the first stage of planet formation, *Astrophysical Journal, 671*, 181–183.

Mukhopadhyay, S. (2012), Early differentiation and volatile accretion in deep mantle neon and xenon, *Nature, 486*, 101–104.

Murthy, V.-R. (1991), Early differentiation of the Earth and the problem of mantle siderophile elements: A new approach, *Science, 253*, 303–306.

Myers, J. S. (1988), Early Archaean Narryer Gneiss Complex, Yilgarn Craton, Western Australia, *Precambrian Research, 38*, 297–307.

Nimmo, F. and C. Agnor (2006), Isotopic outcomes of N-body accretion simulations: Constraints on equilibration processes during large impacts from Hf/W observations, *Earth and Planetary Science Letters, 243*(1–2), 26–43, doi:10.1016/j.epsl.2005.12.009.

Nomura, R., H. Ozawa, S. Tateno, K. Hirose, J. Hernlund, S. Muto, H. Ishii, and N. Hiraoka (2011), Spin crossover and iron-rich silicate melt in the Earth's deep mantle, *Nature, 473*, 199–202.

Norman, M. D., V. C. Bennett, and G. Ryder (2002), Targeting the impactors: siderophile element signatures of lunar impact melts from Serenitatis, *Earth and Planetary Science Letters, 202*, 217–228.

Norman, M. D., R. A. Duncan, and J. J. Huard (2010), Imbrium provenance for the Apollo 16 Descartes terrain: argon ages and geochemistry of lunar breccias 67016 and 67455, *Geochimica et Cosmochimca Acta., 74*, 763–783.

Nutman, A. P., V. C. Bennett, C. R. L. Friend, H. Hidaka, K. Yi, S. R. Lee, and T. Kamiichi (2013), The Itsaq gneiss complex of Greenland: episodic 3900–3660 Ma juvenile crust formation and recycling in the 3660 to 3600 Ma Isukasian orogeny, *American Journal of Science, 313*, 877–911.

Nutman, A. P., V. R. McGregor, C. R. L. Friend, V. C. Bennett, and P. D. Kinny (1996), The Itsaq Gneiss Complex of southern West Greenland; the world's most extensive record of early crustal evolution (3900–3600 Ma), *Precambrian Research, 78*, 1–39.

Nyquist, L., D. Bogard, A. Yamaguchi, C. Y. Shih, Y. Karouji, M. Ebihara, Y. Reese, D. Garrison, G. McKay, and H. Takeda (2006), Feldspathic clasts in Yamato-86032: Remnants of the lunar crust with implications for its formation and impact history, *Geochimica et Cosmochimica Acta., 70*(24), 5990–6015, doi:10.1016/j.gca.2006.07.042.

Nyquist, L. E., B. M. Bansal, H. Wiesmann, and B.-M. Jahn (1974), Taurus-Littrow chronology: some constraints on early lunar crustal development, *Proceedings of the 5th Lunar Science Conference*, 1515–1539.

Nyquist, L. E., C.-Y. Shih, Y. D. Reese, J. Park, D. D. Bogard, D. H. Garrison, and A. Yamaguchi (2010), Lunar crustal history recorded in lunar anorthosites, *Lunar and Planetary Science, 41*, 1383.

Nyquist, L. E., H. Wiesmann, B. Bansal, C.-Y. Shih, J. E. Keith, and C. L. Harper (1995), ^{146}Sm-^{142}Nd formation interval for the lunar mantle, *Geochimica et Cosmochimica Acta., 59*, 2817–2837.

O'Neil, J., M. Boyet, R. W. Carlson, and J.-L. Paquette (2013), Half a billion years of reworking of Hadean mafic crust to produce the Nuvvuagittuq Eoarchean felsic crust, *Earth and Planetary Science Letters, 379*, 13–25.

O'Neil, J., R. W. Carlson, D. Francis, and R. K. Stevenson (2008), Neodymium-142 evidence for Hadean mafic crust, *Science, 321*, 1828–1831.

O'Neil, J., R. W. Carlson, J.-L. Paquette, and D. Francis (2012), Formation age and metamorphic history of the Nuvvuagittuq greenstone belt, *Precambrian Research, 220–221*, 23–44.

O'Neil, J., D. Francis, and R. W. Carlson (2011), Implications of the Nuvvuagittuq greenstone belt for the formation of Earth's early crust, *Journal of Petrology, 52*, 985–1009.

O'Neill, H. S. C. and H. Palme (2008), Collisional erosion and the non-chondritic composition of the terrestrial planets, *Philosophical Transactions Royal Society of London A, 366*, 4205–4238.

Ozima, M. and F. A. Podosek (1983), *Noble Gas Geochemistry*, 367 pp., Cambridge University Press, Cambridge.

Ozima, M. and F. A. Podosek (1999), Formation age of Earth from ^{129}I/^{127}I and ^{244}Pu/^{238}U systematics and the missing Xe, *Journal of Geophysical Research, 104*, 25493–25499.

Palme, H. and H. S. C. O'Neill (2014), Cosmochemical estimates of mantle composition, in *Treatise on Geochemistry*, edited by R. W. Carlson, pp. 1–39, Elsevier, Amsterdam.

Paniello, R. C., J. M. D. Day, and F. Moynier (2012), Zinc isotopic evidence for the origin of the Moon, *Nature, 490*, 376–380.

Patterson, C. (1956), Age of meteorites and the Earth, *Geochimica et Cosmochimica Acta., 10*, 230.

Pepin, R. O. and D. Porcelli (2002), Origin of noble gases in the terrestrial planets, *Reviews in Mineralogy and Geochemistry, 47*, 191–246.

Pepin, R. O. and D. Porcelli (2006), Xenon isotope systematics, giant impacts, and mantle degassing on the early Earth, *Earth and Planetary Science Letters, 250*(3–4), 470–485, doi:10.1016/j.epsl.2006.08.014.

Pidgeon, R. T. and S. A. Wilde (1998), The interpretation of complex zircon U-Pb systems in Archaean granitoids and gneisses from the Jack Hills, Narryer Gneiss terrane, Western Australia, *Precambrian Research, 91*, 309–332.

Podosek, F. A., E. K. Zinner, G. J. MacPherson, L. L. Lundberg, J. C. Brannon, and A. J. Fahey (1991), Correlated study of initial ^{87}Sr/^{86}Sr and Al/Mg isotopic systematics and petrologic properties in a suite of refractory inclusions from the Allende meteorite, *Geochimica et Cosmochimica Acta., 55*, 1083–1110.

Poitrasson, F., A. N. Halliday, D.-C. Lee, S. Levasseur, and N. Teutsch (2004), Iron isotope differences between Earth, Moon, Mars and Vesta as possible records of contrasted accretion mechanisms, *Earth and Planetary Science Letters, 223*, 253–266.

Polat, A., P. W. U. Appel, and B. J. Fryer (2011), An overview of the geochemistry of Eoarchean to Mesoarchean ultramafic to mafic volcanic rocks, SW Greenland: Implications for mantle depletion and petrogenetic processes at subduction zones in the early Earth, *Gondwana Research, 20*, 255–283.

Pujol, M., B. Marty, and R. Burgess (2011), Chondrite-like xenon trapped in Archean rocks: a possible signature of the ancient atmosphere, *Earth and Planetary Science Letters, 308*, 298–306.

Qin, L., C. M. O. D. Alexander, R. W. Carlson, M. F. Horan, and T. Yokoyama (2010), Contributors to chromium isotope variation in meteorites, *Geochimica et Cosmochimica Acta.*, *74*, 1122–1145.

Qin, L., R. W. Carlson, and C. M. O. D. Alexander (2011), Correlated nucleosynthetic isotopic variability in Cr, Sr, Ba, Sm, Nd and Hf in Murchison and QUE 97008, *Geochimica et Cosmochimca Acta.*, *75*, 7806–7828.

Reufer, A., M. M. M. Meier, W. Benz, and R. Wieler (2012), A hit-and-run giant impact scenario, *Icarus*, *221*, 296–299.

Reynolds, J. H. (1960), Isotopic composition of primordial xenon, *Physical Review Letters*, *4*, 351–354.

Richter, S., R. Eykens, H. Kuhn, Y. Aregbe, A. Verbruggen, and S. Weyer (2010), New average values for the n(^{238}U)/n(^{235}U) isotope ratios of natural uranium standards, *International Journal of Mass Spectrometry*, *295*, 94–97.

Righter, K. (2011), Prediction of metal-silicate partition coefficients for siderophile elements: An update and assessment of PT conditions for metal-silicate equilibrium during accretion of the Earth, *Earth and Planetary Science Letters*, *304*, 158–167.

Righter, K., L. Danielson, M. J. Drake, and K. Domanik (2014), Partition coefficients at high pressure and temperature, in *Treatise on Geochemistry*, edited by R. W. Carlson, pp. 449–477, Elsevier, Amsterdam.

Righter, K., M. Humayun, and L. R. Danielson (2008), Partitioning of palladium at high pressures and temperatures during core formation, *Nature Geoscience*, *1*, 321–323.

Righter, K. and C. K. Shearer (2003), Magmatic fractionation of Hf and W: constraints on the timing of core formation and differentiation in the Moon and Mars, *Geochimica et Cosmochimca Acta.*, *67*, 2497–2507.

Ringwood, A. E. (1977), Composition of the core and implications for origin of the earth, *Geochemical Journal*, *11*, 111–135.

Rizo, H., M. Boyet, J. Blichert-Toft, and M. Rosing (2011), Combined Nd and Hf isotope evidence for deep-seated source of Isua lavas, *Earth and Planetary Science Letters*, *312*(3–4), 267–279, doi:10.1016/j.epsl.2011.10.014.

Rizo, H., M. Boyet, J. Blichert-Toft, and M. T. Rosing (2013), Early mantle dynamics inferred from ^{142}Nd variations in Archean rocks from southwest Greenland, *Earth and Planetary Science Letters*, *377–378*, 324–335.

Rizo, H., R.J. Walker, R.W. Carlson, M. Touboul, M.F. Horan, I.S. Puchtel, and M. Boyet (2015), Early Earth differentiation investigated through ^{142}Nd, ^{182}W and highly siderophile element abundances in samples from Isua, Greenland, *Geochimica et Cosmochimica Acta*, submitted.

Roth, A. S. G., B. Bourdon, S. J. Mojzsis, J. F. Rudge, M. Guitreau, and J. Blichert-Toft (2014), Combined ^{147,146}Sm-^{143,142}Nd constraints on the longevity and residence time of early terrestrial crust, *Geochemistry Geophysics Geosystems*, *15*, 2329–2345.

Roth, A. S. G., B. Bourdon, S. J. Mojzsis, M. Touboul, P. Sprung, M. Guitreau, and J. Blichert-Toft (2013), Inherited ^{142}Nd anomalies in Eoarchean protoliths, *Earth and Planetary Science Letters*, *361*, 50–57.

Rubie, D. C., D. J. Frost, U. Mann, Y. Asahara, F. Nimmo, K. Tsuno, P. Kegler, A. Holzheid, and H. Palme (2011), Heterogeneous accretion, composition and core-mantle differentiation of the Earth, *Earth and Planetary Science Letters*, *301*, 31–42.

Rudge, J. F., T. Kleine, and B. Bourdon (2010), Broad bounds on Earth's accretion and core formation constrained by geochemical models, *Nature Geoscience*, *3*, 439–443.

Saal, A. E., E. H. Hauri, C. H. Langmuir, and M. R. Perfit (2002), Vapor undersaturation in primitive mid-ocean-ridge basalt and the volatile content of Earth's upper mantle, *Nature*, *419*, 451–455.

Salters, V. J. M. and A. Stracke (2004), Composition of the depleted mantle, *Geochemistry Geophysics Geosystems*, *5*(5), doi:10.1029/2003gc000597.

Sanborn, M. E., R. W. Carlson, and M. Wadhwa (2014), ^{147,146}Sm-^{143,142}Nd, ^{176}Lu-^{176}Hf, and ^{87}Rb-^{87}Sr systematics in the angrites: implications for chronology and processes on the angrite parent body, *Geochimica et Cosmochimca Acta*, in revision.

Shahar, A., K. Ziegler, E. D. Young, A. Ricolleau, E. A. Schauble, and Y. Fei (2009), Experimentally determined Si isotope fractionation between silicate and Fe metal and implications for Earth's core formation, *Earth and Planetary Science Letters*, *288*(1–2), 228–234, doi:10.1016/j.epsl.2009.09.025.

Shirey, S. B. and R. J. Walker (1998), The Re-Os isotope system in cosmochemistry and high-temperature geochemistry, *Annual Reviews of Earth and Planetary Science*, *26*, 423–500.

Siebert, J., J. Badro, D. Antonangeli, and F. J. Ryerson (2013), Terrestrial accretion under oxidizing conditions, *Science*, *339*, 1194–1197.

Simon, J. I. and D. J. DePaolo (2010), Stable calcium isotopic composition of meteorites and rocky planets, *Earth and Planetary Science Letters*, *289*, 457–466.

Simon, J. I., D. J. DePaolo, and F. Moynier (2009), Calcium isotope composition of meteorites, Earth and Mars, *The Astrophysical Journal*, *702*, 707–715.

Solomatov, V. S. (2000), Fluid dynamics of a terrestrial magma ocean, in *Origin of the Earth and Moon*, edited by R. M. Canup and K. Righter, pp. 323–338, The University of Arizona Press, Tucson.

Song, B., A. P. Nutman, D. Liu, and J. Wu (1996), 3800 to 2500 Ma crustal evolution in the Anshan area of Liaoning Province, northeastern China, *Precambrian Research*, *78*, 79–94.

Sprung, P., T. Kleine, and E. E. Scherer (2013), Isotopic evidence for chondritic Lu/Hf and Sm/Nd of the Moon, *Earth and Planetary Science Letters*, *380*, 77–87.

Staudacher, T. and C. J. Allegre (1982), Terrestrial xenology, *Earth and Planetary Science Letters*, *60*, 389–406.

Steele, R. C. J., T. Elliot, C. D. Coath, and M. Regelous (2011), Confirmation of mass-independent Ni isotopic variability in iron meteorites, *Geochimica et Cosmochimca Acta.*, *75*, 7906–7925.

Stixrude, L. and C. Lithgow-Bertelloni (2012), Geophysics of chemical heterogeneity in the mantle, *Annual Reviews of Earth and Planetary Science*, *40*, 569–595.

Tera, F. and G. J. Wasserburg (1974), U-Th-Pb systematics on lunar rocks and inferences about lunar evolution and the age of the moon, *Proceedings of the 5th Lunar Science Conference*, 1571–1599.

Tera, F., D. A. Papanastassiou, and G. J. Wasserburg (1974), Isotopic evidence for a terminal lunar cataclysm, *Earth and Planetary Science Letters*, *22*, 1–21.

Thiemens, M., and J. E. Heidenreich (1983), Mass independent fractionation of oxygen: A novel isotopic effect and its possible cosmochemical implications, *Science*, *219*, 1073–1075.

Tolstikhin, I. N. and R. K. O'Nions (1994), The Earth's missing xenon—a combination of early degassing and of rare-gas loss from the atmosphere, *Chemical Geology*, *115*, 1–6.

Touboul, M., T. Kleine, B. Bourdon, H. Palme, and R. Wieler (2007), Late formation and prolonged differentiation of the Moon inferred from W isotopes in lunar metals, *Nature*, *450*(7173), 1206–1209, doi:10.1038/nature06428.

Touboul, M., J. Liu, J. O'Neil, I. S. Puchtel, and R. J. Walker (2014), New insights into the Hadean mantle revealed by [182]W and highly siderophile element abundances in supracrustal rocks from the Nuvvuagittuq greenstone belt, Quebec, Canada, *Chemical Geology*, *383*, 63–75.

Touboul, M., I. S. Puchtel, and R. J. Walker (2012), [182]W evidence for long-term preservation of early mantle differentiation products, *Science*, *335*, 1065–1069.

Trinquier, A., J.-L. Birck, and C. J. Allègre (2007), Widespread [54]Cr heterogeneity in the inner solar system, *The Astrophysical Journal*, *655*, 1179–1185.

Trinquier, A., J. L. Birck, C. J. Allègre, C. Göpel, and D. Ulfbeck (2008), [53]Mn–[53]Cr systematics of the early Solar System revisited, *Geochimica et Cosmochimica Acta.*, *72*(20), 5146–5163, doi:10.1016/j.gca.2008.03.023.

Trinquier, A., T. Elliott, D. Ulfbeck, C. Coath, A. N. Krot, and M. Bizzarro (2009), Origin of nucleosynthetic isotope heterogeneity in the solar protoplanetary disk, *Science*, *324*(5925), 374–376, doi:10.1126/science.1168221.

Tucker, J. M. and S. Mukhopadhyay (2014), Evidence for multiple magma ocean outgassing and atmospheric loss episodes from mantle noble gases, *Earth and Planetary Science Letters*, *393*, 254–265.

Turner, S., T. Rushmer, M. Reagan, and J.-F. Moyen (2014), Heading down early on? Start of subduction on Earth, *Geology*, *42*, 139–142.

Valdes, M. C., M. Moreira, J. Foriel, and F. Moynier (2014), The nature of Earth's building blocks as revealed by calcium isotopes, *Earth and Planetary Science Letters*, *394*, 135–145.

Valley, J. W., et al. (2014), Hadean age for a post-magma-ocean zircon confirmed by atom-probe tomography, *Nature Geoscience*, *7*, 219–223.

Villanueva, G. L., M. J. Mumma, B. P. Bonev, M. A. DiSanti, E. L. Gibb, H. Bohnhardt, and M. Lippi (2009), A sensitive search for deuterated water in comet 8P/Tuttle, *The Astrophysical Journal*, *690*, doi:10.1088/0004-637X/690/1/L5.

Wade, J. and B. J. Wood (2005), Core formation and the oxidation state of the Earth, *Earth and Planetary Science Letters*, *236*, 78–95.

Walker, R. J. (2009), Highly siderophile elements in the Earth, Moon and Mars: Update and implications for planetary accretion and differentiation, *Chemie der Erde*, *69*, 101–125.

Walker, R. J., M. F. Horan, J. W. Morgan, H. Becker, J. N. Grossman, and A. E. Rubin (2002), Comparative [187]Re-[187]Os

systematics of chondrites: implications regarding early solar system processes, *Geochimica et Cosmochimica Acta.*, *66*, 4187–4201.

Walker, R. J., M. F. Horan, C. K. Shearer, and J. J. Papike (2004), Depletion of highly siderophile elements in the lunar mantle: evidence for prolonged late accretion, *Earth and Planetary Science Letters*, *224*, 399–413.

Wang, Z. and H. Becker (2013), Ratios of S, Se and Te in the silicate Earth require a volatile-rich late veneer, *Nature*, *499*, 328–331.

Wanke, H., G. Driebus, and E. Jagoutz (1984), Mantle chemistry and accretion history of the Earth, in *Archaean Geochemistry*, edited by A. Kroner, G. N. Hanson and A. M. Goodwin, pp. 1–24, Springer-Verlag, Berlin.

Warren, P. H. (2011), Stable-isotope anomalies and the accretionary assemblage of the Earth and Mars: A subordinate role for carbonaceous chondrites, *Earth and Planetary Science Letters*, *311*, 93–100.

Warren, P. H. and J. T. Wasson (1979), The origin of KREEP, *Rev. Geophys. Space Phys.*, *17*(1), 73–88.

Wetherill, G. W. (1975), Radiometric chronology of the early Solar System, *Annual Reviews of Nuclear Science*, *25*, 283–328.

Weyer, S., A. D. Anbar, A. Gerdes, G. W. Gordon, T. J. Algeo, and E. A. Boyle (2008), Natural fractionation of $^{238}U/^{235}U$, *Geochimica et Cosmochimca Acta.*, *72*, 345–359.

Wheeler, K. T., D. Walker, and W. F. McDonough (2010), Pd and Ag metal-silicate partitioning applied to Earth differentiation and core-mantle exchange, *Meteoritics and Planetary Science*, *46*, 199–217.

Wiechert, U., A. N. Halliday, D.-C. Lee, G. A. Snyder, L. A. Taylor, and D. A. Rumble (2001), Oxygen isotopes and the Moon-forming giant impact, *Science*, *294*, 345–348.

Wilde, S. A., J. W. Valley, W. H. Peck, and C. M. Graham (2001), Evidence from detrital zircons for the existence of continental crust and oceans on Earth 4.4 Gyr ago, *Nature*, *409*, 175–178.

Willbold, M., T. Elliott, and S. Moorbath (2011), The tungsten isotopic composition of the Earth's mantle before the terminal bombardment, *Nature*, *477*, 195–198.

Williams, H. M., A. Markowski, G. Quitté, A. N. Halliday, N. Teutsch, and S. Levasseur (2006), Fe isotope fractionation in iron meteorites: New insights into metal-sulphide segregation and planetary accretion, *Earth and Planetary Science Letters*, *250*(3–4), 486–500, doi:10.1016/j.epsl.2006.08.013.

Willis, J. and J. I. Goldstein (1982), The effects of C, P, and S on trace element partitioning during solidification of Fe-Ni alloys, *Journal of Geophysical Research*, *87*, A435–A445.

Wood, B. J. and A. N. Halliday (2005), Cooling of the Earth and core formation after the giant impact, *Nature*, *437*, 1345–1348.

Wood, B. J. and A. N. Halliday (2010), The lead isotopic age of the Earth can be explained by core formation alone, *Nature*, *465*, 767–770.

Wood, J. A., J. S. D. Jr., U. B. Marvin, and B. N. Powell (1970), Lunar Anorthosites and a geophysical model of the Moon, *Proceedings of the Apollo 11 Lunar Science Conference*, 965–988.

Workman, R. K. and S. R. Hart (2005), Major and trace element composition of the depleted MORB mantle (DMM),
Earth and Planetary Science Letters, *231*(1–2), 53–72,
doi:10.1016/j.epsl.2004.12.005.

Yin, Q., S. B. Jacobsen, K. Yamashita, J. Blichert-Toft, P. Telouk,
and F. Albarede (2002), A short timescale for terrestrial planet
formation from Hf-W chronometry of meteorites, *Nature*,
418, 949–952.

Yin, Q.-Z., and S. B. Jacobsen (2006), Geochemistry: Does
U-Pb date Earth's core formation?, *Nature*, *444*, doi:10.1038/
nature05358.

Zhang, J., N. Dauphas, A. M. Davis, I. Leya, and A. Fedkin
(2012), The proto-Earth as a significant source of lunar material, *Nature Geoscience*, *5*, 251–255.

Zindler, A. and S. Hart (1986), Chemical geodynamics, *Ann.
Rev. Earth Planet. Sci.*, *14*, 493–571.

INDEX

The Early Earth: Accretion and Differentiation, Geophysical Monograph 212, First Edition.
Edited by James Badro and Michael Walter.
© 2015 American Geophysical Union. Published 2015 by John Wiley & Sons, Inc.